D0056202

A Mathematician Plays the Stock Market

***Also by* John Allen Paulos**

Mathematics and Humor (1980)

I Think Therefore I Laugh (1985)

Innumeracy: Mathematical Illiteracy and its Consequences (1988)

Beyond Numeracy: Ruminations of a Numbers Man (1991)

A Mathematician Reads the Newspaper (1995)

Once Upon a Number: The Hidden Mathematical Logic of Stories (1998)

A Mathematician Plays the Stock Market

John Allen Paulos

A Member of the Perseus Books Group

Published by Basic Books,
A Member of the Perseus Books Group
 For information, address Basic Books, 387 Park Avenue South, New York, NY 10016–8810.

Designed by Trish Wilkinson
Set in 11-point Sabon by the Perseus Books Group

Library of Congress Cataloging-in-Publication Data
Paulos, John Allen.
A mathematician plays the stock market / John Allen Paulos.
p. cm.
Includes bibliographical references and index.
ISBN 0–465–05480–3 (alk. paper)
1. Investments—Psychological aspects. 2. Stock exchanges—Psychological aspects. 3. Stock exchanges—Mathematical models. 4. Investment analysis. 5. Stocks. I. Title.

HG4515.15.P38 2003
332.63'2042—dc21 2002156215

03 04 05 / 10 9 8 7 6 5 4 3 2 1

To my father, who never played
the market and knew little about probability,
yet understood one of the prime lessons of both.
"Uncertainty," he would say, "is the only
certainty there is, and knowing how to live
with insecurity is the only security."

Contents

1 | Anticipating Others' Anticipations

It was early 2000, the market was booming, and my investments in various index funds were doing well but not generating much excitement. Why investments should generate excitement is another issue, but it seemed that many people were genuinely enjoying the active management of their portfolios. So when I received a small and totally unexpected chunk of money, I placed it into what Richard Thaler, a behavioral economist I'll return to later, calls a separate mental account. I considered it, in effect, "mad money."

Nothing distinguished the money from other assets of mine except this private designation, but being so classified made my modest windfall more vulnerable to whim. In this case it entrained a series of ill-fated investment decisions that, even now, are excruciating to recall. The psychological ease with which such funds tend to be spent was no doubt a factor in my using the unexpected money to buy some shares of WorldCom (abbreviated WCOM), "the pre-eminent global communications company for the digital generation," as its ads boasted, at $47 per share. (Hereafter I'll generally use WCOM to refer to the stock and WorldCom to refer to the company.)

Today, of course, WorldCom is synonymous with business fraud, but in the halcyon late 1990s it seemed an irrepressibly successful devourer of high-tech telecommunications companies. Bernie Ebbers, the founder and former CEO, is now viewed by many as a pirate, but then he was seen as a swashbuckler. I had read about the company, knew that high-tech guru George Gilder had been long and fervently singing its praises, and was aware that among its holdings were MCI, the huge long-distance telephone company, and UUNet, the "backbone" of the Internet. I spend a lot of time on the net (home is where you hang your @) so I found Gilder's lyrical writings on the "telecosm" and the glories of unlimited bandwidth particularly seductive.

I also knew that, unlike most dot-com companies with no money coming in and few customers, WorldCom had more than $25 billion in revenues and almost 25 million customers, and so when several people I knew told me that WorldCom was a "strong buy," I was receptive to their suggestion. Although the stock had recently fallen a little in price, it was, I was assured, likely to soon surpass its previous high of $64.

If this was all there was to it, there would have been no important financial consequences for me, and I wouldn't be writing about the investment now. Alas, there was something else, or rather a whole series of "something elses." After buying the shares, I found myself idly wondering, why not buy more? I don't think of myself as a gambler, but I willed myself *not* to think, willed myself simply to act, willed myself to buy more shares of WCOM, shares that cost considerably more than the few I'd already bought. Nor were these the last shares I would buy. Usually a hardheaded fellow, I was nevertheless falling disastrously in love.

Although my particular heartthrob was WCOM, almost all of what I will say about my experience is unfortunately applicable to many other stocks and many other investors. Wherever WCOM appears, you may wish to substitute the symbols for Lucent, Tyco, Intel, Yahoo, AOL-Time Warner, Global Crossing, Enron, Adelphia, or, perhaps, the generic symbols WOE or BANE. The time frame of the book—in the midst of a market collapse after a heady, nearly decade-long surge—may also appear rather more specific and constraining than it is. Almost all the points made herein are rather general or can be generalized with a little common sense.

Falling in Love with WorldCom

John Maynard Keynes, arguably the greatest economist of the twentieth century, likened the position of short-term investors in a stock market to that of readers in a newspaper beauty contest (popular in his day). The ostensible task of the readers is to pick the five prettiest out of, say, one hundred contestants, but their real job is more complicated. The reason is that the newspaper rewards them with small prizes only if they pick the five contestants who receive the most votes from readers. That is, they must pick the contestants that they think are most likely to be picked by the other readers, and the other readers must try to do the same. They're not to become enamored of any of the contestants or otherwise give undue weight to their own taste. Rather they must, in Keynes' words, anticipate "what average opinion expects the average opinion to be" (or, worse, anticipate what the average opinion expects the average opinion expects the average opinion to be).

Thus it may be that, as in politics, the golden touch derives oddly from being in tune with the brass masses. People might dismiss rumors, for example, about "Enronitis" or "WorldComism" affecting the companies in which they've invested, but if they believe others will believe the rumors, they can't afford to ignore them.

BWC (before WorldCom) such social calculations never interested me much. I didn't find the market particularly inspiring or exalted and viewed it simply as a way to trade shares in businesses. Studying the market wasn't nearly as engaging as doing mathematics or philosophy or watching the Comedy Network. Thus, taking Keynes literally and not having much confidence in my judgment of popular taste, I refrained from investing in individual stocks. In addition, I believed that stock movements were entirely random and that trying to outsmart dice was a fool's errand. The bulk of my money therefore went into broad-gauge stock index funds.

AWC, however, I deviated from this generally wise course. Fathoming the market, to the extent possible, and predicting it, if at all possible, suddenly became live issues. Instead of snidely dismissing the business talk shows' vapid talk, sportscaster-ish attitudes, and empty prognostication, I began to search for what of substance might underlie all the commentary about the market and slowly changed my mind about some matters. I also sought to account for my own sometimes foolish behavior, instances of which will appear throughout the book, and tried to reconcile it with my understanding of the mathematics underlying the market.

Lest you dread a cloyingly personal account of how I lost my shirt (or at least had my sleeves shortened), I should stress that my primary purpose here is to lay out, elucidate, and explore the basic conceptual mathematics of the market. I'll examine—largely via vignettes and stories rather than formulas and equations—various approaches to investing as well as a number of problems, paradoxes, and puzzles, some old, some new, that encapsulate issues associated with the market. Is it efficient? Random? Is there anything to technical analysis, fundamental analysis? How can one quantify risk? What is the role of cognitive illusion? Of common knowledge? What are the most common scams? What are options, portfolio theory, short-selling, the efficient market hypothesis? Does the normal bell-shaped curve explain the market's occasional extreme volatility? What about fractals, chaos, and other non-standard tools? There will be no explicit investment advice and certainly no segments devoted to the ten best stocks for the new millennium, the five smartest ways to jump-start your 401(k), or the three savviest steps you can take right now. In short, there'll be no financial pornography.

Often inseparable from these mathematical issues, however, is psychology, and so I'll begin with a discussion of the no-man's land between this discipline and mathematics.

Being Right Versus Being Right About the Market

There's something very reductive about the stock market. You can be right for the wrong reasons or wrong for the right reasons, but to the market you're just plain right or wrong. Compare this to the story of the teacher who asks if anyone in the class can name two pronouns. When no one volunteers, the teacher calls on Tommy who responds, "Who, me?" To the market, Tommy is right and therefore, despite being unlikely to get an A in English, he's rich.

Guessing right about the market usually leads to chortling. While waiting to give a radio interview at a studio in Philadelphia in June 2002, I mentioned to the security guard that I was writing this book. This set him off on a long disquisition on the market and how a couple of years before he had received two consecutive statements from his 401(k) administrator indicating that his retirement funds had declined. (He took this to be what in chapter 3 is called a technical sell signal.) "The first one I might think was an accident, but two in a row, no. Do you know I had to argue with that pension person there about getting out of stocks and into those treasury bills? She told me not to worry because I wasn't going to retire for years, but I insisted 'No, I want out now.' And I'm sure glad I did get out." He went on to tell me about "all the big shots at the station who cry like babies every day about how much money they lost. I warned them that two down statements and you get out, but they didn't listen to me."

I didn't tell the guard about my ill-starred WorldCom experience, but later I did say to the producer and sound man that the guard had told me about his financial foresight in response to my mentioning my book on the stock market. They both assured me that he would have told me no matter what. "He tells everyone," they said, with the glum humor of big shots who didn't take his advice and now cry like babies.

Such anecdotes bring up the question: "If you're so smart, why ain't you rich?" Anyone with a modicum of intelligence and an unpaid bill or two is asked this question repeatedly. But just as there is a distinction between being smart and being rich, there is a parallel distinction between being right and being right about the market.

Consider a situation in which the individuals in a group must simultaneously choose a number between 0 and 100. They are further directed to pick the number that they think will be closest to 80 percent of the average number chosen by the group. The one who comes closest will receive $100 for his efforts. Stop for a bit and think what number you would pick.

Some in the group might reason that the average number chosen is likely to be 50 and so these people would guess 40, which is 80 percent of this. Others might anticipate that people will guess 40 for this reason and so they would guess 32, which is 80 percent of 40. Still others might anticipate that people will guess 32 for this reason and so they would guess 25.6, which is 80 percent of 32.

If the group continues to play this game, they will gradually learn to engage in ever more iterations of this meta-reasoning about others' reasoning until they all reach the optimal response, which is 0. Since they all want to choose a number equal to 80 percent of the average, the only way they can all do this is by choosing 0, the only number equal to 80 percent of itself. (Choosing 0 leads to what is called the Nash equilibrium of this game. It results when individuals modify their actions until they can no longer benefit from changing them given what the others' actions are.)

The problem of guessing 80 percent of the average guess is a bit like Keynes's description of the investors' task. What makes it tricky is that anyone bright enough to cut to the heart of the problem and guess 0 right away is almost certain to be wrong, since different individuals will engage in different degrees of meta-reasoning about others' reasoning. Some, to increase their chances, will choose numbers a little above or a little below the natural guesses of 40 or 32 or 25.6 or 20.48. There will be some random guesses as well and some guesses of 50 or more. Unless the group is very unusual, few will guess 0 initially.

If a group plays this game only once or twice, guessing the average of all the guesses is as much a matter of reading the others' intelligence and psychology as it is of following an idea to its logical conclusion. By the same token, gauging investors is often as important as gauging investments. And it's likely to be more difficult.

My Pedagogical Cruelty

Other situations, as well, require anticipating others' actions and adapting yours to theirs. Recall, for example, the television show on which contestants had to guess how their spouses would guess they would answer a particular question.

There was also a show on which opposing teams had to guess the most common associations the studio audience had made with a collection of words. Or consider the game in which you have to pick the location in New York City (or simply the local shopping mall) that others would most likely look for you first. You win if the location you pick is chosen by most of the others. Instances of Keynes's beauty contest metaphor are widespread.

As I've related elsewhere, a number of years ago I taught a summer probability course at Temple University. It met every day and the pace was rapid, so to induce my students to keep up with the material I gave a short quiz every day. Applying a perverse idea I'd experimented with in other classes, I placed a little box at the bottom of each exam sheet and a notation next to it stating that students who crossed the box (placed an X in it) would have ten extra points added to their exam scores. A further notation stated that the points would be added only if less than half the class crossed the box. If more than half crossed the box, those crossing it would lose ten points on their exam scores. This practice, I admit, bordered on pedagogical cruelty.

A few brave souls crossed the box on the first quiz and received ten extra points. As the summer wore on, more and more students did so. One day I announced that more than half the students had crossed the box and that those who did had therefore been penalized ten points. Very few students crossed the box on the next exam. Gradually, however, the number crossing it edged up to around 40 percent of the class and stayed there. But it was always a different 40 percent, and it struck me that the calculation a student had to perform to decide whether to cross the box was quite difficult. It was especially so since the class was composed largely of foreign students who, despite my best efforts (which included this little game), seemed to have developed little camaraderie. Without any collusion that I could discern, the students had to anticipate other students' anticipations of their anticipations in a convoluted and very skittish self-referential tangle. Dizzying.

I've since learned that W. Brian Arthur, an economist at the Santa Fe Institute and Stanford University, has long used an essentially identical scenario to describe the predicament of bar patrons deciding whether or not to go to a popular bar, the experience being pleasant only if the bar is not thronged. An equilibrium naturally develops whereby the bar rarely becomes too full. (This almost seems like a belated scientific justification for Yogi Berra's quip about Toots Shor's restaurant in New York: "Nobody goes there any more. It's too crowded.") Arthur proposed the model to clarify the behavior of market investors who, like my students and the bar patrons, must anticipate others' anticipations of them (and so on). Whether one buys or sells, crosses the box or doesn't cross, goes to the bar or doesn't go, depends upon one's beliefs about others' possible actions and beliefs.

The Consumer Confidence Index, which measures consumers' propensity to consume and their confidence in their own economic future, is likewise subject to a flighty, reflexive sort of consensus. Since people's evaluation of their own economic prospects is so dependent on what they perceive others' prospects to be, the CCI indirectly surveys people's beliefs about other people's beliefs. ("Consume" and "consumer" are, in this context, common but unfortunate terms. "Buy," "purchase," "citizen," and "household" are, I think, preferable.)

Common Knowledge, Jealousy, and Market Sell-Offs

Sizing up other investors is more than a matter of psychology. New logical notions are needed as well. One of them, "common knowledge," due originally to the economist Robert Aumann, is crucial to understanding the complexity of the stock market and the importance of transparency. A bit of information is common knowledge among a group of people if all parties know it, know that the others know it, know that the others know they know it, and so on. It is much more than "mutual knowledge," which requires only that the parties know the particular bit of information, not that they be aware of the others' knowledge.

As I'll discuss later, this notion of common knowledge is essential to seeing how "subterranean information processing" often underlies sudden bubbles or crashes in the markets, changes that seem to be precipitated by nothing at all and therefore are almost impossible to foresee. It is also relevant to the recent market sell-offs and accounting scandals, but before we get to more realistic accounts of the market, consider the following parable from my book *Once Upon a Number*, which illustrates the power of common knowledge. The story takes place in a benightedly sexist village of uncertain location. In this village there are many married couples and each woman immediately knows when another woman's husband has been unfaithful but not when her own has. The very strict feminist statutes of the village require that if a woman can *prove* her husband has been unfaithful, she must kill him that very day. Assume that the women are statute-abiding, intelligent, aware of the intelligence of the other women, and, mercifully, that they never inform other women of their philandering husbands. As it happens, twenty of the men have been unfaithful, but since no woman can prove her husband has been so, village life proceeds merrily and warily along. Then one morning the tribal matriarch comes to visit from the far side of the forest. Her honesty is acknowledged by all and her word is taken as truth. She warns the assembled villagers that there is at least one philandering husband among them. Once this fact, already known to everyone, becomes *common* knowledge, what happens?

The answer is that the matriarch's warning will be followed by nineteen peaceful days and then, on the twentieth day, by a massive slaughter in which twenty women kill their husbands. To see this, assume there is only one unfaithful husband, Mr. A. Everyone except Mrs. A already knows about him, so when the matriarch makes her announcement, only she learns something new from it. Being intelligent, she realizes that she would know if any other husband were unfaithful. She thus infers that Mr. A is the philanderer and kills him that very day.

Now assume there are two unfaithful men, Mr. A and Mr. B. Every woman except Mrs. A and Mrs. B knows about both these cases of infidelity. Mrs. A knows only of Mr. B's, and Mrs. B knows only of Mr. A's. Mrs. A thus learns nothing from the matriarch's announcement, but when Mrs. B fails to kill Mr. B the first day, she infers that there must be a second philandering husband, who can only be Mr. A. The same holds for Mrs. B who infers from the fact that Mrs. A has not killed her husband on the first day that Mr. B is also guilty. The next day Mrs. A and Mrs. B both kill their husbands.

If there are exactly three guilty husbands, Mr. A, Mr. B, and Mr. C, then the matriarch's announcement would have no visible effect the first day or the second, but by a reasoning process similar to the one above, Mrs. A, Mrs. B, and Mrs. C would each infer from the inaction of the other two of them on the first two days that their husbands were also guilty and kill them on the third day. By a process of mathematical induction we can conclude that if twenty husbands are unfaithful, their intelligent wives would finally be able to prove it on the twentieth day, the day of the righteous bloodbath.

Now if you replace the warning of the matriarch with that provided by, say, an announcement by the Securities and Exchange Commission, the nervousness of the wives with the nervousness of investors, the wives' contentment as long as their own husbands weren't straying with the investors' contentment as long their own companies weren't cooking the books, killing husbands with selling stocks, and the gap between the warning and the killings with the delay between announcement of an investigation and big sell-offs, you can understand how this parable of common knowledge applies to the market.

Note that in order to change the logical status of a bit of information from mutually known to commonly known, there must be an independent arbiter. In the parable it was the matriarch; in the market analogue it was the SEC. If there is no one who is universally respected and believed, the motivating and cleansing effect of warnings is lost.

Happily, unlike the poor husbands, the market is capable of rebirth.

2 | Fear, Greed, and Cognitive Illusions

You don't need to have been a temporarily besotted investor to realize that psychology plays an important and sometimes crucial role in the market, but it helps. By late summer 2000, WCOM had declined to $30 per share, inciting me to buy more. As "inciting" may suggest, my purchases were not completely rational. By this I don't mean that there wasn't a rational basis for investing in WCOM stock. If you didn't look too closely at the problems of overcapacity and the long-distance phone companies' declining revenue streams, you could find reasons to keep buying. It's just that my reasons owed less to an assessment of trends in telecommunications or an analysis of company fundamentals than to an unsuspected gambling instinct and a need to be right. I suffered from "confirmation bias" and searched for the good news, angles, and analyses about the stock while avoiding the less sanguine indications.

Averaging Down or Catching a Falling Knife?

After an increasingly intense, albeit one-sided courtship of the stock (the girl never even sent me a dividend), I married it. As its share price fell, I continued to see only opportunities for gains. Surely, I told myself, the stock had reached its bottom and it was now time to average down by buying the considerably cheaper shares. Of course, for every facile invitation I extended myself to "average down," I ignored an equally facile warning about not attempting to "catch a falling knife." The stale, but prudent adage about not putting too many of one's eggs in the same basket never seemed to push itself very forcefully into my consciousness.

I was also swayed by Salomon Smith Barney's Jack Grubman (possessor, incidentally, of a master's degree in mathematics from Columbia) and other analysts, who ritualistically sprinkled their "strong buys" over the object of my affections. In fact, most brokerage houses in early 2000 rated WCOM a "strong buy," and those that didn't had it as a "buy." It required no great perspicacity to notice that at the time, almost no stock ever received a "sell," much less a "strong sell," and that even "holds" were sparingly bestowed. Maybe, I thought, only environmental companies that manufactured solar-powered flashlights qualified for these latter ratings. Accustomed to grade inflation and to movie, book, and restaurant review inflation, I wasn't taken in by the uniformly positive ratings. Still, just as you can be moved by a television commercial whose saccharine dialogue you are simultaneously ridiculing, part of me gave credence to all those "strong buys."

I kept telling myself that I'd incurred only paper losses and had lost nothing real unless I sold. The stock would come back, and if I didn't sell, I couldn't lose. Did I really believe this? Of course not, but I acted as if I did, and "averaging down" continued to seem like an irresistible opportunity. I believed in the company, but greed and fear were already doing their usual two-step in my head and, in the process, stepping all over my critical faculties.

Emotional Overreactions and Homo Economicus

Investors can become (to borrow a phrase Alan Greenspan and Robert Shiller made famous) irrationally exuberant, or, changing the arithmetical sign, irrationally despairing. Some of the biggest daily point gains and declines in Nasdaq's history occurred in a single month in early 2000, and the pattern has continued unabated in 2001 and 2002, the biggest point gain since 1987 occurring on July 24, 2002. (The increase in volatility, although substantial, is a little exaggerated since our perception of gains and losses have been distorted by the rise in the indices. A 2 percent drop in the Dow when the market is at 9,000 is 180 points, whereas not too long ago when it was at 3,000, the same percentage drop was only 60 points.) The volatility has come about as the economy has hovered near a recession, as accounting abuses have come to light, as CEO malfeasance has mounted, as the bubble has fizzled, and as people have continued to trade on their own, influenced no doubt by capricious lists of the fifty most beautiful (er . . . , undervalued) stocks.

As with beautiful people and, for that matter, distinguished universities, emotions and psychology are imponderable factors in the market's jumpy variability. Just as beauty and academic quality don't change as rapidly as ad hoc lists and magazine rankings do, so, it seems, the fundamentals of companies don't change as quickly as our mercurial reactions to news about them do.

It may be useful to imagine the market as a fine race car whose exquisitely sensitive steering wheel makes it impossible to drive in a straight line. Tiny bumps in our path cause us to swerve wildly, and we zigzag from fear to greed and back again, from unreasonable gloom to irrational exuberance and back.

Our overreactions are abetted by the all-crisis-all-the-time business media, which brings to mind a different analogy: the reigning theory in cosmology. The inflationary universe hypothesis holds—very, very roughly—that shortly after the Big Bang the primordial universe inflated so fast that all of our visible universe derives from a tiny part of it; we can't see the rest. The metaphor is strained (in fact I just developed carpal tunnel syndrome typing it), but it seems reminiscent of what happens when the business media (as well as the media in general) focus unrelentingly on some titillating but relatively inconsequential bit of news. Coverage of the item expands so fast as to distort the rest of the global village and render it invisible.

Our responses to business news are only one of the ways in which we fail to be completely rational. More generally, we simply don't always behave in ways that maximize our economic well-being. "Homo economicus" is not an ideal toward which many people strive. My late father, for example, was distinctly uneconimicus. I remember him sitting and chuckling on the steps outside our house one autumn night long ago. I asked what was funny and he told me that he had been watching the news and had heard Bob Buhl, a pitcher for the then Milwaukee Braves, answer a TV reporter's question about his off-season plans. "Buhl said he was going to help his father up in Saginaw, Michigan, during the winter." My father laughed again and continued. "And when the reporter asked Buhl what his father did up in Saginaw, Buhl said, 'Nothing at all. He does nothing at all.' "

My father liked this kind of story and his crooked grin lingered on his face. This memory was jogged recently when I was straightening out my office and found a cartoon he had sent me years later. It showed a bum sitting happily on a park bench as a line of serious businessmen traipsed by him. The bum calls out "Who's winning?" Although my father was a salesman, he always seemed less intent on making a sale than on schmoozing with his customers, telling jokes, writing poetry (not all of it doggerel), and taking innumerable coffee breaks.

Everyone can tell such stories, and you would be hard-pressed to find a novel, even one with a business setting, where the characters are all actively pursuing their economic self-interest. Less anecdotal evidence of the explanatory limits of the homo economicus ideal is provided by so-called "ultimatum games." These generally involve two players, one of whom is given a certain amount of money, say $100, by an experimenter, and the other of whom is given a kind of veto. The first player may offer any non-zero fraction of the $100 to the second player, who can either accept or reject it. If he accepts it, he is given whatever amount the first player has offered, and the first player keeps the balance. If he rejects it, the experimenter takes the money back.

Viewing this in rational game-theoretic terms, one would argue that it's in the interest of the second player to accept whatever is offered since any amount, no matter how small, is better than nothing. One would also suspect that the first player, knowing this, would make only tiny offers to the second player. Both suppositions are false. The offers range up to 50 percent of the money involved, and, if deemed too small and therefore humiliating, they are sometimes rejected. Notions of fairness and equality, as well as anger and revenge, seem to play a role.

Behavioral Finance

People's reactions to ultimatum games may be counterproductive, but they are at least clear-eyed. A number of psychologists in recent years have pointed out the countless ways in which we're all subject to other sorts of counterproductive behavior that spring from cognitive blind spots that are analogues, perhaps, of optical illusions. These psychological illusions and foibles often make us act irrationally in a variety of disparate endeavors, not the least of which is investing.

Amos Tversky and Daniel Kahneman are the founders of this relatively new field of study, many of whose early results are reported upon in the classic book *Judgment Under Uncertainty*, edited by them and Paul Slovic. (Kahneman was awarded the 2002 Nobel Prize in economics, and Tversky almost certainly would have shared it had he not died.) Others who have contributed to the field include Thomas Gilovich, Robin Dawes, J. L. Knetschin, and Baruch Fischhoff. Economist Richard Thaler (mentioned in the first chapter) is one of the leaders in applying these emerging insights to economics and finance, and his book *The Winner's Curse*, as well as Gilovich's *How We Know What Isn't So,* are very useful compendiums of recent results.

What makes these results particularly intriguing is the way they illuminate the tactics used, whether consciously or not, by people in everyday life. For example, a favorite ploy of activists of all ideological stripes is to set the terms of a debate by throwing out numbers, which need have little relation to reality to be influential. If you are appalled at some condition, you might want to announce that more than 50,000 deaths each year are attributable to it. By the time people catch up and realize that the number is a couple of orders of magnitude smaller, your cause will be established.

Unfounded financial hype and unrealistic "price targets" have the same effect. Often, it seems, an analyst cites a "price target" for a stock in order to influence investors by putting a number into their heads. (Since the targets are so often indistinguishable from wishes, shouldn't they always be infinite?)

The reason for the success of this hyperbole is that most of us suffer from a common psychological failing. We credit and easily become attached to any number we hear. This tendency is called the "anchoring effect" and it's been demonstrated to hold in a wide variety of situations.

If an experimenter asks people to estimate the population of Ukraine, the size of Avogadro's number, the date of an historical event, the distance to Saturn, or the earnings of XYZ Corporation two years from now, their guesses are likely to be fairly close to whatever figure the experimenter first suggests as a possibility. For example, if he prefaces his request for an estimate of the population of Ukraine with the question—"Is it more or less than 200 million people?"—the subjects' estimates will vary and generally be a bit less than this figure, but still average, say, 175 million people. If he prefaces his request for an estimate with the question—"Is the population of Ukraine more or less than 5 million people?"—the subjects' estimates will vary and this time be a bit more than this figure, but still average, say, 10 million people. The subjects usually move in the right direction from whatever number is presented to them, but nevertheless remain anchored to it.

You might think this is a reasonable strategy for people to follow. They might realize they don't know much about Ukraine, chemistry, history, or astronomy, and they probably believe the experimenter is knowledgeable, so they stick close to the number presented. The astonishing strength of the tendency comes through, however, when the experimenter obtains his preliminary number by some chance means, say by spinning a dial that has numbers around its periphery—300 million, 200 million, 50 million, 5 million, and so on. Say he spins the dial in front of the subjects, points out where it has stopped, and then asks them if the population of Ukraine is more or less than the number at which the dial has stopped.

The subjects' guesses are still anchored to this number even though, one presumes, they don't think the dial knows anything about Ukraine!

Financial numbers are also vulnerable to this sort of manipulation, including price targets and other uncertain future figures like anticipated earnings. The more distant the future the numbers describe, the more it's possible to postulate a huge figure that is justified, say, by a rosy scenario about the exponentially growing need for bandwidth or online airline tickets or pet products. People will discount these estimates, but usually not nearly enough. Some of the excesses of the dot-coms are probably attributable to this effect. On the sell side too, people can paint a dire picture of ballooning debt or shrinking markets or competing technology. Once again, the numbers presented, this time horrific, need not have much to do with reality to have an effect.

Earnings and targets are not the only anchors. People often remember and are anchored to the fifty-two-week high (or low) at which the stock had been selling and continue to base their deliberations on this anchor. I unfortunately did this with WCOM. Having first bought the stock when it was in the forties, I implicitly assumed it would eventually right itself and return there. Later, when I bought more of it in the thirties, twenties, and teens, I made the same assumption.

Another, more extreme form of anchoring (although there are other factors involved) is revealed by investors' focus on whether the earnings that companies announce quarterly meet the estimates analysts have established for them. When companies' earnings fall short by a penny or two per share, investors sometimes react as if this were tantamount to near-bankruptcy. They seem to be not merely anchored to earnings estimates but fetishistically obsessed with them.

Not surprisingly, studies have shown that companies' earnings are much more likely to come in a penny or two above the analysts' average estimate than a penny or two below it. If earnings were figured without regard to analysts' expectations, they'd come in below the average estimate as often as above it. The reason for the asymmetry is probably that companies sometimes "back in" to their earnings. Instead of determining revenues and expenses and subtracting the latter from the former to obtain earnings (or more complicated variants of this), companies begin with the earnings they need and adjust revenues and expenses to achieve them.

Psychological Foibles, A List

The anchoring effect is not the only way in which our faculties are clouded. The "availability error" is the inclination to view any story, whether political, personal, or financial, through the lens of a superficially similar story that is psychologically available. Thus every recent American military involvement is inevitably described somewhere as "another Vietnam." Political scandals are immediately compared to the Lewinsky saga or Watergate, misunderstandings between spouses reactivate old wounds, normal accounting questions bring the Enron-Andersen-WorldCom fiasco to mind, and any new high-tech firm has to contend with memories of the dot-com bubble. As with anchoring, the availability error can be intentionally exploited.

The anchoring effect and availability error are exacerbated by other tendencies. "Confirmation bias" refers to the way we check a hypothesis by observing instances that confirm it and ignoring those that don't. We notice more readily and even diligently search for whatever might confirm our beliefs, and we don't notice as readily and certainly don't look hard for what disconfirms them. Such selective thinking reinforces the anchoring effect: We naturally begin to look for reasons that the arbitrary number presented to us is accurate. If we succumb completely to the confirmation bias, we step over the sometimes fine line separating flawed rationality and hopeless closed-mindedness.

Confirmation bias is not irrelevant to stock-picking. We tend to gravitate toward those people whose take on a stock is similar to our own and to search more vigorously for positive information on the stock. When I visited WorldCom chatrooms, I more often clicked on postings written by people characterizing themselves as "strong buys" than I did on those written by "strong sells." I also paid more attention to WorldCom's relatively small deals with web-hosting companies than to the larger structural problems in the telecommunications industry.

The "status quo bias" (these various biases are generally not independent of each other) also applies to investing. If subjects are told, for example, that they've inherited a good deal of money and then asked which of four investment options (an aggressive stock portfolio, a more balanced collection of equities, a municipal bond fund, or U.S. Treasuries) they would prefer to invest it in, the percentages choosing each are fairly evenly distributed.

Surprisingly, however, if the subjects are told that they've inherited the money but it is already in the form of municipal bonds, almost half choose to keep it in bonds. It's the same with the other three investment options: Almost half elect to keep the money where it is. This inertia is part of the reason so many people sat by while not only their inheritances but their other investments dwindled away. The "endowment effect," another kindred bias, is an inclination to endow one's holdings with more value than they have simply because one holds them. "It's my stock and I love it."

Related studies suggest that passively endured losses induce less regret than losses that follow active involvement. Someone who sticks with an old investment that then declines by 25 percent is less upset than someone who switches into the same investment before it declines by 25 percent. The same fear of regret underlies people's reluctance to trade lottery tickets with friends. They imagine how they'll feel if their original ticket wins.

Minimizing possible regret often plays too large a role in investors' decisionmaking. A variety of studies by Tversky, Kahneman, and others have shown that most people tend to assume less risk to obtain gains than they do to avoid losses. This isn't implausible: Other research suggests that people feel considerably more pain after incurring a financial loss than they do pleasure after achieving an equivalent gain. In the extreme case, desperate fears about losing a lot of money induce people to take enormous risks with their money.

Consider a rather schematic outline of many of the situations studied. Imagine that a benefactor gives $10,000 to everyone in a group and then offers each of them the following choice. He promises to a) give them an additional $5,000 or else b) give them an additional $10,000 or $0, depending on the outcome of a coin flip. Most people choose to receive the additional $5,000. Contrast this with the choice people in a different group make when confronted with a benefactor who gives them each $20,000 and then offers the following choice to each of them. He will a) take from them $5,000 or else b) will take from them $10,000 or $0, depending on the flip of a coin. In this case, in an attempt to avoid any loss, most people choose to flip the coin. The punchline, as it often is, is that the choices offered to the two groups are the same: a sure $15,000 or a coin flip to determine whether they'll receive $10,000 or $20,000.

Alas, I too took more risks to avoid losses than I did to obtain gains. In early October 2000, WCOM had fallen below $20, forcing the CEO, Bernie Ebbers, to sell 3 million shares to pay off some of his investment debts. The WorldCom chatrooms went into one of their typical frenzies and the price dropped further. My reaction, painful to recall, was, "At these prices I can finally get out of the hole." I bought more shares even though I knew better. There was apparently a loose connection between my brain and my fingers, which kept clicking the buy button on my Schwab online account in an effort to avoid the losses that loomed.

Outside of business, loss aversion plays a role as well. It's something of a truism that the attempt to cover up a scandal often leads to a much worse scandal. Although most people know this, attempts to cover up are still common, presumably because, here too, people are much more willing to take risks to avoid losses than they are to obtain gains.

Another chink in our cognitive apparatus is Richard Thaler's notion of "mental accounts," mentioned in the last chapter. "The Legend of the Man in the Green Bathrobe" illustrates this notion compellingly. It is a rather long shaggy dog story, but the gist is that a newlywed on his honeymoon in Las Vegas wakes up in bed and sees a $5 chip left on the dresser. Unable to sleep, he goes down to the casino (in his green bathrobe, of course), bets on a particular number on the roulette wheel, and wins. The 35 to 1 odds result in a payout of $175, which the newlywed promptly bets on the next spin. He wins again and now has more than $6,000. He bets everything on his number a couple more times, continuing until his winnings are in the millions and the casino refuses to accept such a large bet. The man goes to a bigger casino, wins yet again, and now commands hundreds of millions of dollars. He hesitates and then decides to bets it all one more time. This time he loses. In a daze, he stumbles back up to his hotel room where his wife yawns and asks how he did. "Not too bad. I lost $5."

It's not only in casinos and the stock market that we categorize money in odd ways and treat it differently depending on what mental account we place it in. People who lose a $100 ticket on the way to a concert, for example, are less likely to buy a new one than are people who lose $100 in cash on their way to buy the ticket. Even though the amounts are the same in the two scenarios, people in the former one tend to think $200 is too large an expenditure from their entertainment account and so don't buy a new ticket, while people in the latter tend to assign $100 to their entertainment account and $100 to their "unfortunate loss" account and buy the ticket.

In my less critical moments (although not only then) I mentally amalgamate the royalties from this book, whose writing was prompted in part by my investing misadventure, with my WCOM losses. Like corporate accounting, personal accounting can be plastic and convoluted, perhaps even more so since, unlike corporations, we are privately held.

These and other cognitive illusions persist for several reasons. One is that they lead to heuristic rules of thumb that can save time and energy. It's often easier to go on automatic pilot and respond to events in a way that requires little new thinking, not just in scenarios involving eccentric philanthropists and sadistic experimenters. Another reason for the illusions' persistence is that they have, to an extent, become hardwired over the eons. Noticing a rustle in the bush, our primitive ancestors were better off racing away than they were plugging into Bayes' theorem on conditional probability to determine if a threat was really likely.

Sometimes these heuristic rules lead us astray, again not just in business and investing but in everyday life. Early in the fall 2002 Washington, D.C., sniper case, for example, the police arrested a man who owned a white van, a number of rifles, and a manual for snipers. It was thought at the time that there was one sniper and that he owned all these items, so for the purpose of this illustration let's assume that this turned out to be true. Given this and other reasonable assumptions, which is higher—a) the probability that an innocent man would own all these items, or b) the probability that a man who owned all these items would be innocent? You may wish to pause before reading on.

Most people find questions like this difficult, but the second probability would be vastly higher. To see this, let me make up some plausible numbers. There are about 4 million innocent people in the suburban Washington area and, we're assuming, one guilty one. Let's further estimate that ten people (including the guilty one) own all three of the items mentioned above. The first probability—that an innocent man owns all these items—would be 9/4,000,000 or 1 in 400,000. The second probability—that a man owning all three of these items is innocent—would be 9/10. Whatever the actual numbers, these probabilities usually differ substantially. Confusing them is dangerous (to defendants).

Self-Fulfilling Beliefs and Data Mining

Taken to extremes, these cognitive illusions may give rise to closed systems of thought that are immune, at least for a while, to revision and refutation. (Austrian writer and satirist Karl Kraus once remarked, "Psychoanalysis is that mental illness for which it regards itself as therapy.") This is especially true for the market, since investors' beliefs about stocks or a method of picking them can become a self-fulfilling prophecy. The market sometimes acts like a strange beast with a will, if not a mind, of its own. Studying it is not like studying science and mathematics, whose postulates and laws are (in quite different senses) independent of us. If enough people suddenly wake up believing in a stock, it will, for that reason alone, go up in price and justify their beliefs.

A contrived but interesting illustration of a self-fulfilling belief involves a tiny investment club with only two investors and ten possible stocks to choose from each week. Let's assume that each week chance smiles at random on one of the ten stocks the investment club is considering and it rises precipitously, while the week's other nine stocks oscillate within a fairly narrow band.

George, who believes (correctly in this case) that the movements of stock prices are largely random, selects one of the ten stocks by rolling a die (say an icosehedron—a twenty-sided solid—with two sides for each number). Martha, let's assume, fervently believes in some wacky theory, Q analysis. Her choices are therefore dictated by a weekly Q analysis newsletter that selects one stock of the ten as most likely to break out. Although George and Martha are equally likely to pick the lucky stock each week, the newsletter-selected stock will result in big investor gains more frequently than will any other stock.

The reason is simple but easy to miss. Two conditions must be met for a stock to result in big gains for an investor: It must be smiled upon by chance that week and it must be chosen by one of the two investors. Since Martha always picks the newsletter-selected stock, the second condition in her case is always met, so whenever chance happens to favor it, it results in big gains for her. This is not the case with the other stocks. Nine-tenths of the time, chance will smile on one of the stocks that is not newsletter-selected, but chances are George will not have picked that particular one, and so it will seldom result in big gains for him. One must be careful in interpreting this, however. George and Martha have equal chances of pulling down big gains (10 percent), and each stock of the ten has an equal chance of being smiled upon by chance (10 percent), but the newsletter-selected stock will achieve big gains much more often than the randomly selected ones.

Reiterated more numerically, the claim is that 10 percent of the time the newsletter-selected stock will achieve big gains for Martha, whereas each of the ten stocks has only a 1 percent chance of both achieving big gains *and* being chosen by George. Note again that two things must occur for the newsletter-selected stock to achieve big gains: Martha must choose it, which happens with probability 1, and it must be the stock that chance selects, which happens with probability 1/10th. Since one multiplies probabilities to determine the likelihood that several independent events occur, the probability of both these events occurring is 1 × 1/10, or 10 percent. Likewise, two things must occur for any particular stock to achieve big gains via George: George must choose it, which occurs with probability 1/10th, and it must be the stock that chance selects, which happens with probability 1/10th. The product of these two probabilities is 1/100th or 1 percent.

Nothing in this thought experiment depends on there being only two investors. If there were one hundred investors, fifty of whom slavishly followed the advice of the newsletter and fifty of whom chose stocks at random, then the newsletter-selected stocks would achieve big gains for their investors eleven times as frequently as any particular stock did for its investors. When the newsletter-selected stock is chosen by chance and happens to achieve big gains, there are fifty-five winners, the fifty believers in the newsletter and five who picked the same stock at random. When any of the other nine stocks happens to achieve big gains, there are, on average, only five winners.

In this way a trading strategy, if looked at in a small population of investors and stocks, can give the strong illusion that it is effective when only chance is at work.

"Data mining," the scouring of databases of investments, stock prices, and economic data for evidence of the effectiveness of this or that strategy, is another example of how an inquiry of limited scope can generate deceptive results. The problem is that if you look hard enough, you will always find some seemingly effective rule that resulted in large gains over a certain time span or within a certain sector. (In fact, inspired by the British economist Frank Ramsey, mathematicians over the last half century have proved a variety of theorems on the inevitability of some kind of order in large sets.) The promulgators of such rules are not unlike the believers in bible codes. There, too, people searched for coded messages that seemed to be meaningful, not realizing that it's nearly impossible for there not to be some such "messages." (This is trivially so if you search in a book that has a chapter 11, conveniently foretelling many companies' bankruptcies.)

People commonly pore over price and trade data attempting to discover investment schemes that have worked in the past. In a reductio ad absurdum of such unfocused fishing for associations, David Leinweber in the mid–90s exhaustively searched the economic data on a United Nations CD-ROM and found that the best predictor of the value of the S&P 500 stock index was—a drum roll here—butter production in Bangladesh. Needless to say, butter production in Bangladesh has probably not remained the best predictor of the S&P 500. Whatever rules and regularities are discovered within a sample must be applied to new data if they're to be accorded any limited credibility. You can always arbitrarily define a class of stocks that in retrospect does extraordinarily well, but will it continue to do so?

I'm reminded of a well-known paradox devised (for a different purpose) by the philosopher Nelson Goodman. He selected an arbitrary future date, say January 1, 2020, and defined an object to be "grue" if it is green and the time is before January 1, 2020, or if it is blue and the time is after January 1, 2020. Something is "bleen," on the other hand, if it is blue and the time is before that date or if it is green and the time is after that date. Now consider the color of emeralds. All emeralds examined up to now (2002) have been green. We therefore feel confident that all emeralds are green. But all emeralds so far examined are also grue. It seems that we should be just as confident that all emeralds are grue (and hence blue beginning in 2020). Are we?

A natural objection is that these color words grue and bleen are very odd, being defined in terms of the year 2020. But were there aliens who speak the grue-bleen language, they could make the same charge against us. "Green," they might argue, is an arbitrary color word, being defined as grue before 2020 and bleen afterward. "Blue" is just as odd, being bleen before 2020 and grue from then on. Philosophers have not convincingly shown what exactly is wrong with the terms grue and bleen, but they demonstrate that even the abrupt failure of a regularity to hold can be accommodated by the introduction of new weasel words and ad hoc qualifications.

In their headlong efforts to discover associations, data miners are sometimes fooled by "survivorship bias." In market usage this is the tendency for mutual funds that go out of business to be dropped from the average of all mutual funds. The average return of the surviving funds is higher than it would be if all funds were included. Some badly performing funds become defunct, while others are merged with better-performing cousins. In either case, this practice skews past returns upward and induces greater investor optimism about future returns. (Survivorship bias also applies to stocks, which come and go over time, only the surviving ones making the statistics on performance. WCOM, for example, was unceremoniously replaced on the S&P 500 after its steep decline in early 2002.)

The situation is rather like that of schools that allow students to drop courses they're failing. The grade point averages of schools with such a policy are, on average, higher than those of schools that do not allow such withdrawals. But these inflated GPAs are no longer a reliable guide to students' performance.

Finally, taking the meaning of the term literally, survivorship bias makes us all a bit more optimistic about facing crises. We tend to see only those people who survived similar crises. Those who haven't are gone and therefore much less visible.

Rumors and Online Chatrooms

Online chatrooms are natural laboratories for the observation of illusions and distortions, although their psychology is more often brutally basic than subtly specious. While spellbound by WorldCom, I would spend many demoralizing, annoying, and engaging hours compulsively scouring the various WorldCom discussions at Yahoo! and RagingBull. Only a brief visit to these sites is needed to see that a more accurate description of them would be rantrooms.

Once someone dons a screen name, he (the masculine pronoun, I suspect, is almost always appropriate) usually dispenses with grammar, spelling, and most conventional standards of polite discourse. Other people become morons, idiots, and worse. A poster's references to the stock, if he's shorting it (selling shares he doesn't have in the hope that he can buy them back when the price goes down), put a burden on one's ability to decode scatological allusions and acronyms. Any expression of pain at one's losses is met with unrelenting scorn and sarcasm; ostensibly genuine musings about suicide are no exception. A suicide threat in April 2002, lamenting the loss of house, family, and job because of WCOM, drew this response: "You sad sack loser. Die. You might want to write a note too in case the authorities and your wife don't read the Yahoo! chatrooms."

People who characterize themselves as sellers are generally (but not always) more vituperative than those claiming to be buyers. Some of the regulars appear genuinely interested in discussing the stock rationally, imparting information, and exchanging speculation. A few seem to know a lot, many are devotees of various outlandish conspiracy theories, including the usual anti-Semitic sewage, and even more are just plain clueless, asking, for example, why they "always put that slash between the P and the E in P/E, and is P price or profit." There were also many discussions that had nothing directly to do with the stock. One that I remember fondly was about someone who called a computer help desk because his computer didn't work. It turned out that he had plugged the computer and all his peripheral devices into his surge protector, which he then plugged into itself. The connection with whatever company was being discussed I've forgotten.

Taking advice from such an absurdly skewed sample of posters is silly, of course, but the real-time appeal of the sites is akin to overhearing gossip about a person you're interested in. It's likely to be false, spun, or overstated, but it still holds a certain fascination. Another analogy is to listening to police radio and getting a feel for the raw life and death on the streets.

Chatroom denizens form little groups that spend a lot of time excoriating, but not otherwise responding to, opposing groups. They endorse each other's truisms and denounce those of the others. When WorldCom purchased a small company or had a reversal in its Brazilian operations, this was considered big news. It was not nearly as significant, however, as an analyst changing his recommendation from a strong buy to a buy or vice versa. If you filter out the postings drenched in anger and billingsgate, you find most of the biases mentioned above demonstrated on a regular basis. The posters are averse to risk, anchored to some artificial number, addicted to circular thought, impressed by data mining, or all of the above.

Most boards I visited had a higher percentage of rational posters than did WorldCom's. I remember visiting the Enron board and reading rumors of the bogus deals and misleading accounting practices that eventually came to light. Unfortunately, since there are always rumors of every conceivable and contradictory sort (sometimes posted by the same individual), one cannot conclude anything from their existence except that they're likely to contribute to feelings of hope, fear, anger, and anxiety.

Pump and Dump, Short and Distort

The rumors are often associated with market scams that exploit people's normal psychological reactions. Many of these reactions are chronicled in Edwin Lefevre's 1923 classic novel, *Reminiscences of a Stock Operator,* but the standard "pump and dump" is an illegal practice that has gained new life on the Internet. Small groups of individuals buy a stock and tout it in a misleading hyperbolic way (that is, pump it). Then when its price rises in response to this concerted campaign, they sell it at a profit (dump it). The practice works best in bull markets when people are most susceptible to greed. It is also most effective when used on thinly traded stocks where a few buyers can have a pronounced effect.

Even a single individual with a fast Internet connection and a lot of different screen names can mount a pump and dump operation. Just buy a small stock from an online broker, then visit the chatroom where it's discussed. Post some artful innuendoes or make some outright phony claims and then back yourself up with one of your pseudonyms. You can even maintain a "conversation" among your various screen names, each salivating over the prospects of the stock. Then just wait for it to move up and sell it quickly when it does.

A fifteen-year-old high school student in New Jersey was arrested for successfully pumping and dumping after school. It's hard to gauge how widespread the practice is since the perpetrators generally make themselves invisible. I don't think it's rare, especially since there are gradations in the practice, ranging from organized crime telephone banks to conventional brokers inveigling gullible investors.

In fact, the latter probably constitute a vastly bigger threat. Being a stock analyst used to be a thoroughly respectable profession, and for most practitioners no doubt it still is. Unfortunately, however, there seem to be more than a few whose fervent desire to obtain the investment banking fees associated with underwriting, mergers, and the other quite lucrative practices induces them to shade their analyses—and "shade" may be a kind verb—so as not to offend the companies they're both analyzing and courting. In early 2002, there were well-publicized stories of analysts at Merrill Lynch exchanging private emails deriding a stock that they were publicly touting. Six other brokerage houses were accused of similar wrongdoing.

Even more telling were records from Salomon Smith Barney subpoenaed by Congress indicating that executives at companies generating large investment fees often *personally* received huge dollops of companies' initial public offerings. Not open to ordinary investors, these hot, well-promoted offerings quickly rose in value and their quick sale generated immediate profits. Bernie Ebbers was reported to have received, between 1996 and 2000, almost a million shares of IPOs worth more than $11 million. The $1.4 billion settlement between several big brokerage houses and the government announced in December 2002 left little doubt that the practice was not confined to Ebbers and Salomon.

In retrospect it now seems that some analysts' ratings weren't much more credible than the ubiquitous email invitations from people purporting to be Nigerian government officials in need of a little seed money. The usual claim is that the money will enable them and their gullible respondents to share in enormous, but frozen foreign accounts.

The bear market analogue to pumping and dumping is shorting and distorting. Instead of buying, touting, and selling on the jump in price, shorters and distorters sell, lambast, and buy on the decline in price.

They first short-sell the stock in question. As mentioned, that is the practice of selling shares one doesn't own in the hope that the price of the shares will decline when it comes time to pay the broker for the borrowed shares. (Short-selling is perfectly legal and also serves a useful purpose in maintaining markets and limiting risk.) After short-selling the stock, the scamsters lambast it in a misleading hyperbolical way (that is, distort its prospects). They spread false rumors of writedowns, unsecured debts, technology problems, employee morale, legal proceedings. When the stock's price declines in response to this concerted campaign, they buy the shares at the lower price and keep the difference.

Like its bull-market counterpart, shorting and distorting works best on thinly traded stocks. It is most effective in a bear market when people are most susceptible to fear and anxiety. Online practitioners, like pumpers and dumpers, use a variety of screen names, this time to create the illusion that something catastrophic is about to befall the company in question. They also tend to be nastier toward investors who disagree with them than are pumpers and dumpers, who must maintain a sunny, confident air. Again there are gradations in the practice and it sometimes seems indistinguishable from some fairly conventional practices of brokerage houses and hedge funds.

Even large stocks like WCOM (with 3 billion outstanding shares) can be affected by such shorters and distorters although they must be better placed than the dermatologically challenged isolates who usually carry on the practice. I don't doubt that there was much shorting of WCOM during its long descent, although given what's come to light about the company's accounting, "short and report" is a more faithful description of what occurred.

Unfortunately, after Enron, WorldCom, Tyco, and the others, even an easily generated whiff of malfeasance can cause investors to sell first and ask questions later. As a result, many worthy companies are unfairly tarred and their investors unnecessarily burned.

3 | Trends, Crowds, and Waves

As a predictor of stock prices, psychology goes only so far. Many investors subscribe to "technical analysis," an approach generally intent on discerning the short-term direction of the market via charts and patterns and then devising rules for pursuing it. Adherents of technical analysis, which is not all that technical and would more accurately be termed "trend analysis," believe that "the trend is their friend," that "momentum investing" makes sense, that crowds should be followed. Whatever the validity of these beliefs and of technical analysis in general (and I'll get to this shortly), I must admit to an a priori distaste for the herdish behavior it often seems to counsel: Figure out where the pack is going and follow it. It was this distaste, perhaps, that prevented me from selling WCOM and that caused me to sputter continually to myself that the company was the victim of bad public relations, investor misunderstanding, media bashing, anger at the CEO, a poisonous business climate, unfortunate timing, or panic selling. In short, I thought the crowd was wrong and hated the idea that it must be obeyed. As I slowly learned, however, disdaining the crowd is sometimes simply hubris.

Technical Analysis: Following the Followers

My own prejudices aside, the justification for technical analysis is murky at best. To the extent there is one, it most likely derives from psychology, perhaps in part from the Keynesian idea of conventionally anticipating the conventional response, or perhaps from some as yet unarticulated systemic interactions. "Unarticulated" is the key word here: The quasi-mathematical jargon of technical analysis seldom hangs together as a coherent theory. I'll begin my discussion of it with one of its less plausible manifestations, the so-called Elliott wave theory.

Ralph Nelson Elliott famously believed that the market moved in waves that enabled investors to predict the behavior of stocks. Outlining his theory in 1939, Elliott wrote that stock prices move in cycles based upon the Fibonacci numbers (1, 2, 3, 5, 8, 13, 21, 34, 59, 93, . . . , each successive number in the sequence being the sum of the two previous ones). Most commonly the market rises in five distinct waves and declines in three distinct waves for obscure psychological or systemic reasons. Elliott believed as well that these patterns exist at many levels and that any given wave or cycle is part of a larger one and contains within it smaller waves and cycles. (To give Elliott his due, this idea of small waves within larger ones having the same structure does seem to presage mathematician Benoit Mandelbrot's more sophisticated notion of a fractal, to which I'll return later.) Using Fibonacci-inspired rules, the investor buys on rising waves and sells on falling ones.

The problem arises when these investors try to identify where on a wave they find themselves. They must also decide whether the larger or smaller cycle of which the wave is inevitably a part may temporarily be overriding the signal to buy or sell. To save the day, complications are introduced into the theory, so many, in fact, that the theory soon becomes incapable of being falsified. Such complications and unfalsifiability are reminiscent of the theory of biorhythmns and many other pseudosciences. (Biorhythm theory is the idea that various aspects of one's life follow rigid periodic cycles that begin at birth and are often connected to the numbers 23 and 28, the periods of some alleged male and female principles, respectively.) It also brings to mind the ancient Ptolemaic system of describing the planets' movements, in which more and more corrections and ad hoc exceptions had to be created to make the system jibe with observation. Like most other such schemes, Elliott wave theory founders on the simple question: Why should anyone expect it to work?

For some, of course, what the theory has going for it is the mathematical mysticism associated with the Fibonacci numbers, any two adjacent ones of which are alleged to stand in an aesthetically appealing relation. Natural examples of Fibonacci series include whorls on pine cones and pineapples; the number of leaves, petals, and stems on plants; the numbers of left and right spirals in a sunflower; the number of rabbits in succeeding generations; and, insist Elliott enthusiasts, the waves and cycles in stock prices.

It's always pleasant to align the nitty-gritty activities of the market with the ethereal purity of mathematics.

The Euro and the Golden Ratio

Before moving on to less barren financial theories, I invite you to consider a brand new instance of financial numerology. An email from a British correspondent apprised me of an interesting connection between the euro-pound and pound-euro exchange rates on March 19, 2002.

To appreciate it, one needs to know the definition of the golden ratio from classical Greek mathematics. (Those for whom the confluence of Greek, mathematics, and finance is a bit much may want to skip to the next section.) If a point on a straight line divides the line so that the ratio of the longer part to the shorter is equal to the ratio of the whole to the longer part, the point is said to divide the line in a golden ratio. Rectangles whose length and width stand in a golden ratio are also said to be golden, and many claim that rectangles of this shape, for example, the facade of the Parthenon, are particularly pleasing to the eye. Note that a 3-by-5 card is almost a golden rectangle since 5/3 (or 1.666 . . .) is approximately equal to (5 + 3)/5 (or 1.6).

The value of the golden ratio, symbolized by the Greek letter phi, is 1.618 . . . (the number is irrational and so its decimal representation never repeats). It is not difficult to prove that phi has the striking property that it is exactly equal to 1 plus its reciprocal (the reciprocal of a number is simply 1 divided by the number). Thus 1.618 . . . is equal to 1 + 1/1.618 . . . .

This odd fact returns us to the euro and the pound. An announcer on the BBC on the day in question, March 19, 2002, observed that the exchange rate for 1 pound sterling was 1 euro and 61.8 cents (1.618 euros) and that, lo and behold, this meant that the reciprocal exchange rate for 1 euro was 61.8 pence (.618 pounds). This constituted, the announcer went on, "a kind of symmetry." The announcer probably didn't realize how profound this symmetry was.

In addition to the aptness of "golden" in this financial context, there is the following well-known relation between the golden ratio and the Fibonacci numbers. The ratio of any Fibonacci number to its predecessor is close to the golden ratio of 1.618 . . . , and the bigger the numbers involved, the closer the two ratios become. Consider again, the Fibonacci numbers, 1, 2, 3, 5, 8, 13, 21, 34, 59, ... The ratios, 5/3, 8/5,

13/8, 21/13, . . . , of successive Fibonacci numbers approach the golden ratio of 1.618 . . . !

There's no telling how an Elliott wave theorist dabbling in currencies at the time of the above exchange rate coincidence would have reacted to this beautiful harmony between money and mathematics. An unscrupulous, but numerate hoaxer might have even cooked up some flapdoodle sufficiently plausible to make money from such a "cosmic" connection.

The story could conceivably form the basis of a movie like *Pi*, since there are countless odd facts about phi that could be used to give various investing schemes a superficial plausibility. (The protagonist of *Pi* was a numerologically obsessed mathematician who thought he'd found the secret to just about everything in the decimal expansion of pi. He was pursued by religious zealots, greedy financiers, and others. The only sane character, his mentor, had a stroke, and the syncopated black-and-white cinematography was anxiety-inducing. Appealing as it was, the movie was mathematically nonsensical.) Unfortunately for investors and mathematicians alike, the lesson again is that more than beautiful harmonies are needed to make money on Wall Street. And *Phi* can't match the cachet of *Pi* as a movie title either.

Moving Averages, Big Picture

People, myself included, sometimes ridicule technical analysis and the charts associated with it in one breath and then in the next reveal how much in (perhaps unconscious) thrall to these ideas they really are. They bring to mind the old joke about the man who complains to his doctor that his wife has for several years believed she's a chicken. He would have sought help sooner, he says, "but we needed the eggs." Without reading too much into this story except that we do sometimes seem to need the notions of technical analysis, let me finally proceed to examine some of these notions.

Investors naturally want to get a broad picture of the movement of the market and of particular stocks, and for this the simple technical notion of a moving average is helpful. When a quantity varies over time (such as the stock price of a company, the noontime temperature in Milwaukee, or the cost of cabbage in Kiev), one can, each day, average its values over, say, the previous 200 days. The averages in this sequence vary and hence the sequence is called a moving average, but the value of such a moving average is that it doesn't move nearly as much as the stock price itself; it might be termed the phlegmatic average.

For illustration, consider the three-day moving average of a company whose stock is very volatile, its closing prices on successive days being: 8, 9, 10, 5, 6, 9. On the day the stock closed at 10, its three-day moving average was (8 + 9 + 10)/3 or 9. On the next day, when the stock closed at 5, its three-day moving average was (9 + 10 + 5)/3 or 8. When the stock closed at 6, its three-day moving average was (10 + 5 + 6)/3 or 7. And the next day, when it closed at 9, its three-day moving average was (5 + 6 + 9)/3 or 6.67.

If the stock oscillates in a very regular way and you are careful about the length of time you pick, the moving average may barely move at all. Consider an extreme case, the twenty-day moving average of a company whose closing stock prices oscillate with metronomic regularity. On successive days they are: 51, 52, 53, 54, 55, 54, 53, 52, 51, 50, 49, 48, 47, 46, 45, 46, 47, 48, 49, **50**, 51, 52, 53, and so on, moving up and down around a price of 50. The twenty-day moving average on the day marked in bold is 50 (obtained by averaging the 20 numbers up to and including it). Likewise, the twenty-day moving average on the next day, when the stock is at 51, is also 50. It's the same for the next day. In fact, if the stock price oscillates in this regular way and repeats itself every twenty days, the twenty-day moving average is always 50.

There are variations in the definition of moving averages (some weight recent days more heavily, others take account of the varying volatility of the stock), but they are all designed to smooth out the day-to-day fluctuations in a stock's price in order to give the investor a look at broader trends. Software and online sites allow easy comparison of the stock's daily movements with the slower-moving averages.

Technical analysts use the moving average to generate buy-sell rules. The most common such rule directs you to buy a stock when it exceeds its X-day moving average. Context determines the value of X, which is usually 10, 50, or 200 days. Conversely, the rule directs you to sell when the stock falls below its X-day moving average. With the regularly oscillating stock above, the rule would not lead to any gains or losses. It would call for you to buy the stock when it moves from 50, its moving average, to 51, and for you to sell it when it moves from 50 to 49. In the previous example of the three-day moving average, the rule would require that you buy the stock at the end of the third day and sell it at the end of the fourth, leading in this particular case to a loss.

The rule can work well when a stock fluctuates about a long-term upward- or downward-sloping course. The rationale for it is that trends should be followed, and that when a stock moves above its X-day moving average, this movement signals that a bullish trend has begun. Conversely, when a stock moves below its X-day moving average, the movement signals a bearish trend. I reiterate that mere upward (downward) movement of the stock is not enough to signal a buy (sell) order; a stock must move above (below) its moving average.

Alas, had I followed any sort of moving average rule, I would have been out of WCOM, which moved more or less steadily downhill for almost three years, long before I lost most of my investment in it. In fact, I never would have bought it in the first place. The security guard mentioned in chapter 1 did, in effect, use such a rule to justify the sale of the stocks in his pension plan.

There are a few studies, which I'll get to later, suggesting that a moving average rule is sometimes moderately effective. Even so, however, there are several problems. One is that it can cost you a lot in commissions if the stock price hovers around the moving average and moves through it many times in both directions. Thus you have to modify the rule so that the price must move above or below its moving average by a non-trivial amount. You must also decide whether to buy at the end of the day the price exceeds the moving average or at the beginning of the next day or later still.

You can mine the voluminous time-series data on stock prices to find the X that has given the best returns for adhering to the X-day moving average buy-sell rule. Or you can complicate the rule by comparing moving averages over different intervals and buying or selling when these averages cross each other. You can even adapt the idea to day trading by using X-minute moving averages defined in terms of the mathematical notion of an integral. Optimal strategies can always be found after the fact. The trick is getting something that will work in the future; everyone's very good at predicting the past. This brings us to the most trenchant criticism of the moving-average strategy. If the stock market is efficient, that is, if information about a stock is almost instantaneously incorporated into its price, then any stock's future moves will be determined by random external events. Its past behavior, in particular its moving average, is irrelevant, and its future movement is unpredictable.

Of course, the market may not be all that efficient. There'll be much more on this question in later chapters.

Resistance and Support and All That

Two other important ideas from technical analysis are resistance and support levels. The argument for them assumes that people usually remember when they've been burned, insulted, or left out; in particular, they remember what they paid, or wish they had paid, for a stock. Assume a stock has been selling for $40 for a while and then drops to $32 before slowly rising again. The large number of people who bought it around $40 are upset and anxious to recoup their losses, so if the stock moves back up to $40, they're likely to sell it, thereby driving the price down again. The $40 price is termed a resistance level and is considered an obstacle to further upward movement of the stock price.

Likewise, investors who considered buying at $32 but did not are envious of those who did buy at that price and reaped the 25 percent returns. They are eager to get these gains, so if the stock falls back to $32, they're likely to buy it, driving the price up again. The $32 price is termed a support level and is considered an obstacle to further downward movement.

Since stocks often seem to meander between their support and resistance levels, one rule followed by technical analysts is to buy the stock when it "bounces" off its support level and sell it when it "bumps" up against its resistance level. The rule can, of course, be applied to the market as a whole, inducing investors to wait for the Dow or the S&P to definitively turn up (or down) before buying (or selling).

Since chartists tend to view support levels as shaky, often temporary, floors and resistance levels as slightly stronger, but still temporary, ceilings, there is a more compelling rule involving these notions. It instructs you to buy the stock if the rising price breaks through the resistance level and to sell it if the falling price breaks through the support level. In both these cases breaking through indicates that the stock has moved out of its customary channel and the rule counsels investors to follow the new trend.

As with the moving-average rules, there are a few studies that indicate that resistance-support rules sometimes lead to moderate increases in returns. Against this there remains the perhaps dispiriting efficient-market hypothesis, which maintains that past prices, trends, and resistance and support levels provide no evidence about future movements.

Innumerable variants of these rules exist and they can be combined in ever more complicated ways. The resistance and support levels can change and trend up or down in a channel or with the moving average, for example, rather than remain fixed. The rules can also be made to take account of variations in a stock's volatility as well.

These variants depend on price patterns that often come equipped with amusing names. The "head and shoulders" pattern, for example, develops after an extended upward trend. It is comprised of three peaks, the middle and highest one being the head, and the smaller left and right ones (earlier and later ones, that is) being the shoulders. After falling below the right shoulder and breaking through the support line connecting the lows on either side of the head, the stock price has, technical chartists aver, reversed direction and a downward trend has begun, so sell.

Similar metaphors describe the double-bottom trend reversal. It develops after an extended downward trend and is comprised of two successive troughs or bottoms with a small peak between them. After bouncing off the second bottom, the stock has, technical chartists again aver, reversed direction and an upward trend has begun, so buy.

These are nice stories, and technical analysts tell them with great earnestness and conviction. Even if everyone told the same stories (and they don't), why should they be true? Presumably the rationale is ultimately psychological or perhaps sociological or systemic, but exactly what principles justify these beliefs? Why not triple or quadruple bottoms? Or two heads and shoulders? Or any of innumerable other equally plausible, equally risible patterns? What combination of psychological, financial, or other principles has sufficient specificity to generate effective investment rules?

As with Elliott waves, scale is an issue. If we go to the level of ticks, we can find small double bottoms and little heads and tiny shoulders all over. We find them also in the movement of broad market indices. And do these patterns mean for the market as whole what they are purported to mean for individual stocks? Is the "double-dip" recession discussed in early 2002 simply a double bottom?

Predictability and Trends

I often hear people swear that they make money using the rules of technical analysis. Do they really? The answer, of course, is that they do. People make money using all sorts of strategies, including some involving tea leaves and sunspots. The real question is: Do they make *more* money than they would investing in a blind index fund that mimics the performance of the market as a whole? Do they achieve excess returns? Most financial theorists doubt this, but there is some tantalizing evidence for the effectiveness of momentum strategies or short-term trend-following. Economists Narasimhan Jegadeesh and Sheridan Titman, for example, have written several papers arguing that momentum strategies result in moderate excess returns and that, having done so over the years, their success is not the result of data mining. Whether this alleged profitability—many dispute it—is due to overreactions among investors or to the short-term persistence of the impact of companies' earnings reports, they don't say. They do seem to point to behavioral models and psychological factors as relevant.

William Brock, Josef Lakonishok, and Blake LeBaron have also found some evidence that rules based on moving averages and the notions of resistance and support are moderately effective. They focus on the simplest rules, but many argue that their results have not been replicated on new stock data.

More support for the existence of technical exploitability comes from Andrew Lo, who teaches at M.I.T., and Craig MacKinlay, from the Wharton School. They argue in their book, *A Non-Random Walk Down Wall Street*, that in the short run overall market returns are, indeed, slightly positively correlated, much like the local weather. A hot, sunny day is a bit more likely to be followed by another one, just as a good week in the market is a bit more likely to be followed by another one. Likewise for rainy days and bad markets. Employing state-of-the-art tools, Lo and MacKinlay also claim that in the long term the prognosis changes: Individual stock prices display a slight negative correlation. Winners are a bit more likely to be losers three to five years hence and vice versa.

They also bring up an interesting theoretical possibility. Weeding out some of the details, let's assume for the sake of the argument (although Lo and MacKinlay don't) that the thesis of Burton Malkiel's classic book, *A Random Walk Down Wall Street*, is true and that the movement of the market as a whole is entirely random. Let's also assume that each stock, when its fluctuations are examined in isolation, moves randomly. Given these assumptions it would nevertheless still be possible that the price movements of, say, 5 percent of stocks accurately predict the price movements of a different 5 percent of stocks one week later.

The predictability comes from cross-correlations over time between stocks. (These associations needn't be causal, but might merely be brute facts.) More concretely, let's say stock

X, when looked at in isolation, fluctuates randomly from week to week, as does stock Y. Yet if X's price this week often predicts Y's next week, this would be an exploitable opportunity and the strict random-walk hypothesis would be wrong. Unless we delved deeply into such possible cross-correlations among stocks, all we would see would be a randomly fluctuating market populated by randomly fluctuating stocks. Of course, I've employed the typical mathematical gambit of considering an extreme case, but the example does suggest that there may be relatively simple elements of order in a market that appears to fluctuate randomly.

There are other sorts of stock price anomalies that can lead to exploitable opportunities. Among the most well-known are so-called calendar effects whereby the prices of stocks, primarily small-firm stocks, rise disproportionately in January, especially during the first week of January. (The price of WCOM rose significantly in January 2001, and I was hoping this rise would repeat itself in January 2002. It didn't.) There has been some effort to explain this by citing tax law concerns that end with the close of the year, but the effect also seems to hold in countries with different tax laws. Moreover, unusual returns (good or bad) occur not only at the turn of the year, but, as Richard Thaler and others have observed, at the turn of the month, week, and day as well as before holidays. Again, poorly understood behavioral factors seem to be involved.

Technical Strategies and Blackjack

Most academic financial experts believe in some form of the random-walk theory and consider technical analysis almost indistinguishable from a pseudoscience whose predictions are either worthless or, at best, so barely discernibly better than chance as to be unexploitable because of transaction costs.

I've always leaned toward this view, but I'll reserve my more nuanced judgment for later in the book. In the meantime, I'd like to point out a parallel between market strategies such as technical analysis in one of its many forms and blackjack strategies. (There are, of course, great differences too.)

Blackjack is the only casino game of chance whose outcomes depend on past outcomes. In roulette, the previous spins of the wheel have no effect on future spins. The probability of red on the next spin is 18/38, even if red has come up on the five previous spins. The same is true with dice, which are totally lacking in memory. The probability of rolling a 7 with a pair of dice is 1/6, even if the four previous rolls have not resulted in a single 7. The probability of six reds in a row is $(18/38)^6$; the probability of five 7s in a row is $(1/6)^5$. Each spin and each roll are independent of the past.

A game of blackjack, on the other hand, is sensitive to its past. The probability of drawing two aces in a row from a deck of cards is not $(4/52 \times 4/52)$ but rather $(4/52 \times 3/51)$. The second factor, 3/51, is the probability of choosing another ace given that the first card chosen was an ace. In the same way the probability that a card drawn from a deck will be a face card (jack, queen, or king) given that only three of the thirty cards drawn so far have been face cards is not 12/52, but a much higher 9/22.

This fact—that (conditional) probabilities change according to the composition of the remaining portion of the deck—is the basis for various counting strategies in blackjack that involve keeping track of how many cards of each type have already been drawn and increasing one's bet size when the odds are (occasionally and slightly) in one's favor. Some of these strategies, followed carefully, do work. This is evidenced by the fact that some casinos supply burly guards free of charge to abruptly escort successful counting practitioners from the premises.

The vast majority of people who try these strategies (or, worse, others of their own devising) lose money. It would make no sense, however, to point to the unrelenting average losses of blackjack players and maintain that this proves that there is no effective betting strategy for playing the game.

Blackjack is much simpler than the stock market, of course, which depends on vastly more factors as well as on the actions and beliefs of other investors. But the absence of conclusive evidence for the effectiveness of various investing rules, technical or otherwise, does not imply that no effective rules exist. If the market's movements are not completely random, then it has a kind of memory within it, and investing rules depending on this memory might be effective. Whether they would remain so if widely known is very dubious, but that is another matter.

Interestingly, if there were an effective technical trading strategy, it wouldn't need any convincing rationale. Most investors would be quite pleased to use it, as most blackjack players use the standard counting strategy, without understanding why it works. With blackjack, however, there is a compelling mathematical explanation for those who care to study it. By contrast an effective technical trading strategy might be found that was beyond the comprehension not only of the people using it but of everyone. It might simply work, at least temporarily. In Plato's allegory of the cave the benighted see only the shadows on the wall of the cave and not the real objects behind them that are causing the shadows. If they were really predictive, investors would be quite content with the shadows alone and would simply take the cave to be a bargain basement.

The next segment is a bit of a lark. It offers a suggestive hint for developing a novel and counterintuitive investment strategy that has a bit of the feel of technical analysis.

Winning Through Losing?

The old joke about the store owner losing money on every sale but making it up in volume may have a kernel of truth to it. An interesting new paradox by Juan Parrondo, a Spanish physicist, brings the joke to mind. It deals with two games, each of which results in steady losses over time. When these games are played in succession in random order, however, the result is a steady gain. Bad bets strung together to produce big winnings—very strange indeed!

To understand Parrondo's paradox, let's switch from a financial to a spatial metaphor. Imagine you are standing on stair 0, in the middle of a very long staircase with 1,001 stairs numbered from –500 to 500 (–500, –499, –498, . . . , –4, –3, –2, –1, 0, 1, 2, 3, 4, . . . , 498, 499, 500). You want to go up rather than down the staircase and which direction you move depends on the outcome of coin flips. The first game—let's call it game S—is very Simple. You flip a coin and move up a stair whenever it comes up heads and down a stair whenever it comes up tails. The coin is slightly biased, however, and comes up heads 49.5 percent of the time and tails 50.5 percent. It's clear that this is not only a boring game but a losing one. If you played it long enough, you would move up and down for a while, but almost certainly you would eventually reach the bottom of the staircase.

The second game—let's continue to wax poetic and call it game C—is more Complicated, so bear with me. It involves *two* coins, one of which, the bad one, comes up heads only 9.5 percent of the time, tails 90.5 percent. The other coin, the good one, comes up heads 74.5 percent of the time, tails 25.5 percent. As in game S, you move up a stair if the coin you flip comes up heads and you move down one if it comes up tails.

But which coin do you flip? If the number of the stair you're on is a multiple of 3 (that is, . . . , –9, –6, –3, 0, 3, 6, 9,

12, . . .), you flip the bad coin. If the number of the stair you're on is not a multiple of 3, you flip the good coin. (Note: Changing these odd percentages and constraints may affect the game's outcome.)

Let's go through game C's dance steps. If you were on stair number 5, you would flip the good coin to determine your direction, whereas if you were on stair number 6, you would flip the bad coin. The same holds for the negatively numbered stairs. If you were on stair number –2 and playing game C, you would flip the good coin, whereas if you were on stair number –9, you would flip the bad coin.

Though less obviously so than in game S, game C is also a losing game. If you played it long enough, you would almost certainly reach the bottom of the staircase eventually. Game C is a losing game because the number of the stair you're on is a multiple of 3 more often than a third of the time and thus you must flip the bad coin more often than a third of the time. Take my word for this or read the next paragraph to get a better feel for why it is.

(Assume that you've just started playing game C. Since you're on stair number 0, and 0 is a multiple of 3, you would flip the bad coin, which lands heads with probability less than 10 percent, and you would very likely move down to stair number –1. Then, since –1 is not a multiple of 3, you would flip the good coin, which lands heads with probability almost 75 percent, and would probably move back up to stair 0. You may move up and down like this for a while. Occasionally, however, after the bad coin lands tails, the good coin, which lands tails almost 25 percent of the time, will land tails twice in succession, and you would move down to stair number –3, where the pattern will likely begin again. This latter downward pattern happens slightly more frequently (with probability $.905 \times .255 \times .255$) than does a rare head on the bad coin being followed by two heads on the good one (with probability .095 × .745 × .745) and your moving up three stairs as a consequence. So-called Markov chains are needed for a fuller analysis.)

So far, so what? Game S is simple and results in steady movement down the staircase to the bottom, and game C is complicated and also results in steady movement down the staircase to the bottom. Parrondo's fascinating discovery is that *if you play these two games in succession in random order (keeping your place on the staircase as you switch between games), you will steadily ascend to the top of the staircase.* Alternatively, if you play two games of S followed by two games of C followed by two games of S and so on, all the while keeping your place on the staircase as you switch between games, you will also steadily rise to the top of the staircase. (You might want to look up M. C. Escher's paradoxical drawing, "Ascending and Descending" for a nice visual analog to Parrondo's paradox.)

Standard stock-market investments cannot be modeled by games of this type, but variations of these games might conceivably give rise to counterintuitive investment strategies. The probabilities might be achieved, for example, by complicated combinations of various financial instruments (options, derivatives, and so on), but the decision which coin (which investment, that is) to flip (to make) in game C above would, it seems, have to depend upon something other than whether one's holdings were worth a multiple of $3.00 (or a multiple of $3,000.00). Perhaps the decision could depend in some way on the cross-correlation between a pair of stocks or turn on the value of some index being a multiple of 3.

If strategies like this could be made to work, they would yield what one day might be referred to as Parrondo profits.

Finally, let's consider a companion paradox of sorts that might be called "losing through winning" and that may help explain why companies often overpaid for small companies they were purchasing during the bubble in the late '90s. Professor Martin Shubik has regularly auctioned off $1 to students in his classes at Yale. The bidding takes place at fifty-four intervals, and the highest bidder gets the dollar, of course, but the second highest bidder is required to pay his bid as well. Thus, if the highest bid is 504 and you are second highest at 454, the leader stands to make 504 on the deal and you stand to lose 454 on it if bidding stops. You have an incentive to up your bid to at least 554, but after you've done so the other bidder has an even bigger incentive to raise his bid as well. In this way a one dollar bill can be successfully auctioned off for two, three, four, or more dollars.

If several companies are bidding on a small company and the cost of the preliminary legal, financial, and psychological efforts required to purchase the company are a reasonable fraction of the cost of the company, the situation is formally similar to Shubik's auction. One or more of the bidding companies might feel compelled to make an exorbitant preemptive offer to avoid the fate of the losing bidder on the $1. WorldCom's purchase of the web-hosting company Digex in 2000 for $6 billion was, I suspect, such an offer. John Sidgmore, the CEO who succeeded Bernie Ebbers, says that Digex was worth no more than $50 million, but that Ebbers was obsessed with beating out Global Crossing for the company.

The purchase is much more bizarre than Parrondo's paradox.

4 | Chance and Efficient Markets

If the movement of stock prices is random or near-random, then the tools of technical analysis are nothing more than comforting blather giving one the illusion of control and the pleasure of a specialized jargon. They can prove especially attractive to those who tend to infuse random events with personal significance.

Even some social scientists don't seem to realize that if you search for a correlation between any two randomly selected attributes in a very large population, you will likely find some small but statistically significant association. It doesn't matter if the attributes are ethnicity and hip circumference, or (some measure of) anxiety and hair color, or perhaps the amount of sweet corn consumed annually and the number of mathematics courses taken. Despite the correlation's statistical significance (its unlikelihood of occurring by chance), it is probably not practically significant because of the presence of so many confounding variables. Furthermore, it will not necessarily support the (often ad hoc) story that accompanies it, the one purporting to explain why people who eat a lot of corn take more math. Superficially plausible tales are always available: Corn-eaters are more likely to be from the upper Midwest, where dropout rates are low.

Geniuses, Idiots, or Neither

Around stock market rises and declines, people are prone to devise just-so stories to satisfy various needs and concerns. During the bull markets of the '90s investors tended to see themselves as "perspicacious geniuses." During the more recent bear markets they've tended toward self-descriptions such as "benighted idiots."

My own family is not immune to the temptation to make up pat after-the-fact stories explaining past financial gains and losses. When I was a child, my grandfather would regale me with anecdotes about topics as disparate as his childhood in Greece, odd people he'd known, and the exploits of the Chicago White Sox and their feisty second baseman "Fox Nelson" (whose real name was Nelson Fox). My grandfather was voluble, funny, and opinionated. Only rarely and succinctly, however, did he refer to the financial reversal that shaped his later life. As a young and uneducated immigrant, he worked in restaurants and candy stores. Over the years he managed to buy up eight of the latter and two of the former. His candy stores required sugar, which led him eventually to speculate in sugar markets and—he was always a bit vague about the details—to place a big bet on several train cars full of sugar. He apparently put everything he had into the deal a few weeks before the sugar market crashed. Another version attributed his loss to underinsurance of the sugar shipment. In any case, he lost it all and never really recovered financially. I remember him saying ruefully, "Johnny, I would have been a very, very rich man. I should have known." The bare facts of the story registered with me then, but my recent less calamitous experience with WorldCom has made his pain more palpable.

This powerful natural proclivity to invest random events with meaning on many different levels makes us vulnerable to people who tell engaging stories about these events. In the

Rorschach blot that chance provides us, we often see what we want to see or what is pointed out to us by business prognosticators, distinguishable from carnival psychics only by the size of their fees. Confidence, whether justified or not, is convincing, especially when there aren't many "facts of the matter." This may be why market pundits seem so much more certain than, say, sports commentators, who are comparatively frank in acknowledging the huge role of chance.

Efficiency and Random Walks

The Efficient Market Hypothesis formally dates from the 1964 dissertation of Eugene Fama, the work of Nobel prize-winning economist Paul Samuelson, and others in the 1960s. Its pedigree, however, goes back much earlier, to a dissertation in 1900 by Louis Bachelier, a student of the great French mathematician Henri Poincare. The hypothesis maintains that at any given time, stock prices reflect all relevant information about the stock. In Fama's words: "In an efficient market, competition among the many intelligent participants leads to a situation where, at any point in time, actual prices of individual securities already reflect the effects of information based both on events that have already occurred and on events which, as of now, the market expects to take place in the future."

There are various versions of the hypothesis, depending on what information is assumed to be reflected in the stock price. The weakest form maintains that all information about past market prices is already reflected in the stock price. A consequence of this is that all of the rules and patterns of technical analysis discussed in chapter 3 are useless. A stronger version maintains that all publicly available information about a company is already reflected in its stock price. A consequence of this version is that the earnings, interest, and other elements of fundamental analysis discussed in chapter 5 are useless. The strongest version maintains that all information of all sorts is already reflected in the stock price. A consequence of this is that even inside information is useless.

It was probably this last, rather ludicrous version of the hypothesis that prompted the joke about the two efficient market theorists walking down the street: They spot a hundred dollar bill on the sidewalk and pass by it, reasoning that if it were real, it would have been picked up already. And of course there is the obligatory light-bulb joke. Question: How many efficient market theorists does it take to change a light bulb? Answer: None. If the light bulb needed changing the market would have already done it. Efficient market theorists tend to believe in passive investments such as broad-gauged index funds, which attempt to track a given market index such as the S&P 500. John Bogle, the crusading founder of Vanguard and presumably a believer in efficient markets, was the first to offer such a fund to the general investing public. His Vanguard 500 fund is unmanaged, offers broad diversification and very low fees, and generally beats the more expensive, managed funds. Investing in it does have a cost, however: One must give up the fantasy of a perspicacious gunslinger/investor outwitting the market.

And why do such theorists believe the market to be efficient? They point to a legion of investors of all sorts all seeking to make money by employing all sorts of strategies. These investors sniff out and pounce upon any tidbit of information even remotely relevant to a company's stock price, quickly driving it up or down. Through the actions of this investing horde the market rapidly responds to the new information, efficiently adjusting prices to reflect it. Opportunities to make an excess profit by utilizing technical rules or fundamental analyses, so the story continues, disappear before they can be fully exploited, and investors who pursue them will see their excess profits shrink to zero, especially after taking into account brokers' fees and other transaction costs. Once again, it's not that subscribers to technical or fundamental analysis won't make money; they generally will. They just won't make more than, say, the S&P 500.

(That exploitable opportunities tend to gradually disappear is a general phenomenon that occurs throughout economics and in a variety of fields. Consider an argument about baseball put forward by Steven Jay Gould in his book *Full House: The Spread of Excellence from Plato to Darwin.* The absence of .400 hitters in the years since Ted Williams hit .406 in 1941, he maintained, was not due to any decline in baseball ability but the reverse: a gradual increase in the athleticism of all players and a consequent decrease in the disparity between the worst and best players. When players are as physically gifted and well trained as they are now, the distribution of batting averages and earned run averages shows less variability. There are few "easy" pitchers for hitters and few "easy" hitters for pitchers. One result is that .400 averages are now very scarce. The athletic prowess of hitters and pitchers makes the "market" between them more efficient.)

There is, moreover, a close connection between the Efficient Market Hypothesis and the proposition that the movement of stock prices is random. If present stock prices already reflect all available information (that is, if the information is common knowledge in the sense of chapter 1), then future stock prices must be unpredictable. Any news that might be relevant in predicting a stock's future price has already been weighed and responded to by investors whose buying and selling have adjusted the present price to reflect the news. Oddly enough, as markets become more efficient, they tend to become less predictable. What will move stock prices in the future are truly new developments (or new shadings of old developments), news that is, by definition, impossible to anticipate. The conclusion is that in an efficient market, stock prices move up and down randomly. Evincing no memory of their past, they take what is commonly called a random walk, each step of which is independent of past steps. There is over time, however, an upward trend, as if the coin being flipped were slightly biased.

There is a story I've always liked that is relevant to the impossibility of anticipating new developments. It concerns a college student who completed a speed-reading course. He noted this fact in a letter to his mother. His mother responded with a long, chatty letter of her own in the middle of which she wrote, "Now that you've taken that speed-reading course, you've probably finished reading this letter by now."

Likewise, true scientific breakthroughs or applications, by definition, cannot be foreseen. It would be preposterous to have expected a newspaper headline in 1890 proclaiming "Only 15 Years Until Relativity." It is similarly foolhardy, the efficient market theorist reiterates, to predict changes in a company's business environment. To the extent these predictions reflect a consensus of opinion, they're already accounted for. To the extent that they don't, they're tantamount to forecasting coin flips.

Whatever your views on the subject, the arguments for an efficient market spelled out in Burton Malkiel's *A Random Walk Down Wall Street* and elsewhere can't be grossly wrong. After all, most mutual fund managers continue to generate average gains less than those of, say, the Vanguard Index 500 fund. (This has always seemed to me a rather scandalous fact.) There is other evidence for a *fairly* efficient market as well. There are few opportunities for risk-free money-making or arbitrage, prices seem to adjust rapidly in response to news, and the autocorrelation of the stock prices from day to day, week to week, month to month, and year to year is small (albeit not zero). That is, if the market has done well (or poorly) over a given time period in the past, there is no strong tendency for it to do well (or poorly) during the next time period.

Nevertheless, in the last few years I have qualified my view of the Efficient Market Hypothesis and random-walk theory. One reason is the accounting scandals involving Enron, Adelphia, Global Crossing, Qwest, Tyco, WorldCom, Andersen, and many others from corporate America's Hall of Infamy, which make it hard to believe that available information about a stock always quickly becomes common knowledge.

Pennies and the Perception of Pattern

The *Wall Street Journal* has famously conducted a regular series of stock-picking contests between a rotating collection of stock analysts, whose selections are a result of their own studies, and dart-throwers, whose selections are determined randomly. Over many six-month trials, the pros' selections have performed marginally better than the darts' selections, but not overwhelmingly so, and there is some feeling that the pros' picks may influence others to buy the same stocks and hence drive up their price. Mutual funds, although less volatile than individual stocks, also display a disregard for analysts' pronouncements, often showing up in the top quarter of funds one year and in the bottom quarter the next.

Whether or not you believe in efficient markets and the random movement of stock prices, the huge element of chance present in the market cannot be denied. For this reason an examination of random behavior sheds light on many market phenomena. (So does study of a standard tome on probability such as that by Sheldon Ross.) Sources for such random behavior are penny stocks or, more accessible and more random, stocks of pennies, so let's imagine flipping a penny repeatedly and keeping track of the sequence of heads and tails. We'll assume the coin and the flip are fair (although, if we wish, the penny can be altered slightly to reflect the small upward bias of the market over time).

One odd and little-known fact about such a series of coin flips concerns the proportion of *time* that the number of heads exceeds the number of tails. It's seldom close to 50 percent!

To illustrate, imagine two contestants, Henry and Tommy, who bet that heads and tails respectively will be the outcome of a daily coin flip, a ritual that goes on for years. (Let's not ask why.) Henry is ahead on any given day if up to that day there have been more heads than tails, and Tommy is ahead if up to that day there have been more tails. The coin is fair, so they're equally likely to be in the lead, but one of them will probably be in the lead during most of their rather stultifying contest.

Stated numerically, the claim is that if there have been 1,000 coin flips, then it's considerably more probable that Henry (or Tommy) has been ahead more than, say, 96 percent of the time than that either one has been ahead between 48 percent and 52 percent of the time.

People find this result hard to believe. Many subscribe to the "gambler's fallacy" and believe that the coin's deviations from a 50–50 split between heads and tails are governed by a probabilistic rubber band: the greater the deviation, the greater the equalizing push toward an even split. But even if Henry were way ahead, with 525 heads to Tommy's 475 tails, his lead would be as likely to grow as to shrink. Likewise, a stock that's fallen on a truly random trajectory is as likely to fall further as it is to rise.

The rarity with which the lead switches sides in no way contradicts the fact that the proportion of heads approaches 1/2 as the number of flips increases. Nor does it contradict the phenomenon of regression to the mean. If Henry and Tommy were to start over and flip their penny another 1,000 times, it's quite likely that the number of heads would be smaller than 525.

Given the relative rarity with which Henry and Tommy overtake one another in their penny-flipping contest, it wouldn't be surprising if one of them came to be known as a "winner" and the other a "loser" despite their complete lack of control over the penny. If one professional stock picker outperformed another by a margin of 525 to 475, he might even be interviewed on Moneyline or profiled in *Fortune* magazine. Yet he might, like Henry or Tommy, owe his success to nothing more than getting "stuck" by chance on the up side of a 50–50 split.

But what about such stellar "value investors" as Warren Buffet? His phenomenal success, like that of Peter Lynch, John Neff, and others, is often cited as an argument against the market's randomness. This assumes, however, that Buffett's choices have no effect on the market. Originally no doubt they didn't, but now his selections themselves and his ability to create synergies among them can influence others. His performance is therefore a bit less remarkable than it first appears.

A different argument points to the near certainty of some stocks, funds, or analysts doing well over an extended period merely by chance. Of 1,000 stocks (or funds or analysts), for example, roughly 500 might be expected to outperform the market next year simply by chance, say by the flipping of a coin. Of these 500, roughly 250 might be expected to do well for a second year. And of these 250, roughly 125 might be expected to continue the pattern, doing well three years in a row simply by chance. Iterating in this way, we might reasonably expect there to be a stock (or fund or analyst) among the thousand that does well for ten consecutive years by chance alone. Once again, some in the business media are likely to go gaga over the performance.

The surprising length and frequency of consecutive runs of heads or tails is yet another lesson of penny flipping. If Henry and Tommy were to continue flipping pennies once a day, then there's a better-than-even chance that within about two months Henry will have won at least five flips in a row, as will Tommy. If they continue flipping for six years, there's a better-than-even chance that each will have won at least ten flips in a row.

When people are asked to write down a series of heads and tails that simulates a series of coin flips, they almost always fail to include enough runs of consecutive heads or consecutive tails. In particular, they fail to include any very long runs of heads or tails, and their series are thus easily distinguishable from a real series of coin flips.

But try telling people that long streaks are due to chance alone, whether the streak is a basketball player's shots, a stock analyst's picks, or a series of coin flips. The fact is that random events can frequently seem quite ordered.

To literally see this, take out a large piece of paper and partition it into little squares in a checkerboard pattern. Flip a coin repeatedly and color the squares white or black depending upon whether the coin lands heads or tails. After the checkerboard has been completely filled in, look it over and see if you can discern any patterns or clusters of similarly colored squares. Chances are you will, and if you felt the need to explain these patterns, you would invent a story that might sound superficially plausible or intriguing, but, given how the colors were determined, would necessarily be false.

The same illusion of pattern would result if you were to graph (with time on the horizontal axis) the results of the coin flips, up one unit for a head, down one for a tail. Some chartists and technicians would no doubt see "head and shoulders," "triple tops," or "ascending channels" patterns in these zigzag, up-and-down movements, and they would expatiate on their significance. (One difference between coin flips and models of random stock movements is that in the latter it is generally assumed that stocks move up or down not by a fixed amount per unit time, but by a fixed percentage.)

Leaving aside, once again, the question whether the market is perfectly efficient or whether stock movements follow a truly random walk, we can nevertheless say that phenomena that are truly random often appear almost indistinguishable from real-market behavior. This should, but probably won't, give pause to commentators who provide a neat post hoc explanation for every rally, every sell-off, and everything in between. Such commentators generally don't make remarks analogous to the observation that the penny happened by chance to land heads a few more times than it did tails. Instead they will refer to Tommy's profit-taking, Henry's increased confidence, labor problems in the copper mines, or countless other factors.

Because so much information is available—business pages, companies' annual reports, earnings expectations, alleged scandals, on-line sites, and commentary—something insightful-sounding can always be said. All we need do is filter the sea of numbers until we catch a plausible nugget of speculation. Like flipping a penny, doing so is a snap.

A Stock-Newsletter Scam

The accounting scandals involving WorldCom, Enron, and others derived from the data being selected, spun, and filtered. A scam I first discussed in my book *Innumeracy* derives instead from the *recipients* of the data being selected, spun, and filtered. It goes like this. Someone claiming to be the publisher of a stock newsletter rents a mailbox in a fancy neighborhood, has expensive stationery made up, and sends out letters to potential subscribers boasting of his sophisticated stock-picking software, financial acumen, and Wall Street connections. He writes also of his amazing track record, but notes that the recipients of his letters needn't take his word for it.

Assume you are one of these recipients and for the next six weeks you receive correct predictions about a certain common stock index. Would you subscribe to the newsletter? What if you received ten consecutive correct predictions?

Here's the scam. The newsletter publisher sends out 64,000 letters to potential subscribers. (Using email would save postage, but might appear to be a "spam scam" and hence be less credible.) To 32,000 of the recipients, he predicts the index in question will rise the following week and to the other 32,000, he predicts it will decline. No matter what happens to the index the next week, he will have made a correct prediction to 32,000 people. To 16,000 of them he sends another letter predicting a rise in the index for the following week, and to the other 16,000 he predicts a decline. Again, no matter what happens to the index the next week, he will have made correct predictions for two consecutive weeks to 16,000 people. To 8,000 of them he sends a third letter predicting a rise for the third week and to the other 8,000 he predicts a decline.

Focusing at each stage on the people to whom he's made only correct predictions and winnowing out the rest, he iterates this procedure a few more times until there are 1,000 people left to whom he's made six straight correct "predictions." To these he sends a different sort of follow-up letter, pointing out his successes and saying that they can continue to receive these oracular pronouncements if they pay the $1,000 subscription price to the newsletter. If they all pay, that's a million dollars for someone who need know nothing about stock, indices, trends, or dividends. If this is done knowingly, it is illegal. But what if it's done unknowingly by earnest, confident, and ignorant newsletter publishers? (Compare the faithhealer who takes credit for any accidental improvements.)

There is so much complexity in the market, there are so many different measures of success and ways to spin a story, that most people can manage to convince themselves that they've been, or are about to be, inordinately successful. If people are desperate enough, they'll manage to find some seeming order in random happenings.

Similar to the newsletter scam, but with a slightly different twist, is a story related to me by an acquaintance who described his father's business and its sad demise. He claimed that his father, years before, had run a large college-preparation service in a South American country whose identity I've forgotten. My friend's father advertised that he knew how to drastically improve applicants' chances of getting into the elite national university. Hinting at inside contacts and claiming knowledge of the various forms, deadlines, and procedures, he charged an exorbitant fee for his service, which he justified by offering a money-back guarantee to students who were not accepted.

One day, the secret of his business model came to light. All the material that prospective students had sent him over the years was found unopened in a trash dump. Upon investigation it turned out that he had simply been collecting the students' money (or rather their parents' money) and doing nothing for it. The trick was that his fees were so high and his marketing so focused that only the children of affluent parents subscribed to his service, and almost all of them were admitted to the university anyway. He refunded the fees of those few who were not admitted. He was also sent to prison for his efforts.

Are stock brokers in the same business as my acquaintance's father? Are stock analysts in the same business as the newsletter publisher? Not exactly, but there is scant evidence that they possess any unusual predictive powers. That's why I thought news stories in November 2002 recounting New York Attorney General Eliot Spitzer's criticism of *Institutional Investor* magazine's analyst awards were a tad superfluous. Spitzer noted that the stock-picking performances of most of the winning analysts were, in fact, quite mediocre. Maybe Donald Trump will hold a press conference pointing out that the country's top gamblers don't do particularly well at roulette.

Decimals and Other Changes

Like analysts and brokers, market makers (who make their money on the spread between the bid and the ask price for a stock) have received more than their share of criticism in recent years. One result has been a quiet reform that makes the market a bit more efficient. Wall Street's surrender to radical "decacrats" occurred a couple of years ago, courtesy of a Congressional mandate and a direct order from the Securities and Exchange Commission. Since then stock prices have been expressed in dollars and cents, and we no longer hear "profit-taking drove XYZ down 2 and 1/8" or "news of the deal sent PQR up 4 and 5/16."

Although there may be less romance associated with declines of 2.13 and rises of 4.31, decimalization makes sense for a number of reasons. The first is that price rises and declines are immediately comparable since we no longer must perform the tiresome arithmetic of, say, dividing 11 by 16. Mentally calculating the difference between two decimals generally requires less time than subtracting 3 5/8 from 5 3/16. Another benefit is global uniformity of pricing, as American securities are now denominated in the same decimal units as those in the rest of the world. Foreign securities no longer need to be rounded to the nearest multiple of 1/16, a perverse arithmetical act if there ever was one.

More importantly, the common spread between the bid and ask prices has shrunk. Once generally 1/16 (.0625, that is), the spread in many cases has become .01 and, by so shriveling, will save investors billions of dollars over the years. Market makers aside, most investors applaud this consequence of decimalization.

The last reason for cheering the change is more mathematical. There is a sense in which the old system of halves, quarters, eighths, and sixteenths is more natural than decimals. It is, after all, only a slightly disguised binary system, based on powers of 2 (2, 4, 8, 16) rather than powers of 10. It doesn't inherit any of the prestige of the binary system, however, because it awkwardly combines the base 2 fractional part of a stock price with the base 10 whole-number part.

Thus it is that Ten extends its imperial reach to Wall Street. From the biblical Commandments to David Letterman's lists, the number 10 is ubiquitous. Not unrelated to the perennial yearning for the simplicity of the metric system, 10 envy has also come to be associated with rationality and efficiency. It is thus fitting that all stocks are now expressed in decimals. Still, I suspect that many market veterans miss those pesky fractions and their role in stories of past killings and baths. Except for generation X-ers (Roman numeral ten-ers), many others will too. Anyway, that's my two cents (.02, 1/50th) worth on the subject.

The replacement of marks, francs, drachmas, and other European currencies by euros on stock exchanges and in stores is another progressive step that nevertheless rouses a touch of nostalgia. The coins and bills from my past travels that are scattered about in drawers are suddenly out of work and will never see the inside of a wallet again.

Yet another vast change in trading practices is the greater self-reliance among investors. Despite the faulty accounting that initially disguised their sickly returns, the ladies of Beardstown, Illinois, helped popularize investment clubs. Even more significant in this regard is the advent of effortless online trading, which has further hastened the decline of the traditional broker. The ease with which I clicked on simple icons to buy and sell (specifically sell reasonably performing funds and buy more WCOM shares) was always a little frightening, and I sometimes felt as if there were a loaded gun on my desk. Some studies have linked online trading and day trading to increased volatility in the late '90s, although it's not clear that they remain factors in the '00s.

What's undeniable is that buying and selling online remains easy, so easy that I think it might not be a bad idea were small pictures of real-world items to pop up before every stock purchase or sale as a reminder of the approximate value of what's being traded. If your transaction were for $35,000, a luxury car might appear; if it were for $100,000, a small cottage; and if it were for a penny stock, a candy bar. Investors can now check stock quotations, the size and the number of the bids and the asks, and megabytes of other figures on so-called level-two screens available in (almost) real-time on their personal computers. Millions of little desktop brokerages! Unfortunately, librarian Jesse Sherra's paraphrase of Coleridge often seems apt: Data, data everywhere, but not a thought to think.

Benford's Law and Looking Out for Number One

I mentioned that people find it very difficult to simulate a series of coin flips. Are there other human disabilities that might allow someone to look at a company's books, say Enron's or WorldCom's, and determine whether or not they had been cooked? There may have been, and the mathematical principle involved is easily stated, but counterintuitive.

Benford's Law states that in a wide variety of circumstances, numbers—as diverse as the drainage areas of rivers, physical properties of chemicals, populations of small towns, figures in a newspaper or magazine, and the half-lives of radioactive atoms—have "1" as their first non-zero digit disproportionately often. Specifically, they begin with "1" about 30 percent of the time, with "2" about 18 percent of the time, with "3" about 12.5 percent, and with larger digits progressively less often. Less than 5 percent of the numbers in these circumstances begin with the digit "9." Note that this is in stark contrast to many other situations where each of the digits has an equal chance of appearing.

Benford's Law goes back one hundred years to the astronomer Simon Newcomb (note the letters WCOM in his name), who noticed that books of logarithm tables were much dirtier near the front, indicating that people more frequently looked up numbers with a low first digit. This odd phenomenon remained a little-known curiosity until it was rediscovered in 1938 by the physicist Frank Benford. It wasn't until 1996, however, that Ted Hill, a mathematician at Georgia Tech, established what sort of situations generate numbers in accord with Benford's Law. Then a mathematically inclined accountant named Mark Nigrini generated considerable buzz when he noted that Benford's Law could be used to catch fraud in income tax returns and other accounting documents.

The following example suggests why collections of numbers governed by Benford's Law arise so frequently:

Imagine that you deposit $1,000 in a bank at 10 percent compound interest per year. Next year you'll have $1,100, the year after that $1,210, then $1,331, and so on. (Compounding is discussed further in chapter 5.) The first digit of your account balance remains a "1" for a long time. When your account grows to over $2,000, the first digit will remain a "2" for a shorter period. And when your deposit finally grows to over $9,000, the 10 percent growth will result in more than $10,000 in your account the following year and a long return to "1" as the first digit. If you record your account balance each year for many years, these numbers will thus obey Benford's Law.

The law is also "scale-invariant" in that the dimensions of the numbers don't matter. If you expressed your $1,000 in euros or pounds (or the now defunct francs or marks) and watched it grow at 10 percent per year, about 30 percent of the yearly values would begin with a "1," about 18 percent with a "2," and so on.

More generally, Hill showed that such collections of numbers arise whenever we have what he calls a "distribution of distributions," a random collection of random samples of data. Big, motley collections of numbers will follow Benford's Law.

This brings us back to Enron, WorldCom, accounting, and Mark Nigrini, who reasoned that the numbers on accounting forms, which often come from a variety of company operations and a variety of sources, should be governed by Benford's Law. That is, these numbers should begin disproportionately with the digit "1," and progressively less often with bigger digits, and if they don't, that is a sign that the books have been cooked. When people fake plausible-seeming numbers, they generally use more "5s" and "6s" as initial digits, for example, than Benford's Law would predict.

Nigrini's work has been well publicized and has surely been noted by accountants and by prosecutors. Whether the Enron, WorldCom, and Anderson people have heard of it is unknown, but investigators might want to check if the distribution of leading digits in the Enron documents accords with Benford's Law. Such checks are not foolproof and sometimes lead to false-positive results, but they provide an extra tool that might be useful in certain situations.

It would be amusing if, in looking out for number one, the culprits forgot to look out for their "1s." Imagine the Anderson accountants muttering anxiously that there weren't enough leading "1s" on the documents they were feeding into the shredders. A 1-derful fantasy!

The Numbers Man—A Screen Treatment

An astonishing amount of attention has been paid recently to fictional and narrative treatments of mathematical topics. The movies *Good Will Hunting*, *Pi,* and *The Croupier* come to mind; so do plays such as *Copenhagen*, *Arcadia*, and *The Proof,* the two biographies of Paul Erdos, *A Beautiful Mind*, the biography of John Nash (with its accompanying Academy Award-winning movie), TV specials on Fermat's Last Theorem, and other mathematical topics, as well as countless books on popular mathematics and mathematicians. The plays and movies, in particular, prompted me to expand the idea in the stock-newsletter scam discussed above (I changed the focus, however, from stocks to sports) into a sort of abbreviated screen treatment that highlights the relevant mathematics a bit more than has been the case in the productions just cited. Yet another instance of what columnist Charles Krauthammer has dubbed "Disturbed Nerd Chic," the treatment might even be developed into an intriguing and amusing film. In fact, I rate it a "strong buy" for any studio executive or independent filmmaker.

Rough Idea: Math nerd runs a clever sports-betting scam and accidentally nets an innumerate mobster.

Act One

Louis is a short, lecherous, somewhat nerdy man who dropped out of math graduate school about ten years ago (in the late '80s) and now works at home as a technical consultant. He looks and acts a bit like the young Woody Allen. He's playing cards with his pre-teenage kids and has just finished telling them a funny story. His kids are smart and they ask him how it is that he always knows the right story to tell. His wife, Marie, is uninterested. True to form, he begins telling them the Leo Rosten story about the famous rabbi who was asked by an admiring student how it was that the rabbi always had a perfect parable for any subject. Louis pauses to make sure they see the relevance.

When they smile and his wife rolls her eyes again, he continues. He tells them that the rabbi replied to his students with a parable. It was about a recruiter in the Tsar's army who was riding through a small town and noticed dozens of chalked circular targets on the side of a barn, each with a bullet hole through the bull's-eye. The recruiter was impressed and asked a neighbor who this perfect shooter might be. The neighbor responded, "Oh that's Shepsel, the shoemaker's son. He's a little peculiar." The enthusiastic recruiter was undeterred until the neighbor added, "You see, first Shepsel shoots and *then* he draws the chalk circles around the bullet hole." The rabbi grinned. "That's the way it is with me. I don't look for a parable to fit the subject. I introduce only subjects for which I have parables."

Louis and his kids laugh until a distracted, stricken look crosses his face. Closing the book, Louis hurries his kids off to bed, interrupts Marie's prattling about her new pearl necklace and her Main Line parents' nasty neighbors, distractedly bids her good night, and retreats to his study where he starts scribbling, making calls, and performing calculations. The next day he stops by the bank and the post office and a stationery store, does some research online, and then has a long discussion with his friend, a sportswriter on the local suburban New Jersey newspaper. The conversation revolves around the names, addresses, and intelligence of big sports bettors around the country.

The idea for a lucrative con game has taken shape in his mind. For the next several days he sends letters and emails to many thousands of known sports bettors "predicting" the outcome of a certain sporting event. His wife is uncomprehending when Louis mumbles that, Shepsel-like, he can't lose since whatever happens in the sporting event, his prediction is bound to be right for half the bettors. The reason, it will turn out, is that to half of these people he predicts a certain team will win, and to the other half he predicts that it will lose.

Tall, blond, plain, and dim-witted, Marie is left wondering what exactly her sneaky husband is up to now. She finds the new postage meter behind the computer, notes the increasingly frequent secret telephone calls, and nags him about their worsening financial and marital situation. He replies that she doesn't really need three closets full of clothes and a small fortune of jewelry when she spends all her time watching soaps and puts her off with some mathematical mumbo-jumbo about demographic research and new statistical techniques. She still doesn't follow, but she is mollified by his promise that his mysterious endeavor will end up being lucrative.

They go out to eat to celebrate and Louis, intense and cad-like as always, talks up genetically modified food and tells the cute waitress that he wants to order whatever item on the menu has the most artificial ingredients. Much to Marie's chagrin, he then involves the waitress in a classic mathematical trick by asking her to examine his three cards, one black on both sides, one red on both sides, and one black on one side and red on the other. He asks her for her cap, drops the cards into it, and tells her to pick a card, but only to look at one side of it. The side is red, and Louis notes that the card she picked couldn't possibly be the card that was black on both sides, and therefore it must be one of the other two cards—the red-red card or the red-black card. He guesses that it's the red-red card and offers to double her 15 percent tip if it's the red-black card and stiff her if it's the red-red card. He looks at Marie for approbation that is not forthcoming. The waitress accepts and loses.

Tone-deaf to Marie's discomfort, Louis thinks he's making amends with her by explaining the trick. She is less than enthralled. He tells her that it's not an even bet even though at first glance it appears to be one. There are, after all, two cards it could be, and he bet on one, and the waitress bet on the other. The rub is, he gleefully runs on with his mouth full, there are two ways he can win and only one way the waitress can win. The visible side of the card the waitress picked could be the red side of the red-black card, in which case she wins, or it could be one side of the red-red card, in which case he wins, or it could be the other side of the red-red card, in which case he also wins. His chances of winning are thus 2/3, he concludes exultantly, and the average tip he gives is reduced by a third. Marie yawns and checks her Rolex. He breaks to go to the men's room where he calls his girlfriend May Lee to apologize for some vague indiscretion.

The next week he explains the sports-betting con to May Lee, who looks a bit like Lucy Liu and is considerably smarter than Marie and even more materialistic. They're in her apartment. She is interested in the con and asks clarifying questions. He enthuses to her that he needs her secretarial help. He's sending out more letters and making a second prediction in them, but this time just to the half of the people to whom he sent a correct first prediction; the other half he plans to ignore. To half of this smaller group, he will predict a win in a second sporting event, to the other half a loss. Again for half of this group his prediction is going to be right, and so for one-fourth of the original group he's going to be right two times in a row. "And to this one-fourth of the bettors?" she asks knowingly and excitedly. A mathematico-sexual tension develops.

He smiles rakishly and continues. To half of this one-fourth he will predict a win the following week, to the other half a loss; he again will ignore those to whom he's made an incorrect prediction. Once again he will be right—this time for the third straight time—although for only one-eighth of the original population. May Lee helps with the mailings as he continues this process, focusing only on those to whom he's made correct predictions and winnowing out those to whom he's made incorrect ones. There is a sex scene amid all the letters, and they joke about winning whether the teams in question do or not, whether the predictions are right or wrong. Whether up or down, they'll be happy.

As the mailings go on, so does his other life as a bored consultant, cyber-surfer, and ardent sports fan. He continues to extend his string of successful predictions to a smaller and smaller group of people until finally with great anticipation he sends a letter to the small group of people who are left. In it he points to his impressive string of successes and requests a substantial payment to keep these valuable and seemingly oracular "predictions" coming.

Act Two

He receives many payments and makes a further prediction. Again he's right for half of the remaining people and drops the half for which he's wrong. He asks the former group for even more money for another prediction, receives it, and continues. Things improve with Marie and with May Lee as the money rolls in and Louis realizes his plan is working even better than he expected. He takes his kids and, in turn, each woman to sports events or to Atlantic City, where he comments smugly on the losers who, unlike him, bet on iffy propositions. When Marie worries aloud about shark attacks off the beach, Louis tells her that more Americans die from falling airplane parts each year than from shark attacks. He makes similar pronouncements throughout the trip.

He plays a little blackjack and counts cards while doing so. He complains that it requires too much low-level concentration and that, unless one has a lot of money already, the rate at which one makes money is so slow and uneven that one might as well get a job. Still, he goes on, it's the only game where a strategy exists for winning. All the other games are for mentally flabby losers. He goes to one of the casino restaurants where he shows his kids the waitress tip-cheating game. They think it's great.

Back home in suburban New Jersey again, the sports-betting con resumes. Now there are only a few people left among the original thousands of sports-bettors. One of them, a rough underworld type named Otto, tracks him down, follows him to the parking lot of the basketball arena, and, politely at first and then more and more insistently, demands a prediction on an upcoming game on which he plans to bet a lot of money. Louis dismisses him and Otto, who looks a little like Stephen Segal, promptly orders him into his car at gunpoint and threatens to harm his family. He knows where they live.

Not understanding how he could be the recipient of so many consecutive correct predictions, Otto doesn't believe Louis's protestations that this is a con game. Louis makes some mathematical points in an effort to convince Otto of the possible falsity of any particular prediction. But no matter how he tries, he can't quite convince Otto of the fact that there will always be some people who receive many consecutive correct predictions by chance alone.

Marooned in Otto's basement, the math-nerd scam artist and the bald muscled extortionist are a study in contrasts. They speak different languages and have different frames of reference. Otto claims, for example, that every bet is more or less a 50–50 proposition because you either win or lose. Louis talks of his basketball buddies Lewis Carroll and Bertrand Russell and the names go over Otto's head, of course. Oddly, they have similar attitudes toward women and money and also share an interest in cards, which they play to while away the time. Otto proudly shows off his riffle shuffle that he claims completely mixes the cards, while Louis prefers solitaire and silently scoffs at Otto's lottery expenditures and gambling misconceptions. When they forget why they're there they get along well enough, although now and then Otto renews his threats and Louis renews his disavowal of any special sports knowledge and his plea to go home.

Finally granting that he might receive an incorrect prediction occasionally, Otto still insists that Louis give him his take on who's going to win an upcoming football game. In addition to not being too bright, Otto, it appears, is in serious debt. Under extreme duress (with a gun to his head), Louis makes a prediction that happens to be right, and Otto, desperate and still convinced that he is in control of a money tree, now wants to bet funds borrowed from his gambling associates on Louis's next prediction.

Act Three

Louis at last convinces Otto to let him go home and do research for his next big sports prediction. He and May Lee, whose need for money, baubles, and clothes has all along provided the impetus for the scam, discuss his predicament and realize they must exploit Otto's only weaknesses, his stupidity and gullibility, and his only intellectual interests, money and playing cards.

Both go over to Otto's apartment. Otto is charmed by May Lee, who flirts with him and offers him a deal. She wordlessly takes two decks of cards from her purse and asks Otto to shuffle each of them. Otto is pleased to show off to a more appreciative audience. She then gives him one of the decks and asks him to turn over one card at a time as she, keeping pace with him, does the same thing with the other deck. May Lee asks, what does he think is the likelihood that the cards they turn over will ever match, denomination and suit exactly the same? He scoffs but is entranced by May Lee and is amazed when after a tense minute or so that is exactly what happens. She explains that it will happen more often than not and perhaps he can use this fact to make some money. After all, Louis is a mathematical genius and he's proved that it will. Louis smiles proudly.

Otto is puzzled. May Lee tells Otto again that the sports-betting was a scam and that he's more likely to make money with the card tricks that Louis can teach him. Louis steps forward with the same two decks, which he's now arranged so that the cards in each deck alternate colors. In one, it's red-black, red-black, red-black . . . . In the other it's black-red, black-red, black-red . . . . He gives the two decks to Otto and challenges him to do one of his great riffle shuffles of one deck into the other so that the cards will be mixed. Otto does and arrogantly announces that the cards are completely mixed now, whereupon Louis takes the combined two decks, puts them behind his back, pretends to be manipulating them, and brings forward two cards, one black and one red. So, Otto asks? Louis brings forth two more cards, one of each color, and then he does this again and again. I really shuffled them, Otto observes. How'd you do that? Louis explains that it involves no skill; the cards no longer alternate color in the combined deck, but any two from the top on down are always of different color.

There is a collage scene in which Louis explains various card tricks to Otto and the ways in which they can be exploited to make money. There's always some order, some deviation from randomness, that a card man like you can use to get rich, Louis says to Otto. He even explains to him how he avoids paying waitresses tips. The deal, of course, is that Otto releases them, understanding, vaguely at least, how the betting scheme works and, more precisely, how the new card tricks do. Louis promises a one-day crash course on how to exploit the tricks for money.

In the last scene Louis is seen working the same scam but this time with predictions about the movements of a stock market index. Since he doesn't want any more Ottos, but a higher class clientele, he's redefined himself as the publisher of a stock newsletter. The house he's in is more sumptuous and May Lee, to whom he's now married, bustles about in an expensive suit as Louis plays cards with his slightly older children, occasionally doodling little bull's-eyes and targets on an envelope. He excuses himself and goes to his study to make a secret telephone call to an apartment on Central Park West that he's just purchased for his new mistress.

5 | Value Investing and Fundamental Analysis

I was especially smitten with WorldCom's critical Internet division, UUNet. The Internet wasn't going away, and so, I thought, neither was UUNet or WorldCom. During this time of enchantment my sensible wife would say "UUNet, UUNet" and roll her pretty eyes to mock my rhapsodizing about WorldCom's global IP network and related capabilities. The repetition of the word gradually acquired a more general anti-Pollyannish meaning as well. "Maybe the bill is so exorbitant because the plumber ran into something he didn't expect." "Yeah, sure. UUNet, UUNet."

"Smitten," "rhapsodizing," and "Pollyanna" are not words that come naturally to mind when discussing value investing, a major approach to the market that uses the tools of so-called fundamental analysis. Often associated with Warren Buffett's gimlet-eyed no-nonsense approach to trading, fundamental analysis is described by some as the best, most sober strategy for investors to follow. Had I paid more attention to WorldCom fundamentals, particularly its $30 billion in debt, and less attention to WorldCom fairy tales, particularly its bright future role as a "dumb" network (better not to ask), I would no doubt have fared better. In the stock market's enduring tug-of-war between statistics and stories, fundamental analysis is generally on the side of the numbers.

Still, fundamental analysis has always seemed to me slightly at odds with the general ethic of the market, which is based on hope, dreams, vision, and a certain monetarily tinted yet genuine romanticism. I cite no studies or statistics to back up this contention, only my understanding of the investors I've known or read about and perhaps my own infatuation, quite atypical for this numbers man, with WorldCom.

Fundamentals are to investing what (stereotypically) marriage is to romance or what vegetables are to eating—healthful, but not always exciting. Some understanding of them, however, is essential for any investor and, to an extent, for any intelligent citizen. Everybody's heard of people who refrain from buying a house, for example, because of the amount they would have paid in interest over the years. ("Oh my, don't get a mortgage. You'll end up paying four times as much.") Also common are lottery players who insist that the worth of their possible winnings is really the advertised one million dollars. ("In only 20 years, I'll have that million.") And there are many investors who doubt that the opaque pronouncements of Alan Greenspan have anything to do with the stock or bond markets.

These and similar beliefs stem from misconceptions about compound interest, the bedrock of mathematical finance, which is in turn the foundation of fundamental analysis.

e is the Root of All Money

Speaking of bedrocks and foundations, I claim that e is the root of all money. That's e as in e^x as in exponential growth as in compound interest. An old adage (probably due to an old banker) has it that those who understand compound interest are more likely to collect it, those who don't more likely to pay it. Indeed the formula for such growth is the basis for most financial calculations. Happily, the derivation of a related but simpler formula depends only on understanding percentages, powers, and multiplication—on knowing, for example, that 15 percent of 300 is $.15 \times 300$ (or $300 \times .15$) and that 15 percent of 15 percent of 300 is $300 \times (.15)^2$.

With these mathematical prerequisites stated, let's begin the tutorial and assume that you deposit $1,429.73 into a bank account paying 6.9 percent interest compounded annually. No, let's bow to the great Rotundia, god of round numbers, and assume instead that you deposit $1,000 at 10 percent. After one year, you'll have 110 percent of your original deposit—$1,100. That is, you'll have $1{,}000 \times 1.10$ dollars in your account. (The analysis is the same if you buy $1,000 worth of some stock and it returns 10 percent annually.)

Looking ahead, observe that after two years you'll have 110 percent of your first-year balance—$1,211. That is, you'll have $(\$1{,}000 \times 1.10) \times 1.10$. Equivalently, that is $\$1{,}000 \times 1.10^2$. Note that the exponent is 2.

After three years you'll have 110 percent of your second-year balance—$1,331. That is, you'll have $(\$1{,}000 \times 1.10^2) \times 1.10$. Equivalently, that is $\$1{,}000 \times 1.10^3$. Note the exponent is 3 this time.

The drill should be clear now. After four years you'll have 110 percent of your third-year balance—$1,464.10. That is, you'll have $(\$1{,}000 \times 1.10^3) \times 1.10$. Equivalently, that is $\$1{,}000 \times 1.10^4$. Once again, note the exponent is 4.

Let me interrupt this relentless exposition with the story of a professor of mine long ago who, beginning at the left side of a very long blackboard in a large lecture hall, started writing $1 + 1/1! + 1/2! + 1/3! + 1/4! + 1/5! \ldots.$ (Incidentally the expression 5! is read 5 factorial, not 5 with an exclamatory flair, and it is equal to $5 \times 4 \times 3 \times 2 \times 1$. For any whole number N, N! is defined similarly.) My fellow students initially laughed as this professor, slowly and seemingly in a trance, kept on adding terms to this series. The laughter died out, however, by the time he reached the middle of the board and was writing 1/44! + 1/45! + ... I liked him and remember a feeling of alarm as I saw him continue his senseless repetitions. When he came to the end of the board at 1/83!, he turned and faced the class. His hand shook, the chalk dropped to the floor, and he left the room and never returned.

Mindful thereafter of the risks of too many illustrative repetitions, especially when I'm standing at a blackboard in a classroom, I'll end my example with the fourth year and simply note that the amount of money in your account after t years will be $\$1,000 \times 1.10^t$. More generally, if you deposit P dollars into an account earning r percent interest annually, it will be worth A dollars after t years, where $A = P(1 + r)^t$, the promised formula describing exponential growth of money.

You can adjust the formula for interest compounded semiannually or monthly or daily. If money is compounded four times per year, for example, then the amount you'll have after t years is given by $A = P(1 + r/4)^{4t}$. (The quarterly interest rate is r/4, one-fourth the annual rate of r, and the number of compoundings in t years is 4t, four per year for t years.)

If you compound very frequently (say n times per year for a large number n), the formula $A = P(1 + r/n)^{nt}$ can be mathematically massaged and rewritten as $A = Pe^{rt}$, where e, approximately 2.718, is the base of the natural logarithm. This variant of the formula is used for continuous compounding (and is, of course, the source of my comment that e is the root of all money).

The number e plays a critical role in higher mathematics, best exemplified perhaps by the formula $e^{\pi i} + 1 = 0$, which packs the five arguably most important constants in mathematics into a single equation. The number e also arises if we're simply choosing numbers between 0 and 1 at random. If we (or, more likely, our computer) pick these numbers until their sum exceeds 1, the average number of picks we'd need would be e, about 2.718. The ubiquitous e also happens to equal 1 + 1/1! + 1/2! + 1/3! + 1/4! + . . . , the same expression my professor was writing on the board many years ago. (Inspired by a remark by stock speculator Ivan Boesky, Gordon Gecko in the 1987 movie *Wall Street* stated, "Greed is good." He misspoke. He intended to say, "e is good.")

Many of the formulas useful in finance are consequences of these two formulas: $A = P(1 + r)^t$ for annual compounding and, for continuous compounding, $A = Pe^{rt}$. To illustrate how they're used, note that if you deposit \$5,000 and it's compounded annually for 12 years at 8 percent, it will be worth $\$5,000(1.08)^{12}$ or \$12,590.85. If this same \$4,000 is compounded continuously, it will be worth $\$4,000e^{(.08 \times 12)}$ or \$13,058.48.

Using this interest rate and time interval, we can say that the future value of the present \$5,000 is \$12,590.85 and that the present value of the future \$12,590.85 is \$5,000. (If the compounding is continuous, substitute \$13,058.48 in the previous sentence.) The "present value" of a certain amount of future money is the amount we would have to deposit now so that the deposit would grow to the requisite amount in the allotted time. Alternatively stated (repetition may be an occupational hazard of professors; so may self-reference), the idea is that given an interest rate of 8 percent, you should be indifferent between receiving \$5,000 now (the present value) and receiving something near \$13,000 (the future value) in twelve years.

And just as "George is taller than Martha" and "Martha is shorter than George" are different ways to state the same relation, the interest formulas may be written to emphasize either present value, P, or future value, A. Instead of $A = P(1 + r)^t$, we can write $P = A/(1 + r)^t$, and instead of $A = Pe^{rt}$, we can write $P = A/e^{rt}$. Thus, if the interest rate is 12 percent, the present value of \$50,000 five years hence is given by $P = \$50,000/(1.12)^5$ or \$28,371.34. This amount, \$28,371.34, if deposited at 12 percent compounded annually for five years, has a future value of \$50,000.

One consequence of these formulas is that the "doubling time," the time it takes for a sum of money to double in value, is given by the so-called rule of 72: divide 72 by 100 times the interest rate. Thus, if you can get an 8 percent (.08) rate, it will take you 72/8 or nine years for a sum of money to double, eighteen years for it to quadruple, and twenty-seven years for it to grow to eight times its original size. If you're lucky enough to have an investment that earns 14 percent, your money will double in a little more than five years (since 72/14 is a bit more than 5) and quadruple in a bit over ten years. For continuous compounding, you use 70 rather than 72.

These formulas can also be used to determine the so-called internal rate of return and to define other financial concepts. They provide as well the muscle behind common pleas to young people to begin saving and investing early in life if they wish to become the "millionaire next door." (They don't, however, tell the millionaire next door what he should do with his wealth.)

The Fundamentalists' Creed: You Get What You Pay For

The notion of present value is crucial to understanding the fundamentalists' approach to stock valuation. It should also be important to lottery players, mortgagors, and advertisers. That the present value of money in the future is less than its nominal value explains why a nominal \$1,000,000 award for winning a lottery—say \$50,000 per year at the end of each of the next twenty years—is worth considerably less than $1,000,000. If the interest rate is 10 percent annually, for example, the $1,000,000 has a present value of only about $426,000. You can obtain this value from tables, from financial calculators, or directly from the formulas above (supplemented by a formula for the sum of a so-called geometric series).

The process of determining the present value of future money is often referred to as "discounting." Discounting is important because, once you assume an interest rate, it allows you to compare amounts of money received at different times. You can also use it to evaluate the present or future value of an income stream—different amounts of money coming into or going out of a bank or investment account on different dates. You simply "slide" the amounts forward or backward in time by multiplying or dividing by the appropriate power of $(1 + r)$. This is done, for example, when you need to figure out a payment sufficient to pay off a mortgage in a specified amount of time or want to know how much to save each month to have sufficient funds for a child's college education when he or she turns eighteen.

Discounting is also essential to defining what is often called a stock's fundamental value. The stock's price, say investing fundamentalists (fortunately not the sort who wish to impose their moral certitudes on others), should be roughly equal to the discounted stream of dividends you can expect to receive from holding onto it indefinitely. If the stock does not pay dividends or if you plan on selling it and thereby realizing capital gains, its price should be roughly equal to the discounted value of the price you can reasonably expect to receive when you sell the stock plus the discounted value of any dividends. It's probably safe to say that most stock prices are higher than this. During the 1990 boom years, investors were much more concerned with capital gains than they were with dividends. To reverse this trend, finance professor Jeremy Siegel, author of *Stocks for the Long Run*, and two of his colleagues recently proposed eliminating the corporate dividend tax and making dividends deductible.

The bottom line of bottom-line investing is that you should pay for a stock an amount equal to (or no more than) the present value of all future gains from it. Although this sounds very hard-headed and far removed from psychological considerations, it is not. The discounting of future dividends and the future stock price is dependent on your estimate of future interest rates, dividend policies, and a host of other uncertain quantities, and calling them fundamentals does not make them immune to emotional and cognitive distortion. The tango of exuberance and despair can and does affect estimates of stock's fundamental value. As the economist Robert Shiller has long argued quite persuasively, however, the fundamentals of a stock don't change nearly as much or as rapidly as its price.

Ponzi and the Irrational Discounting of the Future

Before returning to other applications of these financial notions, it may be helpful to take a respite and examine an extreme case of undervaluing the future: pyramids, Ponzi schemes, and chain letters. These differ in their details and colorful storylines. A recent example in California took the form of all-women dinner parties whose new members contributed cash appetizers. Whatever their outward appearance, however, almost all these scams involve collecting money from an initial group of "investors" by promising them quick and extraordinary returns. The returns come from money contributed by a larger group of people. A still larger group of people contributes to both of the smaller earlier groups.

This burgeoning process continues for a while. But the number of people needed to keep the pyramid growing and the money coming in increases exponentially and soon becomes difficult to maintain. People drop out, and the easy marks become scarcer. Participants usually lack a feel for how many people are required to keep the scheme going. If each of the initial group of ten recruits ten more people, for example, the secondary group numbers 100. If each of these 100 recruit ten people, the tertiary group numbers 1,000. Later groups number 10,000, then 100,000, then 1,000,000. The system collapses under its own weight when enough new people can no longer be found. If you enter the scheme early, however, you can make extraordinarily quick returns (or could if such schemes were not illegal).

The logic of pyramid schemes is clear, but people generally worry only about what happens one or two steps ahead and anticipate being able to get out before a collapse. It's not irrational to get involved if you are confident of recruiting a "bigger sucker" to replace you. Some would say that the dot-coms' meteoric stock price rises in the late '90s and their subsequent precipitous declines in 2000 and 2001 were attenuated versions of the same general sort of scam. Get in on the initial public offering, hold on as the stock rockets upward, and jump off before it plummets.

Although not a dot-com, WorldCom achieved its all-too-fleeting dominance by buying up, often for absurdly inflated prices, many companies that were (and a good number that weren't). MCI, MFS, ANS Communication, CAI Wireless, Rhythms, Wireless One, Prime One Cable, Digex, and dozens more companies were acquired by Bernie Ebbers, a pied piper whose song seemed to consist of only one entrancing and repetitive note: acquire, acquire, acquire. The regular drumbeat of WorldCom acquisitions had the hypnotic quality of the tinkling bells that accompany the tiniest wins at casino slot machines. As the stock began its slow descent, I'd check the business news every morning and was tranquilized by news of yet another purchase, web hosting agreement, or extension of services.

While corporate venality and fraud played a role in (some of) their falls, the collapses of the dot-coms and WorldCom were not the brainchilds of con artists. Even when entrepreneurs and investors recognized the bubble for what it was, most figured incorrectly that they'd be able to find a chair when the mania-inducing IPO/acquisition music stopped. Alas, the journey from "have-lots" to "have-nots" was all too frequently by way of "have-dots."

Maybe our genes are to blame. (They always seem to get the rap.) Natural selection probably favors organisms that respond to local or near-term events and ignore distant or future ones, which are discounted in somewhat the same way that future money is. Even the ravaging of the environment may be seen as a kind of global Ponzi scheme, the early "investors" doing well, later ones less well, until a catastrophe wipes out all gains.

A quite different illustration of our short-sightedness comes courtesy of Robert Louis Stevenson's "The Imp in the Bottle." The story tells of a genie in a bottle able and willing to satisfy your every romantic whim and financial desire. You're offered the opportunity to buy this bottle and its amazing denizen at a price of your choice. There is a serious limitation, however. When you've finished with the bottle, you have to sell it to someone else at a price strictly less than what you paid for it. If you don't sell it to someone for a lower price, you will lose everything and will suffer excruciating and unrelenting torment. What would you pay for such a bottle?

Certainly you wouldn't pay 1 cent because then you wouldn't be able to sell it for a lower price. You wouldn't pay

2 cents for it either since no one would buy it from you for 1 cent since everyone knows that it must be sold for a price less than the price at which it is bought. The same reasoning shows that you wouldn't pay 3 cents for it since the person to whom you would have to sell it for 2 cents would object to buying it at that price since he wouldn't be able to sell it for 1 cent. Likewise for prices of 4 cents, 5 cents, 6 cents, and so on. We can use mathematical induction to formalize this argument, which proves conclusively that you wouldn't buy the genie in the bottle for any amount of money. Yet you would almost certainly buy it for $1,000. I know I would. At what point does the argument against buying the bottle cease to be compelling? (I'm ignoring the possibility of foreign currencies that have coins worth less than a penny. This is an American genie.)

The question is more than academic since in countless situations people prepare exclusively for near-term outcomes and don't look very far ahead. They myopically discount the future at an absurdly steep rate.

Average Riches, Likely Poverty

Combining time and money can yield unexpected results in a rather different way. Think back again to the incandescent stock market of the late 1990s and the envious feeling many had that everyone else was making money. You might easily have developed that impression from reading about investing in those halcyon days. In every magazine or newspaper you picked up, you were apt to read about IPOs, the initial public offerings of new companies, and the investment gurus who claimed that they could make your $10,000 grow to more than a million in a year's time. (All right, I'm exaggerating their exaggerations.) But in those same periodicals, even then, you also would have read stories about new companies that were stillborn and naysayers' claims that most investors would lose their $10,000 as well as their shirts by investing in such volatile offerings.

Here's a scenario that helps to illuminate and reconcile such seemingly contradictory claims. Hang on for the math that follows. It may be a bit counterintuitive, but it's not difficult to follow and it illustrates the crucial difference between the arithmetic mean and the geometric mean of a set of returns. (For the record: The arithmetic mean of N different rates of return is what we normally think of as their average; that is, their sum divided by N. The geometric mean of N different rates of return is equal to that rate of return that, if received N times in succession, would be equivalent to receiving the N different rates of return in succession. We can use the formula for compound interest to derive the technical definition. Doing so, we would find that the geometric mean is equal to the Nth root of the product [(1 + first return) × (1 + second return) × (1 + third return) × . . . (1 + Nth return)] – 1.)

Hundreds of IPOs used to come out each year. (Pity that this is only an illustrative flashback.) Let's assume that the first week after the stock comes out, its price is usually extremely volatile. It's impossible to predict which way the price will move, but we'll assume that for half of the companies' offerings the price will rise 80 percent during the first week and for half of the offerings the price will fall 60 percent during this period.

The investing scheme is simple: Buy an IPO each Monday morning and sell it the following Friday afternoon. About half the time you'll earn 80 percent in a week and half the time you'll lose 60 percent in a week for an average gain of 10 percent per week: [(80%) + (–60%)]/2, the arithmetic mean.

Ten percent a week is an amazing average gain, and it's not difficult to determine that after a year of following this strategy, the average worth of an initial $10,000 investment is more than $1.4 million! (Calculation below.) Imagine the newspaper profiles of happy day traders, or week traders in this case, who sold their old cars and turned the proceeds into almost a million and a half dollars in a year.

But what is the most likely outcome if you were to adopt this scheme and the assumptions above held? The answer is that your $10,000 would likely be worth all of $1.95 at the end of a year! Half of all investors adopting such a scheme would have less than $1.95 remaining of their $10,000 nest egg. This same $1.95 is the result of your money growing at a rate equal to the geometric mean of 80 percent and –60 percent over the 52 weeks. (In this case that's equal to the square root (the Nth root for N = 2) of the product [(1 + 80%) × (1 + (–60%))] minus 1, which is the square root of [1.8 × .4] minus 1, which is .85 minus 1, or –.15, a loss of approximately 15 percent each week.)

Before walking through this calculation, let's ask for the intuitive reason for the huge disparity between $1.4 million and $1.95. The answer is that the typical investor will see his investment rise by 80 percent for approximately 26 weeks and decline by 60 percent for 26 weeks. As shown below, it's not difficult to calculate that this results in $1.95 of your money remaining after one year.

The lucky investor, by contrast, will see his investment rise by 80 percent for considerably more than 26 weeks. This will result in astronomical returns that pull the average up. The investments of the unlucky investors will decline by 60 percent for considerably more than 26 weeks, but their losses cannot exceed the original $10,000.

In other words, the enormous returns associated with disproportionately many weeks of 80 percent growth skew the average way up, while even many weeks of 60 percent shrinkage can't drive an investment's value below $0.

In this scenario the stock gurus and the naysayers are both right. The average worth of your $10,000 investment after one year is $1.4 million, but its most likely worth is $1.95.

Which results are the media likely to focus on?

The following example may help clarify matters. Let's examine what happens to the $10,000 in the first two weeks. There are four equally likely possibilities. The investment can increase both weeks, increase the first week and decrease the second, decrease the first week and increase the second, or decrease both weeks. (As we saw in the section on interest theory, an increase of 80 percent is equivalent to multiplying by 1.8. A 60 percent fall is equivalent to multiplying by 0.4.) One-quarter of investors will see their investment increase by a factor of 1.8 × 1.8, or 3.24. Having increased by 80 percent two weeks in a row, their $10,000 will be worth $10,000 × 1.8 × 1.8, or $32,400 in two weeks. One-quarter of investors will see their investment rise by 80 percent the first week and decline by 60 percent the second week. Their investment changes by a factor of 1.8 × 0.4, or 0.72, and will be worth $7,200 after two weeks. Similarly, $7,200 will be the outcome for one-quarter of investors who will see their investment decline the first week and rise the second week, since 0.4 × 1.8 is the same as 1.8 × 0.4. Finally, the unlucky one-quarter of investors whose investment loses 60 percent of its worth for two weeks in a row will have 0.4 × 0.4 × $10,000, or $1,600 after two weeks.

Adding $32,400, $7,200, $7,200, and $1,600 and dividing by 4, we get $12,100 as the average worth of the investments after the first two weeks. That's an average return of 10 percent weekly, since $10,000 × 1.1 × 1.1 = $12,100. More generally, the stock rises an average of 10 percent every week (the average of an 80 percent gain and a 60 percent loss, remember). Thus after 52 weeks, the average value of the investment is $10,000 × $(1.10)^{52}$, which is $1,420,000.

The most *likely* result is that the companies' stock offerings will rise during 26 weeks and fall during 26 weeks. This means that the most likely worth of the investment is $10,000 $\times (1.8)^{26} \times (.4)^{26}$, which is only $1.95. And the geometric mean of 80 percent and –60 percent? Once again, it is the square root of the product of $[(1 + .8) \times (1 - .6)]$ minus 1, which equals approximately –.15. Every week, on average, your portfolio loses 15 percent of its value, and $10,000 $\times (1 - .15)^{52}$ equals approximately $1.95.

Of course, by varying these percentages and time frames, we can get different results, but the principle holds true: The arithmetic mean of the returns far outstrips the geometric mean of the returns, which is also the median (middle) return as well as the most common return. Another example: If half of the time your investment doubles in a week, and half of the time it loses half its value in a week, the most likely outcome is that you'll break even. But the arithmetic mean of your returns is 25 percent per week—[100% + (–50%)]/2, which means that your initial stake will be worth $10,000 $\times 1.25^{52}$, or more than a billion dollars! The geometric mean of your returns is the square root of $(1 + 1) \times (1 - .5)$ minus 1, which is a 0 percent rate of return, indicating that you'll probably end up with the $10,000 with which you began.

Although these are extreme and unrealistic rates of return, these example have much more general importance than it might appear. They explain why a majority of investors receive worse-than-average returns and why some mutual fund companies misleadingly stress their average returns. Once again, the reason is that the average or arithmetic mean of different rates of return is always greater than the geometric mean of these rates of return, which is also the median rate of return.

Fat Stocks, Fat People, and P/E

You get what you pay for. As noted, fundamentalists believe that this maxim extends to stock valuation. They argue that a company's stock is worth only what it returns to its holder in dividends and price increases. To determine what that value is, they try to make reasonable estimates of the amount of cash the stock will generate over its lifetime, and then they discount this stream of payments to the present. And how do they estimate these dividends and stock price increases? Value investors tend to use the company's stream of earnings as a reasonable substitute for the stream of dividends paid to them since, the reasoning goes, the earnings are, or eventually will be, paid out in dividends. In the meantime, earnings may be used to grow the company or retire debt, which also increases the company's value. If the earnings of the company are good and promise to get better, and if the economy is growing and interest rates stay low, then high earnings justify paying a lot for a stock. And if not, not.

Thus we have a shortcut for determining a reasonable price for a stock that avoids complicated estimations and calculations: the stock's so-called P/E ratio. You can't look at the business section of a newspaper or watch a business show on TV without hearing constant references to it. The ratio is just that—a ratio or fraction. It's determined by dividing the price P of a share of the company's stock by the company's earnings per share E (usually over the past year). Stock analysts discuss countless ratios, but the P/E ratio, sometimes called simply the multiple, is the most common.

The share price, P, is discovered simply by looking in a newspaper or online, and the earnings per share, E, is obtained by taking the company's total earnings over the past year and dividing it by the number of shares outstanding. (Unfortunately, earnings are not nearly as cut-and-dried as many once thought. All sorts of dodges, equivocations, and outright lies make it a rather plastic notion.)

So how does one use this information? One very common way to interpret the P/E ratio is as a measure of investors' expectations of future earnings. A high P/E indicates high expectations about the company's future earnings, and a low one low expectations. A second way to think of the ratio is simply as the price you must pay to receive (indirectly via dividends and price appreciation) the company's earnings. The P/E ratio is thus both a sort of prediction and an appraisal of the company.

A company with a high P/E must perform to maintain its high ratio. If its earnings don't continue to grow, its price will decline. Consider Microsoft, whose P/E was somewhere north of 100 a few years ago. Today its P/E is under 50, although it's one of the larger companies in Redmond, Washington. Still a goliath, it's nevertheless growing more slowly than it did in its early days. This shrinking of the P/E ratio occurs naturally as start-ups become blue-chip pillars of the business community.

(The pattern of change in a company's growth rate brings to mind a mathematical curve—the S-shaped or logistic curve. This curve seems to characterize a wide variety of phenomena, including the demand for new items of all sorts. Its shape can most easily be explained by imagining a few bacteria in a petri dish. At first the number of bacteria will increase slowly, then at a more rapid exponential rate because of the rich nutrient broth and the ample space in which to expand. Gradually, however, as the bacteria crowd each other, their rate of increase slows and their number stabilizes, at least until the dish is enlarged.

The curve appears to describe the growth of entities as disparate as a composer's symphony production, the rise of airline traffic, highway construction, mainframe computer installations, television ownership, even the building of Gothic cathedrals. Some have speculated that there is a kind of universal principle governing many natural and human phenomena, including the growth of successful businesses.)

Of course, the P/E ratio by itself does not prove anything. A high P/E does not necessarily indicate that a stock is overvalued (too expensive for the cash flow it's likely to generate) and a candidate for selling, nor does a low one indicate that a stock is undervalued and a candidate for buying. A low P/E might mean that a company is in financial hot water despite its earnings.

As WorldCom approached bankruptcy, for example, it had an extremely low P/E ratio. A constant stream of postings in the chatrooms compared it to the P/Es of SBC, AT&T, Deutsche Telekom, Bell South, Verizon, and other comparable companies, which were considerably higher. The stridency of the postings increased when they failed to have their desired effect: Investors hitting their foreheads with the sudden realization that WCOM was a great buy. The posters did have a point, however. One should compare a company's P/E to its value in the past, to that of similar companies, and to the ratios for the sector and the market as a whole. The average P/E for the entire market ranges somewhere between 15 and 25, although there are difficulties with computing such an average. Companies that are losing money, for example, have negative P/Es although they're generally not reported as such; they probably should be. Despite the recent market sell-offs in 2001–2002, some analysts believe that stocks are still too expensive for the cash flow they're likely to generate.

Like other tools that fundamental analysts employ, the P/E ratio seems to be precise, objective, and quasi-mathematical. But, as noted, it too is subject to events in the economy as a whole, strong economies generally supporting higher P/Es. As bears reiteration (verb appropriate), the P in the numerator is not invulnerable to psychological factors nor is the E in the denominator invulnerable to accountants' creativity.

The P/E ratio does provide a better measure of a company's financial health than does stock price alone, just as, for example, the BMI or body mass index (equal to your weight divided by the square of your height in appropriate units) gives a better measure of somatic health than does weight alone. The BMI also suggests other ratios, such as the P/E^2 or, in general, the P/E^X, whose study might exercise analysts to such a degree that their BMIs would fall.

(The parallel between diet and investment regimens is not that far-fetched. There are a bewildering variety of diets and market strategies, and with discipline you can lose weight or make money on most of them. You can diet or invest on your own or pay a counselor who charges a fee and offers no guarantee. Whether the diet or strategy is optimal or not is another matter, as is whether the theory behind the diet or strategy makes sense. Does the diet result in faster, more easily sustained weight loss than the conventional counsel of more exercise and a smaller but balanced intake? Does the market strategy make any excess returns, over and above what you would earn with a blind index fund? Unfortunately, most Americans' waistlines in recent years have been expanding, while their portfolios have been getting slimmer.

Numerical comparisons of the American economy to the world economy are common, but comparisons of our collective weight to that of others are usually just anecdotal. Although we constitute a bit under 5 percent of the world's population, we make up, I suspect, a significantly greater percentage of the world's human biomass.)

There is one refinement of the P/E ratio that some find very helpful. It's called the PEG ratio and it is the P/E ratio divided by (100 times) the expected annual growth rate of earnings. A low PEG is usually taken to mean that the stock is undervalued, since the growth rate of earnings is high relative to the P/E. High P/E ratios are fine if the rate of growth of the company is sufficiently rapid. A high-tech company with a P/E ratio of 80 and annual growth of 40 percent will have a PEG of

2 and may sound promising, but a stodgier manufacturing company with a P/E of 7 and an earnings growth rate of 14 percent will have a more attractive PEG of .5. (Once again, negative values are excluded.)

Some investors, including the Motley Fool and Peter Lynch, recommend buying stocks with a PEG of .5 or lower and selling stocks with PEG of 1.50 or higher, although with a number of exceptions. Of course, finding stocks having such a low PEG is no easy task.

Contrarian Investing and the *Sports Illustrated* Cover Jinx

As with technical analysis, the question arises: Does it work? Does using the ideas of fundamental analysis enable you to do better than you would by investing in a broad-gauged index fund? Do stocks deemed undervalued by value investors constitute an exception to the efficiency of the markets? (Note that the term "undervalued" itself contests the efficient market hypothesis, which maintains that all stocks are always valued just right.)

The evidence in favor of fundamental analysis is a bit more compelling than that supporting technical analysis. Value investing does seem to yield moderately better rates of return. A number of studies have suggested, for example, that stocks with low P/E ratios (undervalued, that is) yield better returns than do those with high P/E ratios, the effect's strength varying with the type and size of the company. The notion of risk, discussed in chapter 6, complicates the issue.

Value investing is frequently contrasted with growth investing, the chasing of fast-growing companies with high P/Es. It brings better returns, according to some of its supporters, because it benefits from investors' overreactions. Investors sign on too quickly to the hype surrounding fast-growing companies and underestimate the prospects of solid, if humdrum companies of the type that Warren Buffett likes—Coca-Cola, for instance. (I write this in a study littered with empty cans of Diet Coke.)

The appeal of value investing tends to be contrarian, and many of the strategies derived from fundamental analysis reflect this. The "dogs of the Dow" strategy counsels investors to buy the ten Dow stocks (among the thirty stocks that go into the Dow-Jones Industrial Average) whose price-to-dividend, P/D, ratios are the lowest. Dividends are not earnings, but the strategy corresponds very loosely to buying the ten stocks with the lowest P/E ratios. Since the companies are established organizations, the thinking goes, they're unlikely to go bankrupt and thus their relatively poor performance probably indicates that they're temporarily undervalued. This strategy, again similar to one promoted by the Motley Fool, became popular in the late '80s and early '90s and did result in greater gains than those achieved by, say, the broad-gauged S&P 500 average. As with all such strategies, however, the increased returns tended to shrink as more people adopted it.

A ratio that seems to be more strongly related to increased returns than price-to-dividends or price-to-earnings is the price-to-book ratio, P/B. The denominator B is the company's book value per share—its total assets minus the sum of total liabilities and intangible assets. The P/B ratio changes less over time than does the P/E ratio and has the further virtue of almost always being positive. Book value is meant to capture something basic about a company, but like earnings it can be a rather malleable number.

Nevertheless, a well-known and influential study by the economists Eugene Fama and Ken French has shown P/B to be a useful diagnostic device. The authors focused on the period from 1963 to 1990 and divided almost all the stocks on the

New York Stock Exchange and the Nasdaq into ten groups: the 10 percent of the companies with the highest P/B ratios, the 10 percent with the next highest, on down to the 10 percent with the lowest P/B ratios. (These divisions are called deciles.) Once again a contrarian strategy achieved better than average rates of return. Without exception, every decile with lower P/B ratios outperformed the deciles with higher P/B ratios. The decile with lowest P/B ratios had an average return of 21.4 percent versus 8 percent for the decile with the highest P/B ratios. Other studies' findings have been similar, although less pronounced. Some economists, notably James O'Shaughnessy, claim that a low price to sales ratio, P/S, is an even stronger predictor of better-than-average returns.

Concern with the fundamental ratios of a company is not new. Finance icons Benjamin Graham and David Dodd, in their canonical 1934 text *Security Analysis,* stressed the importance of low P/E and P/B ratios in selecting stocks to buy. Some even stipulate that low ratios constitute the definition of "value stocks" and that high ratios define "growth stocks." There are more nuanced definitions, but there is a consensus that value stocks typically include most of those in oil, finance, utilities, and manufacturing, while growth stocks typically include most of those in computers, telecommunications, pharmaceuticals, and high technology.

Foreign markets seem to deliver value investors the same excessive returns. Studies that divide a country's stocks into fifths according to the value of their P/E and P/B ratios, for example, have generally found that companies with low ratios had higher returns than those with high ratios. Once again, over the next few years, the undervalued, unpopular stocks performed better.

There are other sorts of contrarian anomalies. Richard Thaler and Werner DeBondt examined the thirty-five stocks on the New York Stock Exchange with the highest rates of returns and the thirty-five with the lowest rates for each year from the 1930s until the 1970s. Three to five years later, the best performers had average returns lower than those of the NYSE, while the worst performers had averages considerably higher than the index. Andrew Lo and Craig MacKinlay, as mentioned earlier, came to similar contrarian conclusions more recently, but theirs were significantly weaker, reflecting perhaps the increasing popularity and hence decreasing effectiveness of contrarian strategies.

Another result with a contrarian feel derives from management guru Tom Peters's book *In Search of Excellence,* in which he deemed a number of companies "excellent" based on various fundamental measures and ratios. Using these same measures a few years after Peters's book, Michelle Clayman compiled a list of "execrable" companies (my word, not hers) and compared the fates of the two groups of companies. Once again there was a regression to the mean, with the execrable companies doing considerably better than the excellent ones five years after being so designated.

All these contrarian findings underline the psychological importance of a phenomenon I've only briefly mentioned: regression to the mean. Is the decline of Peters's excellent companies, or of other companies with good P/E and P/B ratios the business analogue of the *Sports Illustrated* cover jinx?

For those who don't follow sports (a field of endeavor where the numbers are usually more trustworthy than in business), a black cat stared out from the cover of the January 2002 issue of *Sports Illustrated* signaling that the lead article was about the magazine's infamous cover jinx. Many fans swear that getting on the cover of the magazine is a prelude to a fall from grace, and much of the article detailed instances of an athlete's or a team's sudden decline after appearing on the cover.

There were reports that St. Louis Rams quarterback Kurt Warner turned down an offer to pose with the black cat on the issue's cover. He wears No. 13 on his back, so maybe there's a limit to how much bad luck he can withstand. Besides, a couple of weeks after gracing the cover in October 2000, Warner broke his little finger and was sidelined for five games.

The sheer number of cases of less than stellar performance or worse following a cover appearance is impressive at first. The author of the jinx story, Alexander Wolff, directed a team of researchers who examined almost all of the magazine's nearly 2,500 covers dating back to the first one, featuring Milwaukee Braves third baseman Eddie Mathews in August 1954. Mathews was injured shortly after that. In October 1982, Penn State was unbeaten and the cover featured its quarterback, Todd Blackledge. The next week Blackledge threw four interceptions against Alabama and Penn State lost big. The jinx struck Barry Bonds in late May 1993, seeming to knock him into a dry spell that reduced his batting average forty points in just two weeks.

I'll stop. The article cited case after case. More generally, the researchers found that within two weeks of a cover appearance, over a third of the honorees suffered injuries, slumps, or other misfortunes. Theories abound on the cause of the cover jinx, many having to do with players or teams choking under the added performance pressure.

A much better explanation is that no explanation is needed. It's what you would expect. People often attribute meaning to phenomena governed only by a regression to the mean, the mathematical tendency for an extreme value of an at least partially chance-dependent quantity to be followed by a value closer to the average. Sports and business are certainly chancy enterprises and thus subject to regression. So is genetics to an extent, and so very tall parents can be expected to have offspring who are tall, but probably not as tall as they are. A similar tendency holds for the children of very short parents.

If I were a professional darts player and threw one hundred darts at a target (or a list of companies in a newspaper's business section) during a tournament and managed to hit the bull's-eye (or a rising stock) a record-breaking eighty-three times, the next time I threw one hundred darts, I probably wouldn't do nearly as well. If featured on a magazine cover (*Sports Illustrated* or *Barron's*) for the eighty-three hits, I'd probably be adjudged a casualty of the jinx too.

Regression to the mean is widespread. The sequel to a great CD is usually not as good as the original. The same can be said of the novel after the best-seller, the proverbial sophomore slump, Tom Peters's excellent companies faring relatively badly after a few good years, and, perhaps, the fates of Bernie Ebbers of WorldCom, John Rigas of Adelphia, Ken Lay of Enron, Gary Winnick of Global Crossing, Jean-Marie Messier of Vivendi (to throw in a European), Joseph Nacchio of Qwest, and Dennis Kozlowski of Tyco—all CEOs of large companies who received adulatory coverage before their recent plunges from grace. (Satirewire.com refers to these publicity-fleeing, company-draining executives as the CEOnistas.)

There is a more optimistic side to regression. I suggest that *Sports Illustrated* consider featuring an established player who has had a particularly bad couple of months on its *back* cover. Then they could run feature stories on the *boost* associated with such appearances. *Barron's* could do the same thing with its back cover.

An expectation of a regression to the mean is not the whole story, of course, but there are dozens of studies suggesting that value investing, generally over a three-to-five year period, does result in better rates of return than, say, growth investing. It's important to remember, however, that the size of the effect varies with the study (not surprisingly, some studies find zero or a negative effect), transaction costs can eat up some or all of it, and competing investors tend to shrink it over time.

In chapter 6 I'll consider the notion of risk in general, but there is a particular sort of risk that may be relevant to value stocks. Invoking the truism that higher risks bring greater returns even in an efficient market, some have argued that value companies are risky because they're so colorless and easily ignored that their stock prices must be lower to compensate! Using "risky" in this way is risky, however, since it seems to explain too much and hence nothing at all.

Accounting Practices, WorldCom's Problems

Even if value investing made better sense than investing in broad-gauged index funds (and that is certainly not proved) a big problem remains. Many investors lack a clear understanding of the narrow meanings of the denominators in the P/E, P/B, and P/D ratios, and an uncritical use of these ratios can be costly.

People are easily bamboozled about numbers and money even in everyday circumstances. Consider the well-known story of the three men attending a convention at a hotel. They rent a booth for $30, and after they go to their booth, the manager realizes that it costs only $25 and that he's overcharged them. He gives $5 to the bellhop and directs him to give it back to the three men. Not knowing how to divide the $5 evenly, the bellhop decides to give $1 to each of the three men and pockets the remaining $2 for himself. Later that night the bellhop realizes that the men each paid $9 ($10 minus the $1 they received from him). Thus, since the $27 the men paid (3 × $9 = $27) plus the $2 that he took for himself sums to $29, the bellhop wonders what happened to the missing dollar. What did happen to it?

The answer, of course, is that there is no missing dollar. You can see this more easily if we assume that the manager originally made a bigger mistake, realizing after charging the men $30 that the booth costs only $20 and that he's overcharged them $10. He gives $10 to the bellhop and directs him to give it back to the three men. Not knowing how to divide the $10 evenly, the bellhop decides to give $3 to each of the three men and pockets the remaining $1 for himself. Later that night the bellhop realizes that the men each paid $7 ($10 minus the $3 they received from him). Thus, since the $21 the men paid (3 × $7 = $21) plus the $1 he took for himself sums to $22, the bellhop wonders what happened to the missing $8. In this case there's less temptation to think that there's any reason the sum should be $30.

If people are baffled by these "disappearances," and many are, what makes us so confident that they understand the accounting intricacies on the basis of which they may be planning to invest their hard-earned (or even easily earned) dollars? As the recent accounting scandals make clear, even a good understanding of these notions is sometimes of little help in deciphering the condition of a company's finances. Making sense of accounting documents and seeing how balance sheets, cash flow statements, and income statements feed into each other is not something investors often do. They rely instead on analysts and auditors, and this is why conflating the latter roles with those of investment bankers and consultants causes such concern.

If an accounting firm auditing a company also serves as a consultant to the company, there is a troubling conflict of interest. (A similar crossing of professional lines that is more upsetting to me has been curtailed by Eliot Spitzer, New York attorney general. One typical instance involved Jack Grubman, arguably the most influential analyst of telecommunications companies such as WorldCom, who was incestually entangled in the investments and underwriting of the very companies he was supposed to be dispassionately analyzing.)

A student's personal tutor who is paid to improve his or her performance should not also be responsible for grading the student's exams. Nor should an athlete's personal coach be the referee in a game in which the athlete competes. The situation may not be exactly the same since, as accounting firms have argued, different departments are involved in auditing and consulting. Nevertheless, there is at least the appearance of impropriety, and often enough the reality too.

Such improprieties come in many flavors. Enron's accounting feints and misdirections involving off-shore entities and complicated derivatives trading were at least subtle and almost elegant. WorldCom's moves, by contrast, were so simple and blunt that Arthur Andersen's seeming blindness is jaw-dropping. Somehow Andersen's auditors failed to note that WorldCom had classified \$3.8 billion in corporate expenses as capital investments. Since expenses are charged against profits as they are incurred, while capital investments are spread out over many years, this accounting "mistake" allowed WorldCom to report profits instead of losses for at least two years and probably longer. After this revelation, investigators learned that earnings were increased another \$3.3 billion by some combination of the same ruse and the shifting of funds from exaggerated one-time charges against earnings (bad debts and the like) back into earnings as the need arose, creating, in effect, a huge slush fund. Finally (almost finally?) in November 2002 the SEC charged WorldCom with inflating earnings by an additional \$2 billion, bringing the total financial misstatements to over \$9 billion! (Many comparisons with this sum are possible; one is that \$9 billion is more than twice the gross domestic product of Somalia.)

WorldCom's accounting fraud first came to light in June 2002, long after I had invested a lot of money in the company and passively watched as its value shriveled to almost nothing. Bernie Ebbers and company had not merely made \$1 disappear as in the puzzle above, but had presided over the vanishing of approximately $190 billion, the value of WorldCom's market capitalization in 1999—$64 a share times 3 billion shares. For this and many other reasons it might be argued that both the multi-trillion-dollar boom of the '90s and the comparably sized bust of the early '00s were largely driven by telecommunications. (With such gargantuan numbers it's important to remember the fundamental laws of financial estimation: A trillion dollars plus or minus a few dozen billion is still a trillion, just as a billion dollars plus or minus a few dozen million is still a billion.)

I was a victim, but the primary victimizer, I'm sorry to say, was not WorldCom management but myself. Putting so much money into one stock, failing to place stop-loss orders or to buy insurance puts, and investing on margin (puts and margin will be discussed in chapter 6) were foolhardy and certainly not based by the company's fundamentals. Besides, these fundamentals and other warning signs should have been visible even through the accounting smoke screen.

The primary indication of trouble was the developing glut in the telecommunications industry. Several commentators have observed that the industry's trajectory over the last decade resembled that of the railroad industry after the Civil War. The opening of the West, governmental inducements, and new technology led the railroads to build thousands of miles of unneeded track. They borrowed heavily, each company attempting to be the dominant player; their revenue couldn't keep pace with the rising debt; and the resulting collapse brought on an economic depression in 1873.

Substitute fiber-optic cable for railroad tracks, the opening of global markets for the opening of the West, the Internet for the intercontinental railroad network, and governmental inducements for governmental inducements, and there you have it. Millions of miles of unused fiber-optic cable costing billions of dollars were laid to capture the insufficiently burgeoning demand for online music and pet stores. In a nutshell: Debts increased, competition grew keener, revenue declined, and bankruptcies loomed. Happily, however, no depression, at least as of this writing.

In retrospect, it's clear that the situation was untenable and that WorldCom's accounting tricks and deceptions (as well as Global Crossing's and others') merely papered over what would soon have come to light anyway: These companies were losing a lot of money. Still, anyone can be forgiven for not recognizing the problem of overcapacity or for not seeing through the hype and fraudulent accounting. (Far less blameless, if I may self-flagellate again, were my dumb investing practices, for which WorldCom management and accountants certainly weren't responsible.) The real source of most people's dismay and apprehension, I suspect, derives less from accountants' malfeasance than from the market's continuing to flounder. If it were rising, interest in the various accounting reforms that have been proposed and enacted would rival the public's keen fascination with partial differential equations or Cantor's continuum hypothesis.

Reforms can only accomplish so much. There are countless ways for accountants to dissemble, many of which shade into legitimate moves, and this highlights a different tension running through the accounting profession. The precision and objectivity of its bookkeeping fit uncomfortably alongside the vagueness and subjectivity of many of its practices. Every day accountants must make judgments and determinations that are debatable—about the way to value inventory, the burdens of pensions and health care, the quantification of goodwill, the cost of warranties, or the classification of expenses—but once made, these judgments result in numbers, exact to the nearest penny, that seem indubitable.

The situation is analogous to that in applied mathematics where the appropriateness of a mathematical model is always vulnerable to criticism. Is this model the right one for this situation? Are these assumptions warranted? Once the assumptions are made and the model is adopted, however, the numbers and organizational clarity that result have an irresistible appeal. Responding to this appeal two hundred years ago, the German poet Goethe rapturously described accounting this way: "Double entry bookkeeping is one of the most beautiful discoveries of the human spirit."

Focusing only on the bookkeeping and the numerical output, however, and refusing to examine the legitimacy of the assumptions made, can be disastrous, both in mathematics and accounting. Recall the tribe of bear hunters who became extinct once they became expert in the complex calculations of vector analysis. Before they encountered mathematics, the tribesmen killed, with their bows and arrows, all the bears they could eat. After mastering vector analysis, they starved. Whenever they spotted a bear to the northeast, for example, they would fire, as vector analysis suggested, one arrow to the north and one to the east.

Even more important than the appropriateness of accounting rules and models is the transparency of these practices. It makes compelling sense, for example, for companies to count the stock options given to executives and employees as expenses. Very few do so, but as long as everyone knows this, the damage is not as great as it could be. Everyone knows what's going on and can adapt to it.

If an accounting practice is transparent, then an outside auditor who is independent and trusted can, when necessary, issue a statement analogous to the warning made by the independent and trusted matriarch from chapter 1. By making a bit of information common knowledge, an auditor (or the SEC) can alert everyone involved to a violation and stimulate remedial action. If the auditor is not independent or not trusted (as Harvey Pitt, the recently departed chairman of the SEC, was not), then he is simply another player and violations, although perhaps widely and mutually known, will not become commonly known (everyone knowing that everyone else knows it and knowing that everyone else knows they know it and so on), and no action will result. In a similar way, family secrets take on a different character and have some hope of being resolved when they become common knowledge rather than merely mutual knowledge. Family and corporate "secrets" (such as WorldCom's misclassification of expenses) are often widely known, just not talked about.

Transparency, trust, independence, and authority are all needed to make the accounting system work. They are all in great demand, but sometimes in short supply.

6 | Options, Risk, and Volatility

Consider a rather ugly mathematical physicist who goes to the same bar every evening, always takes the second to the last seat, and seems to speak toward the empty seat next to his as if someone were there. The bartender notes this, and on Valentine's Day when the physicist seems to be especially fervent in his conversation, he asks why he is talking into the air. The physicist scoffs that the bartender doesn't know anything about quantum mechanics. "There is no such thing as a vacuum. Virtual particles flit in and out of existence, and there is a non-zero probability that a beautiful woman will materialize and, when she does, I want to be here to ask her out." The bartender is baffled and asks why the physicist doesn't just ask one of the real women who are in the bar. "You never know. One of them might say yes." The physicist sneers, "Do you know how unlikely that is?"

Being able to estimate probabilities, especially minuscule ones, is essential when dealing with stock options. I'll soon describe the language of puts and calls, and we'll see why the January 2003 calls on WCOM at 15 have as much chance of ending up in the money as Britney Spears has of suddenly materializing before the ugly physicist.

Options and the Calls of the Wild

Here's a thought experiment: Two people (or the same person in parallel universes) have roughly similar lives until each undertakes some significant endeavor. The endeavors are equally worthy and equally likely to result in success, but one endeavor ultimately leads to good things for X and his family and friends, and the other leads to bad things for Y and his family and friends. It seems that X and Y should receive roughly comparable evaluations for their decision, but generally they won't. Unwarranted though it may be, X will be judged kindly and Y harshly. I tell this in part because I'd like to exonerate myself for my investing behavior by claiming status as a faultless Mr. Y, but I don't qualify.

By late January 2002, WCOM had sunk to about $10 per share, and I was feeling not only dispirited but guilty about losing so much money on it. Losing money in the stock market often induces guilt in those who have lost it, whether they've done anything culpable or not. Whatever your views on the randomness of the market, it's indisputable that chance plays a huge role, so it makes no sense to feel guilty about having called heads when a tails comes up. If this was what I'd done, I could claim to be a Mr. Y: It wouldn't have been my fault. Alas, as I mentioned, it does make sense to blame yourself for betting recklessly on a particular stock (or on options for it).

There is a term used on Wall Street to describe traders and others who "blow up" (that is, lose a fortune) and as a result become hollow, sepulchral figures. The term is "ghost" and I have developed more empathy for ghosts than I wanted to have. Often they achieve their funereal status by taking *unnecessary* risks, risks that they could and should have "diversified away." One perhaps counterintuitive way in which to reduce risk is to buy and sell stock options.

Many people think of stock options as slot machines, roulette wheels, or dark horse long shots; that is, as pure gambles. Others think of them as absurdly large inducements for people to stay with a company or as rewards for taking a company public. I have no argument with these characterizations, but much of the time an option is more akin to a boring old insurance policy. Just as one buys an insurance policy in case one's washing machine breaks down, one often buys options in case one's stock breaks down. They lessen risk, which is the bete noire, bugbear, and bane of investors' lives and the topic of this chapter.

How options work is best explained with a few numerical examples. (How they're misused is reserved for the next section.) Assume that you have 1,000 shares of AOL (just to give WCOM a rest), and it is selling at $20 per share. Although you think it's likely to rise in the long term, you realize there's a chance that it may fall significantly in the next six months. You could insure against this by buying 1,000 "put" options at an appropriate price. These would give you the right to sell 1,000 shares of AOL for, say, $17.50 for the next six months. If the stock rises or falls less than $2.50, the puts become worthless in six months (just as your washing machine warranty becomes worthless on its expiration if your machine has not broken down by then). Your right to sell shares at $17.50 is not attractive if the price of the stock is more than that. However, if the stock plunges to, say, $10 per share within the six-month period, your right to sell shares at $17.50 is worth at least $7.50 per share. Buying put options is a hedge against a precipitous decline in the price of the underlying stock.

As I was first writing this, only a few paragraphs and a few days after WCOM had fallen to $10, it fell to under $8 per share, and I wished I had bought a boatload of puts on it months before when they were dirt cheap.

In addition to put options, there are "call" options. Buying them gives you the right to buy a stock at a certain price within a specified period of time. You might be tempted to buy calls when you strongly believe that a stock, say Intel this time (abbreviated INTC), selling at $25 per share, will rise substantially during the next year. Maybe you can't afford to buy many shares of INTC, but you can afford to buy calls giving you the right to buy shares at, say, $30 during the next year. If the stock falls or rises less than $5 during the next year, the calls become worthless. Your right to buy shares at $30 is not attractive if the price of the stock is less than that. But if the stock rises to, say, $40 per share within the year, each call is worth at least $10. Buying call options is a bet on a substantial rise in the price of the stock. It is also a way to insure that you are not left out when a stock, too expensive to buy outright, begins to take off. (The figures $17.50 and $30 in the AOL and INTC examples above are the "strike" prices of the respective options; this is the price of the stock that determines the point at which the option has intrinsic value or is "in the money.")

One of the most alluring aspects of buying puts and calls is that your losses are limited to what you have paid for them, but the potential gains are unlimited in the case of calls and very substantial in the case of puts. Because of these huge potential gains, options probably induce a comparably huge amount of fantasy—countless investors thinking something like "the option for INTC with a $30 strike price costs around a dollar, so if the stock goes to $45 in the next year, I'll make 15 times my investment. And if it goes to $65, I'll make 35 times my investment." The attraction for some speculators is not much different from that of a lottery.

Although I've often quoted approvingly Voltaire's quip that lotteries are a tax on stupidity (or at least on innumeracy), yes, I did buy a boatload of now valueless WCOM calls. In fact, over the two years of my involvement with the stock, I bought many thousands of January 2003 calls on WCOM at $15. I thought that whatever problems the company had were temporary and that by 2003 it would right itself and, in the process, me as well. Call me an ugly physicist.

There is, of course, a market in puts and calls, which means that people sell them as well as buy them. Not surprisingly, the payoffs are reversed for sellers of options. If you sell calls for INTC with a strike price of $30 that expires in a year, then you keep your proceeds from the sale of the calls and pay nothing unless the stock moves above $30. If, however, the stock moves to, say, $35, you must supply the buyer of the calls with shares of INTC at $30. Selling calls is thus a bet that the stock will either decline or rise only slightly in a given time period. Likewise, selling puts is a bet that the stock will either rise or decline only slightly.

One common investment strategy is to buy shares of a stock and simultaneously sell calls on them. Say, for example, you buy some shares of INTC stock at $25 per share and sell six-month calls on them with a strike price of $30. If the stock price doesn't rise to $30, you keep the proceeds from the sale of the calls, but if the stock price does exceed $30, you can sell your own shares to the buyer of the calls, thus limiting the considerable risk in selling calls. This selling of "covered" calls (covered because you own the stock and don't have to buy it at a high price to satisfy the buyer of the call) is one of many hedges investors can employ to maximize their returns and minimize their risks.

More generally, you can buy and sell the underlying stock and mix and match calls and puts with different expiration dates and strike prices to create a large variety of potential profit and loss outcomes. These combinations go by names like "straddles," "strangles," "condors," and "butterflies," but whatever strange and contorted animal they're named for, like all insurance policies, they cost money. A surprisingly difficult question in finance has been "How does one place a value on a put or a call?" If you're insuring your house, some of the determinants of the policy premium are the replacement cost of the house, the length of time the policy is in effect, and the amount of the deductible. The considerations for a stock include these plus others having to do with the rise and fall of stock prices.

Although the practice and theory of insurance have a long history (Lloyd's of London dates from the late seventeenth century), it wasn't until 1973 that a way was found to rationally assign costs to options. In that year Fischer Black and Myron Scholes published a formula that, although much refined since, is still the basic valuation tool for options of all sorts. Their work and that of Robert Merton won the Nobel prize for economics in 1997.

Louis Bachelier, whom I mentioned in chapter 4, also devised a formula for options more than one hundred years ago. Bachelier's formula was developed in connection with his famous 1900 doctoral dissertation in which he was the first to conceive of the stock market as a chance process in which price movements up and down were normally distributed. His work, which utilized the mathematical theory of Brownian motion, was way ahead of its time and hence was largely ignored. His options formula was also prescient, but ultimately misleading. (One reason for its failure is that Bachelier didn't take account of the effect of compounding on stock returns. Over time this leads to what is called a "lognormal" distribution rather than a normal one.)

The Black-Scholes options formula depends on five parameters: the present price of the stock, the length of time until the option expires, the interest rate, the strike price of the option, and the volatility of the underlying stock. Without getting into the mechanics of the formula, we can see that certain general relations among these parameters are commonsensical. For example, a call that expires two years from now has to cost more than one that expires in three months since the later expiration date gives the stock more time to exceed the strike price. Likewise, a call with a strike price a point or two above the present stock price will cost more than one five points above the stock price. And options on a stock whose volatility is high will cost more than options on stocks that barely move from quarter to quarter (just as a short man on a pogo stick is more likely to be able to peek over a nine-foot fence than a tall man who can't jump). Less intuitive is the fact that the cost of an option also rises with the interest rate, assuming all other parameters remain unchanged.

Although there are any number of books and websites on the Black-Scholes formula, it and its variants are more likely to be used by professional traders than by gamblers, who rely on commonsense considerations and gut feel. Viewing options as pure bets, gamblers are generally as interested in carefully pricing them as casino-goers are in the payoff ratios of slot machines.

The Lure of Illegal Leverage

Because of the leverage possible with the purchase, sale, or mere possession of options, they sometimes attract people who aren't content to merely play the slots but wish to stick their thumbs onto the spinning disks and directly affect the outcomes. One such group of people are CEOs and other management personnel who stand to reap huge amounts of money if they can somehow contrive (by hook, crook, or, too often, by cooking the books) to raise their companies' stock price. Even if the rise is only temporary, the suddenly valuable call options can "earn" them tens of millions of dollars. This is the luxury version of "pump and dump" that has animated much of the recent corporate malfeasance.

(Such malfeasance might make for an interesting novel. On public television one sometimes sees a fantasia in which diverse historical figures are assembled for an imaginary conversation. Think, for example, of Leonardo da Vinci, Thomas Edison, and Benjamin Franklin discussing innovation. Sometimes a contemporary is added to the mix or simply paired with an illustrious precursor—maybe Karl Popper and David Hume, Stephen Hawking and Isaac Newton, or Henry Kissinger and Machiavelli. Recently I tried to think with whom I might pair a present-day ace CEO, investor, or analyst. There are a number of books about the supposed relevance to contemporary business practices of Plato, Aristotle, and other ancient wise men, but the conversation I'd be most interested in would be one between a current wheeler-dealer and some accomplished hoaxer of the past, maybe Dennis Koslowski and P. T. Barnum, or Kenneth Lay and Harry Houdini, or possibly Bernie Ebbers and Elmer Gantry.)

Option leverage works in the opposite direction as well, the options-fueled version of "short and distort." One particularly abhorrent example may have occurred in connection with the bombing of the World Trade Center. Just after September 11, 2001, there were reports that Al Qaeda operatives in Europe had bought millions of dollars worth of puts on various stock indices earlier in the month, reasoning that the imminent attacks would lead to a precipitous drop in the value of these indices and a consequent enormous rise in the value of their puts. They may have succeeded, although banking secrecy laws in Switzerland and elsewhere make that unclear.

Much more commonly, people buy puts on a stock and then try to depress its price in less indiscriminately murderous ways. A stockbroker friend of mine tells me, for example, of his fantasy of writing a mystery novel in which speculators buy puts on a company whose senior management is absolutely critical to the success of the company. The imaginary speculators then proceed to embarrass, undermine, and ultimately kill the senior management in order to reap the benefit of the soon-to-be valuable puts. The WorldCom chatroom, home to all sorts of utterly baseless rumors, once entertained a brief discussion about the possibility of WorldCom management having been blackmailed into doing all the ill-considered things they did on pain of having some awful secrets revealed. The presumption was that the blackmailers had bought WCOM puts.

Intricacies abound, but the same basic logic governing stock options is at work in the pricing of derivatives. Sharing only the same name as the notion studied in calculus, derivatives are financial instruments whose value is derived from some underlying asset—the stock of a company, commodities like cotton, pork bellies, and natural gas, or almost anything whose value varies significantly over time. They present the same temptation to directly change, affect, or manipulate conditions, and the opportunities for doing so are more varied and would also make for an intriguing business mystery novel.

The leverage involved in trading options and derivatives brings to mind a classic quote from Archimedes, who maintained that given a fulcrum, a long enough lever, and a place to stand, he could move the earth. The world-changing dreams that created the suggestively named WorldCom, Global Crossing, Quantum Group (George Soros' companies, no stranger to speculation), and others may have been similar in scope. The metaphorical baggage of levers and options is telling.

One can also look at seemingly non-financial situations and discern something like the buying, selling, and manipulating of options. For example, the practice of defraying the medical bills of AIDS patients in exchange for being made the beneficiary of their insurance policies has disappeared with the increased longevity of those with AIDS. However, if the deal were modified so that the parties put a time limit on their agreement, it could bc considered a standard option sale. The "option buyer" would pay a sum of money, and the patient/option seller would make the buyer the beneficiary for an agreed-upon period of time. If the patient happens not to expire within that time, the "option" does. Maybe another mystery novel here?

Less ghoulish variants of option buying, selling, and manipulating play an important role in everyday life from education and family planning to politics. Political options, better known as campaign contributions to relatively unknown candidates, usually expire worthless after the candidate loses the race. If he or she is elected, however, the "call option" becomes very valuable, enabling the contributor to literally call on the new officeholder. There is no problem with that, but direct manipulation of conditions that might increase the value of the political option is generally called "dirty tricks."

For all the excesses options sometimes inspire, they are generally a good thing, a valuable lubricant that enables prudent hedgers and adventurous gamblers to form a mutually advantageous market. It's only when the option holders do something to directly affect the value of the options that the lure of leverage turns lurid.

Short-Selling, Margin Buying, and Familial Finances

An old Wall Street couplet says, "He who sells what isn't his'n must buy it back or go to prison." The lines allude to "short-selling," the selling of stocks one doesn't own in the hope that the price will decline and one can buy the shares back at a lower price in the future. The practice is very risky because the price might rise precipitously in the interim, but many frown upon short-selling for another reason. They consider it hostile or anti-social to bet that a stock will decline. You can bet that your favorite horse wins by a length, not that some other horse breaks its leg. A simple example, however, suggests that short-selling can be a necessary corrective to the sometimes overly optimistic bias of the market.

Imagine that a group of investors has a variety of attitudes to the stock of company X, ranging from a very bearish 1 through a neutral 5 or 6 to a very bullish 10. In general, who is going to buy the stock? It will generally be those whose evaluations are in the 7 to 10 range. Their average valuation will be, let's assume, 8 or 9. But if those investors in the 1 to 4 range who are quite dubious of the stock were as likely to short sell X as those in the 7 to 10 range were to buy it, then the average valuation might be a more realistic 5 or 6.

Another positive way to look at short-selling is as a way to double the number of stock tips you receive. Tips about a bad stock become as useful as tips about a good one, assuming that you believe any tips. Short-selling is occasionally referred to as "selling on margin," and it is closely related to "buying on margin," the practice of buying stock with money borrowed from your broker.

To illustrate the latter, assume you own 5,000 shares of WCOM and it's selling at $20 per share (ah, remembrance of riches past). Since your investment in WCOM is worth $100,000, you can borrow up to this amount from your broker and, if you're very bullish on WCOM and a bit reckless, you can use it to buy an additional 5,000 shares on margin, making the total market value of your WCOM holdings $200,000 ($20 × 10,000 shares). Federal regulations require that the amount you owe your broker be no more than 50 percent of the total market value of your holdings. (Percentages vary with the broker, stock, and type of account.) This is no problem if the price of WCOM rises to $25 per share, since the $100,000 you owe your broker will then constitute only 40 percent of the $250,000 ($25 × 10,000) market value of your WCOM shares. But consider what happens if the stock falls to $15 per share. The $100,000 you owe now constitutes 67 percent of the $150,000 ($15 × 10,000) market value of your WCOM shares, and you will receive a "margin call" to deposit immediately enough money ($25,000) into your account to bring you back into compliance with the 50 percent requirement. Further declines in the stock price will result in more margin calls.

I'm embarrassed to reiterate that my devotion to WCOM (others may characterize my relationship to the stock in less kindly terms) led me to buy it on margin and to make the margin calls on it as it continued its long, relentless decline. Receiving a margin call (which often takes the literal form of a telephone call) is, I can attest, unnerving and confronts you with a stark choice. Sell your holdings and get out of the game now or quickly scare up some money to stay in it.

My first margin call on WCOM is illustrative. Although the call was rather small, I was leaning toward selling some of my shares rather than depositing yet more money in my account. Unfortunately (in retrospect), I needed a book quickly and decided to go to the Borders store in Center City, Philadelphia, to look for it. While doing so, I came across the phrase "staying in the game" while browsing and realized that staying in the game was what I still wanted to do. I realized too that Schwab was very close to Borders and that I had a check in my pocket.

My wife was with me, and though she knew of my investment in WCOM, at the time she was not aware of its extent nor of the fact that I'd bought on margin. (Readily granting that this doesn't say much for the transparency of my financial practices, which would not likely be approved by even the most lax Familial Securities Commission, I plead guilty to spousal deception.) When she went upstairs, I ducked out of the store and made the margin call. My illicit affair with WCOM continued. Occasionally exciting, it was for the most part anxiety-inducing and pleasureless, not to mention costly.

I took some comfort from the fact that my margin buying distantly mirrored that of WorldCom's Bernie Ebbers, who borrowed approximately $400 million to buy WCOM shares. (More recent allegations have put his borrowings at closer to $1 billion, some of it for personal reasons unrelated to WorldCom. Enron's Ken Lay, by contrast, borrowed only $10 to $20 million.) When he couldn't make the ballooning margin calls, the board of directors extended him a very low interest loan that was one factor leading to further investor unrest, massive sell-offs, and more trips to Borders for me.

Relatively few individuals short-sell or buy on margin, but the practice is very common among hedge funds—private, lightly regulated investment portfolios managed by people who employ virtually every financial tool known to man. They can short-sell, buy on margin, use various other sorts of leverage, or engage in complicated arbitrage (the near simultaneous buying and selling of the same stock, bond, commodity, or anything else, in order to profit from tiny price discrepancies). They're called "hedge funds" because many of them try to minimize the risks of wealthy investors. Others fail to hedge their bets at all.

A prime example of the latter is the collapse in 1998 of Long-Term Capital Management, a hedge fund, two of whose founding partners, Robert Merton and Myron Scholes, were the aforementioned Nobel prize winners who, together with Fischer Black, derived the celebrated formula for pricing options. Despite the presence of such seminal thinkers on the board of LTCM, the debacle roiled the world's financial markets and, had not emergency measures been enacted, might have seriously damaged them. (Then again, there is a laissez-faire argument for letting the fund fail.)

I admit I take a certain self-serving pleasure from this story since my own escapades pale by comparison. It's not clear, however, that the LTCM collapse was the fault of the Nobel laureates and their models. Many believe it was a consequence of a "perfect storm" in the markets, a vanishingly unlikely confluence of chance events. (The claim that Merton and Scholes were not implicated is nevertheless a bit disingenuous, since many invested in LTCM precisely because the fund was touting them and their models.)

The specific problems encountered by LTCM concerned a lack of liquidity in world markets, and this was exacerbated by the disguised dependence of a number of factors that were assumed to be independent. Consider, for illustration's sake, the likelihood that 3,000 specific people will die in New York on any given day. Provided that there is no connection among them, this is an impossibly minuscule number—a small probability raised to the 3,000th power. If most of the people work in a pair of buildings, however, the independence assumption that allows us to multiply probabilities fails. The 3,000 deaths are still extraordinarily unlikely, but not impossibly minuscule. Of course, the probabilities associated with possible LTCM scenarios were nowhere near as small and, according to some, could and should have been anticipated.

Are Insider Trading and Stock Manipulation So Bad?

It's natural to take a moralistic stance toward the corporate fraud and excess that have dominated business news the last couple of years. Certainly that attitude has not been completely absent from this book. An elementary probability puzzle and its extensions suggest, however, that some arguments against insider trading and stock manipulation are rather weak. Moral outrage, rather than actual harm to investors, seems to be the primary source of many people's revulsion toward these practices.

Let me start with the original puzzle. Which of the following two situations would you prefer to be in? In the first one you're given a fair coin to flip and are told that you will receive $1,000 if it lands heads and lose $1,000 if it lands tails. In the second you're given a very biased coin to flip and must decide whether to bet on heads or tails. If it lands the way you predict you win $1,000 and, if not, you lose $1,000. Although most people prefer to flip the fair coin, your chances of winning are 1/2 in both situations, since you're as likely to pick the biased coin's good side as its bad side.

Consider now a similar pair of situations. In the first one you are told you must pick a ball at random from an urn containing 10 green balls and 10 red balls. If you pick a green one, you win $1,000, whereas if you pick a red one, you lose $1,000. In the second, someone you thoroughly distrust places an indeterminate number of green and red balls in the urn. You must decide whether to bet on green or red and then choose a ball at random. If you choose the color you bet on, you win $1,000 and, if not, you lose $1,000. Again, your chances of winning are 1/2 in both situations.

Finally, consider a third pair of similar situations. In the first one you buy a stock that is being sold in a perfectly efficient market and your earnings are $1,000 if it rises the next day and –$1,000 if it falls. (Assume that in the short run it moves up with probability 1/2 and down with the same probability.) In the second there is insider trading and manipulation and the stock is very likely to rise or fall the next day as a result of these illegal actions. You must decide whether to buy or sell the stock. If you guess correctly, your earnings are

$1,000 and, if not, –$1,000. Once again your chances of winning are 1/2 in both situations. (They may even be slightly higher in the second situation since you might have knowledge of the insiders' motivations.)

In each of these pairs, the unfairness of the second situation is only apparent. You have the same chance of winning that you do in the first situation. I do not by any means defend insider trading and stock manipulation, which are wrong for many other reasons, but I do suggest that they are, in a sense, simply two among many unpredictable factors affecting the price of a stock.

I suspect that more than a few cases of insider trading and stock manipulation result in the miscreant guessing wrong about how the market will respond to his illegal actions. This must be depressing for the perpetrators (and funny for everyone else).

Expected Value, Not Value Expected

What can we anticipate? What should we expect? What's the likely high, low, and average value? Whether the quantity in question is height, weather, or personal income, extremes are more likely to make it into the headlines than are more informative averages. "Who makes the most money," for example, is generally more attention-grabbing than "what is the average income" (although both terms are always suspect because—surprise—like companies, people lie about how much money they make).

Even more informative than averages, however, are distributions. What, for example, is the distribution of all incomes and how spread out are they about the average? If the average income in a community is $100,000, this might reflect the fact that almost everyone makes somewhere between $80,000 and $120,000, or it might mean that a big majority earns less than $30,000 and shops at Kmart, whose spokesperson, the (too) maligned Martha Stewart, also lives in town and brings the average up to $100,000. "Expected value" and "standard deviation" are two mathematical notions that help clarify these issues.

An expected value is a special sort of average. Specifically, the expected value of a quantity is the average of its values, but weighted according to their probabilities. If, for example, based on analysts' recommendations, our own assessment, a mathematical model, or some other source of information, we assume that 1/2 of the time a stock will have a 6 percent rate of return, that 1/3 of the time it will have a –2 percent rate of return, and that the remaining 1/6 of the time it will have a 28 percent rate of return, then, on average, the stock's rate of return over any given six periods will be 6 percent three times, –2 percent twice, and 28 percent once. The expected value of its return is simply this probabilistically weighted average—(6% + 6% + 6% + (–2%) + (–2%) + 28%)/6, or 7%.

Rather than averaging directly, one generally obtains the expected value of a quantity by multiplying its possible values by their probabilities and then adding up these products. Thus $.06 \times 1/2 + (-.02) \times 1/3 + .28 \times 1/6 = .07$, or 7%, the expected value of the above stock's return. Note that the term "mean" and the Greek letter μ (mu) are used interchangeably with "expected value," so 7% is also the mean return, μ.

The notion of expected value clarifies a minor investing mystery. An analyst may simultaneously and without contradiction believe that a stock is very likely to do well but that, on average, it's a loser. Perhaps she estimates that the stock will rise 1 percent in the next month with probability 95 percent and that it will fall 60 percent in the same time period with probability 5 percent. (The probabilities might come, for example, from an appraisal of the likely outcome of an impending court decision.) The expected value of its price change is thus $(.01 \times .95) + (-.60) \times .05)$, which equals $-.021$ or an expected loss of 2.1%. The lesson is that the expected value, −2.1%, is not the value expected, which is 1%.

The same probabilities and price changes can also be used to illustrate two complementary trading strategies, one that usually results in small gains but sometimes in big losses, and one that usually results in small losses but sometimes in big gains. An investor who's willing to take a risk to regularly make some "easy money" might sell puts on the above stock, puts that expire in a month and whose strike price is a little under the present price. In effect, he's betting that the stock won't decline in the next month. Ninety-five percent of the time he'll be right, and he'll keep the put premiums and make a little money. Correspondingly, the buyer of the puts will lose a little money (the put premiums) 95 percent of the time. Assuming the probabilities are accurate, however, when the stock declines, it declines by 60 percent, and so the puts (the right to sell the stock at a little under the original price) become very valuable 5 percent of the time. The buyer of the puts then makes a lot of money and the seller loses a lot.

Investors can play the same game on a larger scale by buying and selling puts on the S&P 500, for example, rather than on any particular stock. The key to playing is coming up with reasonable probabilities for the possible returns, numbers about which people are as likely to differ as they are in their preferences for the above two strategies. Two exemplars of these two types of investor are Victor Niederhoffer, a well-known futures trader and author of *The Education of a Speculator*, who lost a fortune by selling puts a few years ago, and Nassim Taleb, another trader and the author of *Fooled by Randomness*, who makes his living by buying them.

For a more pedestrian illustration, consider an insurance company. From past experience, it has good reason to believe that each year, on average, one out of every 10,000 of its homeowners' policies will result in a claim of $400,000, one out of 1,000 policies will result in a claim of $60,000, one out of 50 will result in a claim of $4,000, and the remainder will result in a claim of $0. The insurance company would like to know what its average payout will be per policy written. The answer is the expected value, which in this case is ($400,000 × 1/10,000) + ($60,000 × 1/1,000) + ($4,000 × 1/50) + ($0 × 9,979/10,000) = $40 + $60 + $80 + $0 = $180. The premium the insurance company charges the homeowners will no doubt be at least $181.

Combining the techniques of probability theory with the definition of expected value allows for the calculation of more interesting quantities. The rules for the World Series of baseball, for example, stipulate that the series ends when one team wins four games. The rules further stipulate that team A plays in its home stadium for games 1 and 2 and however many of games 6 and 7 are necessary, whereas team B plays in its home stadium for games 3, 4, and, if necessary, game 5. If the teams are evenly matched, you might be interested in the expected number of games that will be played in each team's stadium. Skipping the calculation, I'll simply note that team A can expect to play 2.9375 games and team B 2.875 games in their respective home stadiums.

Almost any situation in which one can calculate (or reasonably estimate) the probabilities of the values of a quantity allows us to determine the expected value of that quantity. An example more tractable than the baseball problem concerns the decision whether to park in a lot or illegally on the street. If you park in a lot, the rate is $10 or $14, depending upon whether you stay for less than an hour, the probability of which you estimate to be 25 percent. You may, however, decide to park illegally on the street and have reason to believe that 20 percent of the time you will receive a simple parking ticket for $30, 5 percent of the time you will receive an obstruction of traffic citation for $100, and 75 percent of the time you will get off for free.

The expected value of parking in the lot is ($10 × .25) + ($14 × .75), which equals $13. The expected value of parking on the street is ($100 × .05) + ($30 × .20) + ($0 × .75), which equals $11. For those to whom this is not already Greek, we might say that μ_L, the mean costs of parking in the lot, and μ_S, the mean cost of parking on the street, are $13 and $11, respectively.

Even though parking in the street is cheaper on average (assuming money was your only consideration), the variability of what you'll have to pay there is much greater than it is with the lot. This brings us to the notion of standard deviation and stock risk.

What's Normal? Not Six Sigma

Risk in general is frightening, and the fear it engenders explains part of the appeal of quantifying it. Naming bogeymen tends to tame them, and chance is one of the most terrifying bogeyman around, at least for adults.

So how might one get at the notion of risk mathematically? Let's start with "variance," one of several mathematical terms for variability. Any chance-dependent quantity varies and deviates from its mean or average; it's sometimes more than the average, sometimes less. The actual temperature, for example, is sometimes warmer than the mean temperature, sometimes cooler. These deviations from the mean constitute risk and are what we want to quantify. They can be positive or negative, just as the actual temperature minus the mean temperature can be positive or negative, and hence they tend to cancel out. If we square them, however, the deviations are all positive, and we come to the definition: the variance of a chance-dependent quantity is the expected value of all its squared deviations from the mean. Before I numerically illustrate this, note the etymological/psychological association of risk with "deviation from the mean." This is a testament, I suspect, to our fear not only of risk but of anything unusual, peculiar, or deviant.

Be that as it may, let's switch from temperature back to our parking scenario. Recall that the mean cost of parking in the lot is \$13, and so $(\$10 - \$13)^2$ and $(\$14 - \$13)^2$, which equal \$9 and \$1, respectively, are the squares of the deviations of the two possible costs from the mean. They don't occur equally frequently, however. The first occurs with probability 25%, and the second occurs with probability 75%, and so the variance, the expected value of these numbers, is $(\$9 \times .25) + (\$1 \times .75)$, or \$3. More commonly used in statistical applications in finance and elsewhere is the square root of the variance, which is usually symbolized by the Greek letter σ (sigma). Termed the "standard deviation," it is in this case the square root of \$3, or approximately \$1.73. The standard deviation is (not exactly, but can be thought of as) the average deviation from the mean, and it is the most common mathematical measure of risk.

Forget the numerical examples if you like, but remember that, for any quantity, the larger the standard deviation, the more spread out its possible values are about the mean; the smaller it is, the more tightly the possible values cluster around the mean. Thus, if you read that in Japan the standard deviation of personal incomes is much less than it is in the United States, you should infer that Japanese incomes vary considerably less than U.S. incomes.

Returning to the street, you may wonder what the variance and standard deviation are of your parking costs there. The mean cost of parking in the street is \$11, and the squares of the deviations of the three possible costs from the mean are $(\$100 - \$11)^2$, $(\$30 - \$11)^2$, and $(\$0 - \$11)^2$, or \$7,921, \$361, and \$121, respectively. The first occurs with probability 5%, the second with probability 20%, and the third with probability 75%, and so the variance, the expected values of these numbers, is (\$7,921 × .05) + (\$361 × .20) + (\$0 × .75), or \$468.25. The square root of this gives us the standard deviation of \$21.64, more than twelve times the standard deviation of parking in the lot.

Despite this blizzard of numbers, I reiterate that all we have done is quantify the obvious fact that the possible outcomes of parking on the street are much more varied and unpredictable than those of parking in the lot. Even though the average cost of parking in the street (\$11) is less than that of parking in the lot (\$13), most would prefer to incur less risk and would therefore park in the lot for prudential reasons, if not moral ones.

This brings us to the market's use of standard deviation (sigma) to measure a stock's volatility. Let's use the same approach to calculate the variance of the returns for our stock that yields a rate of 6% about 1/2 the time, –2% about 1/3 of the time, and 28% the remaining 1/6 of the time. The mean or expected value of its returns is 7%, and so the squares of the deviations from the mean are $(.06 - .07)^2$, $(-.02 - .07)^2$, and $(.28 - .07)^2$ or .0001, .0081, and .0441, respectively. These occur with probabilities 1/2, 1/3, and 1/6, and so the variance, the expected value of the squares of these deviations from the mean, is (.0001 × 1/2) + (.0081 × 1/3) + (.0441 × 1/6), which is .01. The square root of .01 is .10 or 10%, and this is the standard deviation of the returns for this stock.

The Greek lesson again: The expected value of a quantity is its (probabilistically weighted) average and is symbolized by the letter μ (mu), and the standard deviation of a quantity is a measure of its variability and is symbolized by the letter σ (sigma). If the quantity in question is the rate of return on a stock price, its volatility is generally taken to be the standard deviation.

If there are only two or three possible values a quantity might assume, the standard deviation is not that helpful a notion. It becomes very useful, however, when a quantity can assume many different values and these values, as they often do, have an approximately normal bell-shaped distribution—high in the middle and tapering off on the sides. In this case, the expected value is the high point of the distribution. Moreover, approximately 2/3 of the values (68 percent) lie within one standard deviation of the expected value, and 95 percent of the values lie within two standard deviations of the expected value.

Before we go on, let's list a few of the quantities that have a normal distribution: age-specific heights and weights, natural gas consumption in a city for any given winter day, water use between 2 A.M. and 3 A.M. in a given city, thicknesses of a particular machined part coming off an assembly line, I.Q.s (whatever it is that they measure), the number of admissions to a large hospital on any given day, distances of darts from a bull's-eye, leaf sizes, nose sizes, the number of raisins in boxes of breakfast cereal, and possible rates of return for a stock. If we were to graph any of these quantities, we would obtain bell-shaped curves whose values are clustered about the mean.

Take as an example the number of raisins in a large box of cereal. If the expected number of raisins is 142 and the standard deviation is 8, then the high point of the bell-shaped graph would be at 142. About two-thirds of the boxes would contain between 134 and 150 raisins, and 95 percent of the boxes would contain between 126 and 158 raisins.

Or consider the rate of return of a conservative stock. If the possible rates are normally distributed with an expected value of 5.4 percent and a volatility (standard deviation, that is) of only 3.2 percent, then about two-thirds of the time, the rate of return will be between 2.2 percent and 8.6 percent, and 95 percent of the time the rate will be between –1 percent and 11.8 percent. You might prefer this stock to a more risky one with the same expected value but a volatility of, say, 20.2 percent. About two-thirds of the time, the rate of return of this more volatile stock will be between –14.8 percent and 25.6 percent, and 95 percent of the time it will be between –35 percent and 45.8 percent.

In all cases, the more standard deviations from the expected value, the more unusual the result. This fact helps account for the many popular books on management and quality control having the words "six sigma" in their titles. The covers of many of these books suggest that by following their precepts, you can attain results that are six standard deviations above the norm, leading, for example, to a minuscule number of product defects. A six-sigma performance is, in fact, so unlikely that the tables in most statistics texts don't even include values for it. If you look into the books on management, however, you learn that Sigma is usually capitalized and means something other than sigma, the standard deviation of a chance-dependent quantity. A new oxymoron: minor capital offense.

Whether they are defects, nose sizes, raisins, or water use in a city, almost all normally distributed quantities can be thought of as the average or sum of many factors (genetic, physical, social, or financial). This is not an accident: The so-called Central Limit Theorem states that averages and sums of a sufficient number of chance-dependent quantities are always normally distributed.

As we'll see in chapter 8, however, not everyone believes that stocks' rates of return are normally distributed.

7 | Diversifying Stock Portfolios

Long before my children's fascination with Super Mario Brothers, Tetris, and more recent addictive games, I spent interminable hours as a kid playing antediluvian, low-tech Monopoly with my two brothers. The game requires the players to roll dice and move around the board buying, selling, and trading real estate properties. Although I paid attention to the probabilities and expected values associated with various moves (but not to what have come to be called the game's Markov chain properties), my strategy was simple: Play aggressively, buy every property whether it made sense or not, and then bargain to get a monopoly. I always traded away railroads and utilities if I could, much preferring to build hotels on the real estate I owned instead.

A Reminiscence and a Parable

Although the game's get-out-of-jail-free card was one of the few ties to the present-day stock market, I've recently had a tiny epiphany. On some atavistic level I've likened hotel building to stock buying and the railroads and utilities to bonds. Railroads and utilities seemed safe in the short run, but the ostensibly risky course of putting most of one's money into building hotels was ultimately more likely to make one a winner (especially since we occasionally altered the rules to allow unlimited hotel building on a property).

Was my excessive investment in WorldCom a result of a bad generalization from playing Monopoly? I strongly doubt it, but such just-so stories come naturally to mind. Aside from the jail card, a board game called WorldCom would have few features in common with Monopoly (but might more closely resemble Grand Theft Auto). Different squares along players' paths would call for SEC investigations, Eliot Spitzer prosecutions, IPO giveaways, or favorable analyst ratings. If you attained CEO status, you would be allowed to borrow up to $400 million ($1 billion in later versions of the game), whereas if you were reduced to the rank of employee, you would have to pay a coffee fee after each move and invest a certain portion of your savings in company stock. If you were unfortunate enough to become a stockholder, you would be required to remove your shirt while playing, while if you became CFO, you would receive stock options and get to keep the stockholders' shirts. The object of the game would be to make as much money and collect as many of your fellow players' shirts as possible before the company went bankrupt.

The game might be fun with play money; it wasn't with the real thing.

Here's a better analogue for the market. People are milling around a huge labyrinthine bazaar. Occasionally some of the booths in the bazaar attract a swarm of people jostling to buy their wares. Likewise, some booths are occasionally devoid of any prospective customers. At any given time most booths have a few customers. At the intersections of the bazaar's alleys are sales people from some of the bigger booths as well as well-traveled seers. They know the various sections of the bazaar intimately and claim to be able to foretell the fortunes of various booths and collections of booths. Some of these sales people and some of the prognosticators have very large bullhorns and can be heard throughout the bazaar, while others make do by shouting.

In this rather primitive setting, many aspects of the stock market can already be discerned. The forebears of technical traders might be those who buy from booths where crowds are developing, while the forebears of fundamental traders might be those who coolly weigh the worth of the goods on display. The seers are the progenitors of analysts, the sales people progenitors of brokers. The bullhorns are a rudimentary form of business media, and, of course, the goods on sale are companies' stocks. Crooks and swindlers have their ancestors as well with some of the booths hiding their shoddy merchandise under the better goods.

If everyone, not just the booth owners, could sell as well as buy, this would be a better elemental model of an equities market. (I don't intend this as an historical account, but merely as an idealized narrative.) Nevertheless, I think it's clear that stock exchanges are natural economic phenomena. It's not hard to imagine early analogues of options trading, corporate bonds, or diversified holdings developing out of such a bazaar.

Maybe there'd even be some arithmeticians around too, analyzing booths' sales and devising purchasing strategies. In acting on their theories, some might even lose their togas and protractors.

Are Stocks Less Risky Than Bonds?

Perhaps because of Monopoly, certainly because of WorldCom, and for many other reasons, the focus of this book has been the stock market, not the bond market (or real estate, commodities, and other worthy investments). Stocks are, of course, shares of ownership in a company, whereas bonds are loans to a company or government, and "everybody knows" that bonds are generally safer and less volatile than stocks, although the latter have a higher rate of return. In fact, as Jeremy Siegel reports in *Stocks for the Long Run*, the average annual rate of return for stocks between 1802 and 1997 was 8.4 percent; the rate on treasury bills over the same period was between 4 percent and 5 percent. (The rates that follow are before inflation. What's needless to say, I hope, is that an 8 percent rate of return in a year of 15 percent inflation is much worse than a 4 percent return in a year of 3 percent inflation.)

Despite what "everybody knows," Siegel argues in his book that, as with Monopoly's hotels and railroads, stocks are actually less risky than bonds because, over the long run, they have performed so much better than bonds or treasury bills. In fact, the longer the run, the more likely this has been the case. (Comments like "everybody knows" or "they're all doing this" or "everyone's buying that" usually make me itch. My background in mathematical logic has made it difficult for me to interpret "all" as signifying something other than all.) "Everybody" does have a point, however. How can we believe Siegel's claims, given that the standard deviation for stocks' annual rate of return has been 17.5 percent?

If we assume a normal distribution and allow ourselves to get numerical for a couple of paragraphs, we can see how stomach-churning this volatility is. It means that about two-thirds of the time, the rate of return will be between –9.1 percent and 25.9 percent (that is, 8.4 percent plus or minus 17.5 percent), and about 95 percent of the time the rate will be between –26.6 percent and 43.4 percent (that is, 8.4 percent plus or minus two times 17.5 percent). Although the precision of these figures is absurd, one consequence of the last assertion is that the returns will be worse than –26.6 percent about 2.5 percent of the time (and better than 43.4 percent with the same frequency). So about once every forty years (1/40 is 2.5 percent), you will lose more than a quarter of the value of your stock investments and much more frequently than that do considerably worse than treasury bills.

These numbers certainly don't seem to indicate that stocks are less risky than bonds over the long term. The statistical warrant for Siegel's contention, however, is that over time, the returns even out and the deviations shrink. Specifically, the annualized standard deviation for rates of return over a number N of years is the standard deviation divided by the square root of N. The larger N is, the smaller is the standard deviation. (The cumulative standard deviation is, however, greater.) Thus over any given four-year period the annualized standard deviation for stock returns is 17.5%/2, or 8.75%. Likewise, since the square root of 30 is about 5.5, the annualized standard deviation of stock returns over any given thirty-year period is only 17.5%/5.5, or 3.2%. (Note that this annualized thirty-year standard deviation is the same as the annual standard deviation for the conservative stock mentioned in the example at the end of chapter 6.)

Despite the impressive historical evidence, there is no guarantee that stocks will continue to outperform bonds. If you look at the period from 1982 to 1997, the average annual rate of return for stocks was 16.7 percent with a standard deviation of 13.1 percent, while the returns for bonds were between 8 percent and 9 percent. But from 1966 to 1981, the average annual rate of return for stocks was 6.6 percent with a standard deviation of 19.5 percent, while the returns for bonds were about 7 percent.

So is it really the case that, despite the debacles, deadbeats, and doomsday equities like WCOM and Enron, the less risky long-term investment is in stocks? Not surprisingly, there is a counterargument. Despite their volatility, stocks as a whole have proven less risky than bonds over the long run because their average rates of return have been considerably higher. Their rates of return have been higher because their prices have been relatively low. And their prices have been relatively low because they've been viewed as risky and people need some inducement to make risky investments.

But what happens if investors believe Siegel and others, and no longer view stocks as risky? Then their prices will rise because risk-averse investors will need less inducement to buy them; the "equity-risk premium," the amount by which stock returns must exceed bond returns to attract investors, will decline. And the rates of return will fall because prices will be higher. And stocks will therefore be riskier because of their lower returns.

Viewed as less risky, stocks become risky; viewed as risky, they become less risky. This is yet another instance of the skittish, self-reflective, self-corrective dynamic of the market. Interestingly, Robert Shiller, a personal friend of Siegel, looks at the data and sees considerably lower stock returns for the next ten years.

Market practitioners as well as academics disagree. In early October 2002, I attended a debate between Larry Kudlow, a CNBC commentator and Wall Street fixture, and Bob Prechter, a technical analyst and Elliot wave proponent. The audience at the CUNY graduate center in New York seemed affluent and well-educated, and the speakers both seemed very sure of themselves and their predictions. Neither seemed at all affected by the other's diametrically opposed expectations. Prechter anticipated very steep declines in the market, while Kudlow was quite bullish. Unlike Siegel and Shiller, they didn't engage on any particulars and generally talked past each other.

What I find odd about such encounters is how typical they are of market discussions. People with impressive credentials regularly expatiate upon stocks and bonds and come to conclusions contrary to those of other people with equally impressive credentials. An article in the *New York Times* in November 2002 is another case in point. It described three plausible prognoses for the market—bad, so-so, and good—put forth by economic analysts Steven H. East, Charles Pradilla, and Abby Joseph Cohen, respectively. Such stark disagreement happens very rarely in physics or mathematics. (I'm not counting crackpots who sometimes receive a lot of publicity but aren't taken seriously by anybody knowledgeable.)

The market's future course may lie beyond what, in chapter 9, I term the "complexity horizon." Nevertheless, aside from some real estate, I remain fully vested in stocks, which may or may not result in my remaining fully shirted.

The St. Petersburg Paradox and Utility

Reality, like the perfectly ordinary woman in Virginia Woolf's famous essay "Mr. Bennett and Mrs. Brown," is endlessly complex and impossible to capture completely in any model. Expected value and standard deviation seem to reflect the ordinary meanings of average and variability most of the time, but it's not hard to find important situations where they don't.

One such case is illustrated by the so-called St. Petersburg paradox. It takes the form of a game that requires that you flip a coin repeatedly until a tail first appears. If a tail appears on the first flip, you win $2. If the first tail appears on the second flip, you win $4. If the first tail appears on the third flip, you win $8, and, in general, if the first tail appears on the Nth flip, you win 2^N dollars. How much would you be willing to pay to play this game? One could argue that you should be willing to pay any amount to play this game.

To see why this is so, recall that the probability of a sequence of independent events such as coin flips is obtained by multiplying the probabilities of each of the events. Thus the probability of getting the first tail, T, on the first flip is 1/2; of getting a head and then the first tail on the second flip, HT, is $(1/2)^2$ or 1/4; of getting the first tail on the third flip, HHT, is $(1/2)^3$ or 1/8; and so on. Putting these probabilities and the possible winnings associated with them into the formula for expected value, we see that the expected value of the game is $(\$2 \times 1/2) + (\$4 \times 1/4) + (\$8 \times 1/8) + (\$16 \times 1/16) + \ldots (2^N \times (1/2))^N + \ldots$. All of these products are 1, there are infinitely many of them, and so their sum is infinite. The failure of expected value to capture our intuitions becomes clear when you ask yourself why you'd be reluctant to pay even a measly $1,000 for the privilege of playing this game.

The most common resolution is roughly that provided by the eighteenth century mathematician Daniel Bernoulli, who wrote that people's enjoyment of any increase in wealth (or regret at any decrease) is "inversely proportionate to the quantity of goods previously possessed." The fewer dollars you have, the more you appreciate gaining one and the more you fear losing one, and so, for almost everyone, the likely prospect of losing $1,000 more than cancels the remote possibility that you'll win, say, a billion dollars.

What's important is the "utility" to you of the dollars that you receive, and this utility drops off as you receive more of them. (Note that this is not irrelevant to the rationale for progressive taxation.) For this reason people consider not the dollar amount involved in any investment (or game), but the utility of the dollar amount for the individual involved. The St. Petersburg paradox disappears, for example, if we consider a so-called logarithmic utility function, which attempts to reflect the slowly diminishing satisfaction of having more money and which results in the expected value of the game above being finite. Other versions of the game, in which the payoffs increase even faster, require even slower-growing utility functions so that the expected value remains finite.

People do differ in their utility assignments. Some are so acquisitive that the 741,783,219th dollar is almost as dear to them as the first; others are so laid back that their 25,000th dollar is almost worthless to them. There are probably relatively few of the latter, although my father in his later years came close. His attitude suggests that utility functions vary not only across people but also over time. Furthermore, utility may not be so easily described by simple functions since, for example, there may be variations in the utility of money as one approaches a certain age or reaches some financial milestone such as X million dollars. And we're back to Virginia Woolf's essay.

Portfolios: Benefiting from the Hatfields and McCoys

John Maynard Keynes wrote, "Practical men, who believe themselves to be quite exempt from any intellectual influences, are usually the slaves of some defunct economist. Madmen in authority, who hear voices in the air, are distilling their frenzy from some academic scribbler of a few years back." A corollary of this is that fund managers and stock gurus, who slickly dispense their investment ideas and advice, generally derive them from a previous generation's Nobel prize-winning finance professor.

To get a taste of what a couple more of these Nobelists have written, assume you're a fund manager intent on measuring the expected return and volatility (risk) of a portfolio. In stock market contexts a portfolio is simply a collection of different stocks—a mutual fund, for example, or Uncle Jake's ragbag of mysterious picks, or a nightmare inheritance containing a bunch of different stocks, all in telecommunications. Portfolios like the latter that are so lacking in diversification often become portfolios lacking in dollars. How can you more judiciously choose stocks to maximize a portfolio's returns and minimize its risks?

Let's first envision a simple portfolio consisting of only three stocks, Abbey Roads, Barkley Hoops, and Consolidated Fragments. Let's further assume that 40 percent (or $40,000) of a $100,000 portfolio is in Abbey, 25 percent in Barkley, and the remaining 35 percent in Consolidated. Assume further that the expected rate of return from Abbey is 8 percent, from Barkley is 13 percent, and from Consolidated is 7 percent. Using these weights, we compute that the expected return from the portfolio as a whole is $(.40 \times .08) + (.25 \times .13) + (.35 \times .07)$, which is .089 or 8.9 percent.

Why not put all our money in Barkley Hoops since its expected rate of return is the highest of the three stocks? The answer has to do with volatility and the risk of not diversifying, of putting all one's proverbial eggs in one basket. (The result, as was the case with my WorldCom misadventure, may well be egg on one's face and the transformation of one's nest egg into a scrambled egg if not a goose egg. Sorry, but thought of the stock even now sometimes momentarily unhinges me.) If you were indifferent to risk, however, and simply wanted to maximize your returns, you might well put all your money in Barkley Hoops.

So how does one determine the volatility—that is, sigma, the standard deviation—of a portfolio? Does one just weight the volatilities of the companies' stocks as we weighted their returns to get the volatility of the portfolio? In general, we can't do this because the stocks' performances are sometimes not independent of each other. When one goes up in response to some news, the others' chances of going up or down may be affected and this in turn affects their joint volatility.

Let me illustrate with an even simpler portfolio consisting of only two stocks, Hatfield Enterprises and McCoy Productions. They both produce thingamajigs, but history tells us that when one does well, the other suffers and vice versa, and that overall dominance seems to shift regularly back and forth between them. Perhaps Hatfield produces snow shovels and McCoy makes tanning lotion. To be specific, let's say that half the time Hatfield's rate of return is 40 percent and half the time it is –20 percent, so its expected rate of return is $(.50 \times .40) + (.50 \times (-.20))$, which is .10 or 10 percent. McCoy's returns are the same, but again it does well when Hatfield does poorly and vice versa.

The volatility of each company is the same too. Recalling the definition, we first find the squares of the deviations from the mean of 10 percent, or .10. These squares are $(.40 - .10)^2$ and $(-.20 - .10)^2$ or .09 and .09. Since they each occur half the time, the variance is $(.50 \times .09) + (.50 \times .09)$, which is .09. The square root of this is .3 or 30 percent, which is the standard deviation or volatility of each company's returns.

But what if we don't choose one or the other to invest in, but split our investment funds and buy half as much of each stock? Then we're always earning 40 percent from half our investment and losing 20 percent on the other half, and our expected return is still 10 percent. But notice that this 10 percent return is constant. The volatility of the portfolio is zero! The reason is that the returns of these two stocks are not independent, but are perfectly negatively correlated. We get the same average return as if we bought either the Hatfield or the McCoy stock, but with no risk. This is a good thing; we get richer and don't have to worry about who's winning the battle between the Hatfields and the McCoys.

Of course, it's difficult to find stocks that are perfectly negatively correlated, but that is not required. As long as they aren't perfectly positively correlated, the stocks in a portfolio will decrease volatility somewhat. Even a portfolio of stocks from the same sector will be less volatile than the individual stocks in it, while a portfolio consisting of Wal-Mart, Pfizer, General Electric, Exxon, and Citigroup, the biggest stocks in their respective sectors, will provide considerably more protection against volatility. To find the volatility of a portfolio in general, we need what is called the "covariance" (closely related to the correlation coefficient) between any pair of stocks X and Y in the portfolio. The covariance between two stocks is roughly the degree to which they vary together—the degree, that is, to which a change in one is proportional to a change in the other.

Note that unlike many other contexts in which the distinction between covariance (or, more familiarly, correlation) and causation is underlined, the market generally doesn't care much about it. If an increase in the price of ice cream stocks is correlated to an increase in the price of lawn mower stocks, few ask whether the association is causal or not. The aim is to use the association, not understand it—to be right about the market, not necessarily to be right for the right reasons.

Given the above distinction, some of you may wish to skip the next three paragraphs on the calculation of covariance. Go directly to "For example, if we let H be the cost . . . ."

Technically, the covariance is the expected value of the product of the deviation from the mean of one of the stocks and the deviation from the mean of the other stock. That is, the covariance is the expected value of the product $[(X - \mu_X) \times (Y - \mu_Y)]$, where μ_X and μ_Y are the means of X and Y, respectively. Thus, if the stocks vary together, when the price of one is up, the price of the other is likely to be up too, so both deviations from the mean will be positive, and their product will be positive. And when the price of one is down, the price of the other is likely to be down too, so both deviations will be negative, and their product will again be positive. If the stocks vary inversely, however, when the price of one is up (or down), the price of the other is likely to be down (or up), so when the deviation of one stock is positive, that of the other is negative, and the product will be negative. In general and in short, we want negative covariance.

We may now use this notion of covariance to find the *variance* of a two-equity portfolio, p percent of which is in stock X and q percent in stock Y. The mathematics involves nothing more than squaring the sum of two terms. (Remember, however, that $(A + B)^2 = A^2 + B^2 + 2AB$.) By definition, the variance of the portfolio, $(pX + qY)$, is the expected value of the squares of its deviations from its mean, $p\mu_X + q\mu_Y$. That is, the variance of $(pX + qY)$ is the expected value of $[(pX + qY) - (p\mu_X + q\mu_Y)]^2$, which, upon rewriting, is the expected value of $[(pX - p\mu_X) + (qY - q\mu_Y)]^2$, which, using the algebra rule cited above, is the expected value of $[(pX - p\mu_X)^2 + (qY - q\mu_Y)^2 + 2 \times$ the expected value of $[(pX - p\mu_X) \times (qY - q\mu_Y)]$.

Minding (that is, factoring out) our p's and q's, we find that the variance of the portfolio, $(pX + qY)$, equals [($p^2 \times$ the variance of X) + ($q^2 \times$ the variance of Y) + ($2pq \times$ the covariance of X and Y)]. If the stocks vary negatively (that is, have negative covariance), the variance of the portfolio is reduced by the last factor. (In the case of the Hatfield and McCoy stocks, the variance was reduced to zero.) And when they vary positively (that is, have positive covariance), the variance of the portfolio is increased by the last factor, a situation we want to avoid, volatility and risk being bad for our peace of mind and stomach.

For example, if we let H be the cost of a randomly selected homeowner's house in a given community and I be his or her household income, then the variance of $(H + I)$ is greater than the variance of H plus the variance of I. People who live in expensive houses generally have higher incomes than people who don't, so the extremes of the sum, house cost plus personal income, are going to be considerably greater than they would be if house cost and personal income did not have a positive covariance.

Likewise, if C is the number of classes skipped during the year by a randomly selected student in a large lecture and S is his score on the final exam, then the variance of (C + S) is smaller than the variance of C plus the variance of S. Students who miss a lot of classes generally (although certainly not always) achieve a lower score, so the extremes of the sum, number of classes missed plus exam scores, are going to be considerably less that they would be if number of classes missed and exam scores did not have a negative covariance.

When choosing stocks for a diversified portfolio, investors, as noted, generally look for negative covariances. They want to own equities like the Hatfield and the McCoy stocks and not like WCOM, say, and some other telecommunications stock. With three or more stocks in a portfolio, one uses the stocks' weights in the portfolio as well as the definitions just discussed to compute the portfolio's variance and standard deviation. (The algebra is tedious, but easy.) Unfortunately, the covariances between all possible pairs of stocks in the portfolio are needed for the computation, but good software, troves of stock data, and fast computers allow investors to determine a portfolio's risk (volatility, standard deviation) fairly quickly. With care, you can minimize the risk of a portfolio without hurting its expected rate of return.

Diversification and Politically Incorrect Funds

There are countless mutual funds, and many commentators have noted that there are more funds than there are stocks, as if this were a surprising fact. It isn't. In mathematical terms a fund is simply a set of stocks, so, theoretically at least, there are vastly more possible funds than there are stocks. Any set of n stocks (people, books, CDs) has 2^N subsets. Thus, if there were only 20 stocks in the world, there would be 2^{20} or approximately 1 million possible subsets of these stocks—1 million possible mutual funds. Of course, most of these subsets would not have a compelling reason for existence. Something more is needed, and that is the financial balancing act that ensures diversification and low volatility.

We can increase the number of possibilities even further by extending the notion of diversification. Instead of searching for individual stocks or whole sectors that are negatively correlated, we can search for concerns of ours that are negatively correlated. Say, for example, financial and social ones. A number of portfolios purport to be socially progressive and politically correct, but in general their performance is not stellar. Less appealing to many are funds that are socially regressive and politically incorrect but that do perform well. In this latter category many people would place tobacco, alcohol, defense contractors, fast food, or any of several others.

The existence of these politically incorrect funds suggests, for those passionately committed to various causes, a nonstandard strategy that exploits the negative correlation that sometimes exists between financial and social interests. Invest heavily in funds holding shares in companies that you find distasteful. If these funds do well, you make money, money that you could, if you wished, contribute to the political causes you favor. If these funds cool off, you can rejoice that the companies are no longer thriving, and your psychic returns will soar.

Such "diversification" has many applications. People often work for organizations, for example, whose goals or products they find unappealing and use part of their salary to counter the organization's goals or products. Taken to its extreme, diversification is something we do naturally in dealing with the inevitable trade-offs in our daily lives.

Of course, extending the notion of diversification to these other realms is difficult for several reasons. One is that quantifying contributions and payoffs is problematic. How do you place a numerical value on your efforts and their various consequences? The number of possible "funds," subsets of all your possible concerns, also grows exponentially.

Another problem derives from the logic of the notion of diversification. It often makes sense in life, where some combination of work, play, family, personal experiences, study, friends, money, and so forth, seems more likely to lead to satisfaction than, say, all toil or pure hedonism. Nevertheless, diversification may not be appropriate when you are trying to have a personal impact. Take charity, for example.

As the economist Steven Landsburg has argued, you diversify when investing to protect yourself, but when contributing to large charities in which your contributions are a small fraction of the total, your goal is presumably to help as much as possible. Since you incur no personal risk, if you truly think that Mothers Against Drunk Driving is more worthy than the American Cancer Society or the American Heart Association, why would you split your charitable dollars among them? The point isn't to insure that your money will do some good, but to maximize the good it will do. There are other situations too where bulleting one's efforts is preferable to a bland diversification.

Metaphorical extensions of the notion of diversification can be useful, but uncritical use of them can lead you to, in the words of W. H. Auden, "commit a social science."

Beta—Is It Better?

Returning to more quantitative matters, we choose stocks so that when some are down, others are up (or at least not as down), giving us a healthy rate of return with as little risk as possible. More precisely, given any portfolio of stocks, we grind the numbers describing their past performances and come up with estimates for their expected returns, volatilities, and covariances, and then use these to determine the expected returns and volatilities of the portfolio as a whole. We could, if we had the time, the price data, and fast computers, do this for a variety of different portfolios. The Nobel prize-winning economist Harry Markowitz, one of the originators of this approach, developed mathematical techniques for carrying out these calculations in the early 1950s, graphed his results for a few portfolios (computers weren't fast enough to do much more then), and defined what he called the "efficient frontier" of portfolios.

If we were to use these techniques and construct comparable graphs for a wide variety of contemporary portfolios, what would we find? Arraying the (degree of) volatility of these portfolios along the graph's horizontal axis and their expected rates of return along its vertical axis, we would see a swarm of points. Each point would represent a portfolio whose coordinates would be its volatility and expected return, respectively. We'd also notice that among all the portfolios having a given level of risk (that is, volatility, standard deviation), there would be one with the highest expected rate of return. If we single out the portfolio with the highest expected rate of return for each level of risk, we would obtain a curve, Markowitz's efficient frontier of optimal portfolios.

The more risky a portfolio on the efficient frontier curve is, the higher is its expected return. In part, this is because most investors are risk-averse, making risky stocks cheaper. The idea is that investors decide upon a risk level with which they're comfortable and then choose the portfolio with this risk level that has the highest possible return. Call this Variation One of the theory of portfolio selection.

Don't let this mathematical formulation blind you to the generality of the psychological phenomenon. Automobile engineers have noted, for example, that safety advances in automobile design (say anti-lock brakes) often result in people driving faster and turning more sharply. Their driving performance is enhanced rather than their safety. Apparently, people choose a risk level with which they're comfortable and then seek the highest possible return (performance) for it.

Inspired by this trade-off between risk and return, William Sharpe proposed in the 1960s what is now a common measure of the performance of a portfolio. It is defined as the ratio of the excess return of a portfolio (the difference between its expected return and the return on a risk-free treasury bill) to the portfolio's volatility (standard deviation). A portfolio might have a hefty rate of return, but if the volatility the investor must endure to achieve this return is roller coasterish, the portfolio's Sharpe measure won't be very high. By contrast, a portfolio with a moderate rate of return but a less anxiety-inducing volatility will have a higher Sharpe measure.

There are many complications to portfolio selection theory. As the Sharpe measure suggests, an important one is the existence of risk-free investments, such as U.S. treasury bills. These pay a fixed rate of return and have essentially zero volatility. Investors can always invest in such risk-free assets and can borrow at the risk-free rates as well. Moreover, they can combine risk-free investment in treasury bills with a risky stock portfolio.

Variation Two of portfolio theory claims that there is one and only one optimal stock portfolio on the efficient frontier with the property that some combination of it and a risk-free investment (ignoring inflation) constitute a set of investments having the highest rates of return for any given level of risk. If you wish to incur no risk, you put all your money into treasury bills. If you're comfortable with risk, you put all your money into this optimal stock portfolio. Alternatively, if you want to divide your money between the two, you put p% into the risk-free treasury bills and (100 – p)% into the optimal risky stock portfolio for an expected rate of return of [p × (risk-free return) + (1 – p) × (stock portfolio)]. An investor can also invest more money than he has by borrowing at the risk-free rate and putting this borrowed money into the risky portfolio.

In this refinement of portfolio selection, all investors choose the same optimal stock portfolio and then adjust how much risk they're willing to take by increasing or decreasing the percentage, p, of their holdings that they put into risk-free treasury bills.

This is easier said than done. In both variations the required mathematical procedures put enormous pressure on one's computing facilities, since countless calculations must be performed regularly on new data. The expected returns, variances, and covariances are, after all, derived from their values in the recent past. If there are twenty stocks in a portfolio, we would need to compute the covariance of every possible pair of stocks, and there are (20 × 19)/2, or 190, such covariances. If there were fifty stocks, we'd need to compute (50 × 49)/2, or 1,225 covariances. Doing this for each of a wide class of portfolios is not possible without massive computational power.

As a way to avoid much of the computational burden of updating and computing all these covariances, efficient frontiers, and optimal risky portfolios, Sharpe, yet another Nobel Prize winner in economics, developed (with others) what's called the "single index model." This Variation Three relates a portfolio's rate of return not to that of all possible pairs of stocks in the portfolio, but simply to the change in some index representing the market as a whole. If your portfolio or stock is statistically determined to be relatively more volatile than the market as a whole, then changes in the market will bring about exaggerated changes in the stock or portfolio. If it is relatively less volatile than the market as a whole, then changes in the market will bring about attenuated changes in the stock or portfolio.

This brings us to the so-called Capital Asset Pricing Model, which maintains that the expected excess return on one's stock or portfolio (the difference between the expected return on the portfolio, R_p, and the return on risk-free treasury bills, R_f) is equal to the notorious beta, symbolized by β, multiplied by the expected excess return of the general market (the difference between the market's expected return, R_m, and the return on risk-free treasury bills, R_f). In algebraic terms: $(R_p - R_f) = \beta(R_m - R_f)$. Thus, if you can get a sure 4 percent on treasury bills and if the expected return on a broad market index fund is 10 percent and if the relative volatility, beta, of your portfolio is 1.5, then the portfolio's expected return is obtained by solving $(R_p - 4\%) = 1.5(10\% - 4\%)$, which yields 13 percent for R_p. A beta of 1.5 means that your stock or portfolio gains (or loses) an average of 1.5 percent for every 1 percent gain (or loss) in the market as a whole.

Betas for the stocks of high-tech companies like WorldCom are often considerably more than 1, meaning that changes in the market, both up and down, are magnified. These stocks are more volatile and thus riskier. Betas for utility company stocks, by contrast, are often less than 1, which means that changes in the market are muted. If a company has a beta of .5, then its expected return is obtained by solving $(R_p - 4\%) = .5(10\% - 4\%)$, which yields 7% for R_p, the expected return on the portfolio. Note that for short-term treasury bills, whose returns don't vary at all, beta is 0. To reiterate: Beta quantifies the degree to which a stock or a portfolio fluctuates in relation to market fluctuations. It is not the same as volatility.

This all sounds neat and clean, but you beta watch your step with all of these portfolio selection models. Specifically with regard to Variation Three, we might wonder where the number beta comes from. Who says your stock or portfolio will be 40 percent more volatile or 25 percent less volatile than the market as a whole? Here's the rough technique for finding beta. You check the change in the broad market for the last three months—say it's 3 percent—and check the change in the price of your stock or portfolio for the same period—say it's 4.1 percent. You do the same thing for the three months before that—say the numbers this time are 2 percent and 2.5 percent, respectively—and for the three months before that—say –1.2 percent and –3 percent, respectively. You continue doing this for a number of such periods and then on a graph you plot the points (3%, 4.1%), (2%, 2.5%), (–1.2%, –3%), and so on. Most of the time if you squint hard enough, you'll see a sort of linear relationship between changes in the market and changes in your stock or portfolio, and you then use standard mathematical methods for determining the line of closest fit through these points. The slope or steepness of this line is beta.

One problem with beta is that companies change over time, sometimes rather quickly. AT&T, for example, or IBM is not the same company it was twenty years ago or even two years ago. Why should we expect a company's relative volatility, beta, to remain the same? In the opposite direction is a related difficulty. Beta is often of very limited value in the short term and varies with the index chosen for comparison and the time period used in its definition. Still another problem is that beta depends on market returns, and market returns depend on a narrow definition of the market, namely just the stock market rather than stocks, bonds, real estate, and so forth. For all its limitations, however, beta can be a useful notion if it's not turned into a fetish.

You might compare beta to different people's emotional reactivity and expressiveness. Some respond to the slightest good news with outbursts of joy and to the tiniest hardship with wails of despair. At the other end of the emotional spectrum are those who say "ouch" when they accidentally touch a scalding iron and allow themselves an "oh, good" when they win the lottery. The former have high emotional beta, the latter low emotional beta. A zero beta person would have to be unconscious, perhaps from ingesting too many beta-blockers. Unfortunate for the prospect of predicting the behavior of people, however, is the commonplace that people's emotional betas vary depending on the type of stimulus a person faces. I'll leave out the examples, but this may be beta's biggest limitation as a measure of the relative volatility of a portfolio or stock. Betas may vary with the type of stimulus a company faces.

Whatever refinements of portfolio theory are developed, one salient point remains: Portfolios, although often less risky than individual stocks, are still risky (as millions of 401(k) returns attest). Some mathematical manipulation of the notions of variance and covariance and a few reasonable assumptions are sufficient to show that this risk can be partitioned into two parts. There is a systematic part that is related to general movements in the market, and there is a non-systematic part that is idiosyncratic to the stocks in the portfolio. The latter, non-systematic risk, specific to the individual stocks in the portfolio, can be eliminated or "diversified away" by an appropriate choice of thirty or so stocks. An irreducible core, however, remains inherent in the market and cannot be avoided. This systematic risk depends on the beta of one's portfolio.

Or so the story goes. To the criticisms of beta above should be added the problems associated with forcing a non-linear world into a linear mold.

8 | Connectedness and Chaotic Price Movements

Near the end of my involvement with WorldCom, when I was particularly concerned about what the new day would bring, I would sometimes wake up very early, grab a Diet Coke, and check how the stock was faring on the German or English exchanges. As the computer was booting up, I grew more and more apprehensive. The European response to bad overnight news sometimes prefigured Wall Street's response, and I dreaded seeing a steeply downward-sloping graph pop up on my screen. More often the European exchanges treaded water on WCOM until trading began in New York. Occasionally I'd be encouraged when the stock was up there, but I soon learned that the small volume sold on overseas exchanges didn't always mean much.

Whether haunted by a bad investment or not, we're all connected. No investor is an island (or even a peninsula). Stated mathematically, this means that statistical independence often fails; your actions affect mine. Most accounts of the stock market acknowledge in a general way that we learn from and respond to one another, but a better understanding of the market requires that one's models reflect the complexity of investors' interaction. In a sense, the market *is* the interaction. Stocks R Us. Before discussing some of the consequences of this complexity, let me consider three such sources for it: one micro, one macro, and the third mucro (yup, it's a word).

The micro example involves insider trading, which has always struck me as an odd sort of crime. Few people who aren't psychopaths daydream about murder or burglary, but many investors, I suspect, fantasize about coming upon inside information and making a bundle from it. The thought of finding myself on a plane next to Bernie Ebbers and Jack Grubman (assuming they flew economy class on commercial airliners) and overhearing their conversation about an impending merger or IPO offering, for example, did cross my mind a few times. Insider trading seems the limit or culmination of what investors and traders do naturally: getting all the information possible and acting on it before others see and understand what they see and understand.

Insider Trading and Subterranean Information Processing

The kind of insider trading I want to consider is relevant to seemingly unexplained price movements. It's also related to good poker playing, which may explain why the training program of at least one very successful hedge fund has a substantial unit on the game. The strategies associated with poker include learning not only the relevant probabilities but also the bluffing that is a necessary part of the game. Options traders often deal with relatively few other traders, many of whom they recognize, and this gives rise to the opportunity for feints, misdirection, and the exploitation of idiosyncrasies.

The example derives from the notion of common knowledge introduced in chapter 1. Recall that a bit of information is common knowledge among a group of people if they all know it, know that the others know it, know that the others know that they know it, and so on. Robert Aumann, who first defined the notion, proved a theorem that can be roughly paraphrased as follows: Two individuals cannot forever agree to disagree. As their beliefs, formed in rational response to different bits of private information, gradually become common knowledge, the beliefs change and eventually coincide.

When private information becomes common knowledge, it induces decisions and actions. As anyone who has overheard teenagers' gossip with its web of suppositions can attest, this transition to common knowledge sometimes relies on convoluted inferences about others' beliefs. Sergiu Hart, an economist at Hebrew University and one of a number of people who have built on Aumann's result, demonstrates this with an example relevant to the stock market. Superficially complicated, it nevertheless requires no particular background besides an ability to decode gossip, hearsay, and rumor and decide what others really think.

Hart asks us to consider a company that must make a decision. In keeping with the WorldCom leitmotif, let's suppose it to be a small telecommunications company that must decide whether to develop a new handheld device or a cell phone with a novel feature. Assume that the company is equally likely to decide on one or the other of these products, and assume further that whatever decision it makes, the product chosen has a 50 percent chance of being successful, say being bought in huge numbers by another company. Thus there are four equally likely outcomes: Handheld+, Handheld–, Phone+, Phone– (where Handheld+ means the handheld device was chosen for development and it was a success, Handheld– means the handheld was chosen but it turned out to be a failure, and similarly for Phone+ and Phone–).

Let's say there are two influential investors, Alice and Bob. They both decide that at the current stock price, if the chances of success of this product development are better than 50 percent, they should (continue to) buy, and if they're 50 percent or less they should (continue to) sell.

Furthermore, they are each privy to a different piece of information about the company. Because of her inside contacts, Alice knows which product decision was made, Handheld or Phone, but not whether it was successful or not.

Bob, because of his position with another company, stands to get the "rejects" from a failed phone project, so he knows whether or not the cell phone was chosen for development and failed. That is, Bob knows whether Phone– or not.

Let's assume that the handheld device was chosen for development. So the true situation is either Handheld+ or Handheld–. Alice therefore knows Handheld, while Bob knows that the decision is not Phone– (else he would have received the rejects).

After the first period (week, day, or hour), Alice sells since Handheld+ and Handheld– are equally likely, and one sells if the probability of success is 50 percent or less. Bob buys since he estimates that the probability for success is 2/3. With Phone– ruled out, the remaining possibilities are Handheld+, Handheld–, and Phone+, and two out of three of them are successes.

After the second period, it is common knowledge that the true situation is not Phone– since otherwise Bob would have sold in the first period. This is not news to Alice, who continues to sell. Bob continues to buy.

After the third period, it is common knowledge that it is not Phone (neither Phone+ nor Phone–) since otherwise Alice would have bought in the second period. Thus it's either Handheld+ or Handheld–. Both Bob and Alice take the probability of success to be 50 percent, thus both sell, and there is a mini-crash of the stock price. (Selling by both influential investors triggers a general sell-off.)

Note that at the beginning both Alice and Bob know that the true situation is not Phone–, but this knowledge is mutual, not common. Alice knows that Bob knows it is not Phone–, but Bob does not know that Alice knows this. From his position the true situation might be Phone+, in which case Alice would know Phone but not whether the situation is Phone+ or Phone–.

The example can be varied in a number of ways: there needn't be merely three periods before a crash, but an arbitrary number; there may be a bubble (sellers suddenly switching to become buyers) instead of a crash; there may be an arbitrarily large number of investors or investor groups; there may be an issue other than buying or selling under deliberation, perhaps a decision whether to employ one stock-picking approach rather than another.

In all these cases the stock's price can move in response to no external news. Nevertheless, the subterranean information processing leading to common knowledge among the investors eventually leads to precipitous and unexpected movement in the stock's price. Analysts will express surprise at the crash (or bubble) because "nothing happened."

The example is also relevant to what I suspect is a relatively common kind of insider trading, in which "partial insiders" are privy to bits of insider information but not to the whole story.

Trading Strategies, Whim, and Ant Behavior

A more macro-level interaction among investors occurs between technical traders and value traders. Also contributing over time to booms and busts, this interaction comes through clearly in computer models of the following commonsense dynamic.

Let's suppose that value traders perceive individual stocks or the market as a whole to be strongly undervalued. They start buying and, by doing so, raise prices. As prices increase, a trend develops and technical traders, as is their wont, follow it, increasing prices even further. Soon enough, the market is seen as overvalued by value traders, who begin to sell and thereby slow and then reverse the trend. The trend-following technical traders eventually follow suit, and the cycle begins over again. There are, of course, other sources of variation (one being the number of people who are technical traders and value traders at any given time), and the oscillations are irregular.

The bottom line of much of this modeling is that contrarian value traders have a stabilizing effect on the market, whereas technical traders increase volatility. So does computer-generated program trading, which tends to produce buying or selling in lockstep. There are other sorts of interaction among different classes of investors leading to cycles of varying duration, all of which have differential impacts on the others on which they are superimposed.

In addition to these more or less rational interactions among investors I must also note influences inspired by nothing more than whim, where behavior turns on a mucro. I recall many times, for example, reluctantly beginning work on a project when a niggling detail about some utterly irrelevant matter came to mind. It may have concerned the etymology of a word, or the colleague whose paper bag ripped open at a departmental meeting revealing an embarrassing magazine inside, or why caller ID misidentified a friend's telephone number. These in turn brought to mind the next in a train of associations and musings, which ultimately led me to an entirely different project. My impulsively deciding, while browsing in Borders, to make my first margin call on WCOM is another instance.

When this capriousness extends to influential analysts, the effect is more pronounced. In November 2002 the *New York Times* reported on such a case involving Jack Grubman, telecommunications analyst and anxious father. In an email to a friend Grubman allegedly stated that his boss, Sanford Weill, the chairman of Citigroup, helped get Grubman's children into an exclusive nursery school after he raised his rating of AT&T in 1999. Gretchen Morgenson, the article's author, further reported that Weill had his own personal reasons for wanting this upgrade. Whether these particular charges are true or not is immaterial. It's very hard to believe, however, that this sort of influence is rare.

Such episodes strongly suggest to me that there will never be a precise science of finance or economics. Buying and selling must surely partake of a similar iffiness, at least sometimes. *Butterfly Economics,* by the British economic theorist Paul Ormerod, faults these disciplines for not sufficiently taking into account the commonsense fact that people, whether knowledgeable or not, influence each other.

People do not, as chapter 2 demonstrated, have a set of fixed preferences on which they coolly and rationally base their economic decisions. The assumption that investors are sensitive only to price and a few ratios simplifies the mathematical models, but it is not always true to our experience of fads, fashions, and people's everyday monkey-see, monkey-do behavior.

Ormerod tells of an experiment involving not monkeys but ants that provides a useful metaphor. Two identical piles of food are set up at equal distances from a large nest of ants. Each pile is automatically replenished and the ants have no reason to prefer one to the other. Entomologists tell us that once an ant has found food, it usually returns to the same source. Upon returning to the nest, however, it physically stimulates other ants, who might be frequenting the other pile, to follow it to the first pile.

So where do the ants go? It might be speculated that either they would split into two roughly even groups or perhaps a large majority would arbitrarily settle on one or the other pile. Their actual behavior is counterintuitive. The number of ants going to each pile fluctuates wildly and doesn't ever settle down. A graph of these fluctuations looks suspiciously like a graph of the stock market.

And in a way, the ants are like stock traders (or people deciding whether or not to make a margin call). Upon leaving the nest, each ant must make a decision: Go to the pile visited last time, be influenced by another to switch piles, or switch piles of its own volition. This slight openness to the influence of other ants is enough to insure the complicated and volatile fluctuations in the number of ants visiting the two sites.

An astonishingly simple formal model of such influence is provided by Stephen Wolfram in his book *A New Kind of Science*. Imagine a colossally high brick wall wherein each brick rests on parts of two bricks below it and, except for the top row, has parts of two bricks above it. Imagine further that the top row has some red bricks and some green ones. The coloring of the bricks in the top row determines the coloring of the bricks in the second row as follows. Pick a brick in the second row and check the colors of the two bricks above it in the first row. If exactly one of these bricks is green, then the brick in the second row is colored green. If both or neither are green, then the brick is colored red. Do this for every brick in the second row.

The coloring of the bricks in the second row determines the coloring of the bricks in the third row in the same way, and in general, the coloring of the bricks in any row determines the coloring of the bricks in the row below it in the same way. That's it.

Now if we interpret a row of bricks as a collection of investors at any given instant, green ones for buyers and red ones for sellers, then the change from moment to moment of investor sentiment is reflected in the changing color composition of the succeeding rows of bricks. If we let P be the difference between the number of green bricks and the number of red bricks, then P is a rough analogue of a stock's price. Graphing it, we see that it oscillates up and down in a way that looks random.

The model can be made more realistic, but it is significant that even this bare-bones version, like the ant behavior, evinces a kind of internally generated random noise. This suggests that *part of* the oscillation of stock prices is also internally generated and is not a response to anything besides investors' reactions to each other. The theme of Wolfram's book, borne out here, is that complex behavior can result from very simple rules of interaction.

Chaos and Unpredictability

What is the relative importance of private information, investor trading strategies, and pure whim in predicting the market? What is the relative importance of conventional economic news (interest rates, budget deficits, accounting scandals, and trade balances), popular culture fads (in sports, movies, fashions), and germane political and military events (terrorism, elections, war) too disparate even to categorize? If we were to carefully define the problem, predicting the market with any precision is probably what mathematicians call a universal problem, meaning that a complete solution to it would lead immediately to solutions for a large class of other problems. It is, in other words, as hard a problem in social prediction as there is.

Certainly, too little notice is taken of the complicated connections among these variables, even the more clearly defined economic ones. Interest rates, for example, have an impact on unemployment rates, which in turn influence revenues; budget deficits affect trade deficits, which sway interest rates and exchange rates; corporate fraud influences consumer confidence, which may depress the stock market and alter other indices; natural business cycles of various periods are superimposed on one another; an increase in some quantity or index positively (or negatively) feeds back on another, reinforcing or weakening it and being reinforced or weakened in turn.

Few of these associations are accurately described by a straight-line graph and so they bring to a mathematician's mind the subject of nonlinear dynamics, more popularly known as chaos theory. The subject doesn't deal with anarchist treatises or surrealist manifestoes but with the behavior of so-called nonlinear systems. For our purposes these may be thought of as any collection of parts whose interactions and connections are described by nonlinear rules or equations. That is to say, the equations' variables may be multiplied together, raised to powers, and so on. As a consequence the system's parts are not necessarily linked in a proportional manner as they are, for example, in a bathroom scale or a thermometer; doubling the magnitude of one part will not double that of another—nor will outputs be proportional to inputs. Not surprisingly, trying to predict the precise long-term behavior of such systems is often futile.

Let me, in place of a technical definition of such nonlinear systems, describe instead a particular physical instance of one. Picture before you a billiards table. Imagine that approximately twenty-five round obstacles are securely fastened to its surface in some haphazard arrangement. You hire the best pool player you can find and ask him to place the ball at a particular spot on the table and take a shot toward one of the round obstacles. After he's done so, his challenge is to make exactly the same shot from the same spot with another ball.

Even if his angle on this second shot is off by the merest fraction of a degree, the trajectories of these two balls will very soon diverge considerably. An infinitesimal difference in the angle of impact will be magnified by successive hits of the obstacles. Soon one of the balls will hit an obstacle that the other misses entirely, at which point all similarity between the two trajectories ends.

The sensitivity of the billiard balls' paths to minuscule variations in their initial angles is characteristic of nonlinear systems. The divergence of the billiard balls is not unlike the disproportionate effect of seemingly inconsequential events, the missed planes, serendipitous meetings, and odd mistakes and links that shape and reshape our lives.

This sensitive dependence of nonlinear systems on even tiny differences in initial conditions is, I repeat, relevant to various aspects of the stock market in general, in particular its sometimes wildly disproportionate responses to seemingly small stimuli such as companies' falling a penny short of earnings estimates. Sometimes, of course, the differences are more substantial. Witness the notoriously large discrepancies between government economic figures on the size of budget surpluses and corporate accounting statements of earnings and the "real" numbers.

Aspects of investor behavior too can no doubt be better modeled by a nonlinear system than a linear one. This is so despite the fact that linear systems and models are much more robust, with small differences in initial conditions leading only to small differences in final outcomes. They're also easier to predict mathematically, and this is why they're so often employed whether their application is appropriate or not. The chestnut about the economist looking for his lost car keys under the street lamp comes to mind. "You probably lost them near the car," his companion remonstrates, to which the economist responds, "I know, but the light is better over here."

The "butterfly effect" is the term often used for the sensitive dependence of nonlinear systems, a characteristic that has been noted in phenomena ranging from fluid flow and heart fibrillations to epilepsy and price fluctuations. The name comes from the idea that a butterfly flapping its wings someplace in South America might be sufficient to change future weather systems, helping to bring about, say, a tornado in Oklahoma that would otherwise not have occurred. It also explains why long-range precise prediction of nonlinear systems isn't generally possible. This non-predictability is the result not of randomness but of complexity too great to fathom.

Yet another reason to suspect that parts of the market may be better modeled by nonlinear systems is that such systems' "trajectories" often follow a fractal course. The trajectories of these systems, of which the stock price movements may be considered a proxy, turn out to be aperiodic and unpredictable and, when examined closely, evince even more intricacy. Still closer inspection of the system's trajectories reveals yet smaller vortices and complications of the same general kind.

In general, fractals are curves, surfaces, or higher dimensional objects that contain more, but similar, complexity the closer one looks. A shoreline, to cite a classic example, has a characteristic jagged shape at whatever scale we draw it; that is, whether we use satellite photos to sketch the whole coast, map it on a fine scale by walking along some small section of it, or examine a few inches of it through a magnifying glass. The surface of the mountain looks roughly the same whether seen from a height of 200 feet by a giant or close up by an insect. The branching of a tree appears the same to us as it does to birds, or even to worms or fungi in the idealized limiting case of infinite branching.

As the mathematician Benoit Mandelbrot, the discoverer of fractals, has famously written, "Clouds are not spheres, mountains are not cones, coastlines are not circles, and bark is not smooth, nor does lightning travel in a straight line." These and many other shapes in nature are near fractals, having characteristic zigzags, push-pulls, bump-dents at almost every size scale, greater magnification yielding similar but ever more complicated convolutions.

And the bottom line, or, in this case, the bottom fractal, for stocks? By starting with the basic up-down-up and down-up-down patterns of a stock's possible movements, continually replacing each of these patterns' three segments with smaller versions of one of the basic patterns chosen at random, and then altering the spikiness of the patterns to reflect changes in the stock's volatility, Mandelbrot has constructed what he calls multifractal "forgeries." The forgeries are patterns of price movement whose general look is indistinguishable from that of real stock price movements. In contrast, more conventional assumptions about price movements, say those of a strict random-walk theorist, lead to patterns that are noticeably different from real price movements.

These multifractal patterns are so far merely descriptive, not predictive of specific price changes. In their modesty, as well as in their mathematical sophistication, they differ from the Elliott waves mentioned in chapter 3.

Even this does not prove that chaos (in the mathematical sense) reigns in (part of) the market, but it is clearly a bit more than suggestive. The occasional surges of extreme volatility that have always been a part of the market are not as nicely accounted for by traditional approaches to finance, approaches Mandelbrot compares to "theories of sea waves that forbid their swells to exceed six feet."

Extreme Price Movements, Power Laws, and the Web

Humans are a social species, which means we're all connected to each other, some in more ways than others. This is especially so in financial matters. Every investor responds not only to relatively objective economic considerations, but also in varying degrees to the pronouncements of national and world leaders (not least of those Mr. Greenspan), consumer confidence, analysts' ratings (bah), general and business media reports and their associated spin, investment newsletters, the behavior of funds and large institutions, the sentiments of friends, colleagues, and of course the much-derided brother-in-law.

The linkage of changes in stock prices to the varieties of investor responses and interactions suggests to me that communication networks, degrees of connectivity, and so-called small world phenomena ("Oh, you must know my uncle Waldo's third wife's botox specialist") can shine a light on the workings of Wall Street.

First the conventional story. Movements in a stock or index over small units of time are usually slightly positive or slightly negative, less frequently very positive or very negative. A large fraction of the time, the price will rise or fall between 0 percent and 1 percent; a smaller fraction of the time, it will rise or fall between 1 percent and 2 percent; a very small fraction of the time will the movement be more than, say, 10 percent up or down. In general, the movements are well described by a normal bell-shaped curve. The most likely change for a small unit of time is probably a minuscule jot above zero, reflecting the market's long-term (and recently invisible) upward bias, but the fact remains that extremely large price movements, whether positive or negative, are rare.

It's been clear for some time, however (that is, since Mandelbrot made it clear), that extreme movements are not as rare as the normal curve would predict. If you measure commodity price changes, for example, in each of a large number of small time units and make from these measurements a histogram, you will notice that the graph is roughly normal near its middle. The distribution of these price movements, however, seems to have "fatter tails" than the normal distribution, suggesting that crashes and bubbles in a stock, an index, or the entire market are less unlikely than many would like to admit. There is, in fact, some evidence that very large movements in stock prices are best described by a so-called power law (whose definition I'll get to shortly) rather than the tails of the normal curve.

An oblique approach to such evidence is via the notions of connectivity and networks. Everyone's heard people exclaim about how amazed they were to run into someone they knew so far from home. (What I find amazing is how they can be continually amazed at this sort of thing.) Most have heard too of the alleged six degrees of separation between any two people in this country. (Actually, under reasonable assumptions each of us is connected to everyone else by an average of two links, although we're not likely to know who the two intermediate parties are.) Another popular variant of the notion concerns the number of movie links between film actors, say between Marlon Brando and Christina Ricci or between Kevin Bacon and anyone else. If A and B appeared together in X, and B and C appeared together in Y, then A is linked to C via these two movies.

Although they may not know of Kevin Bacon and his movies, most mathematicians are familiar with Paul Erdös and his theorems. Erdös, a prolific and peripatetic Hungarian mathematician, wrote hundreds of papers in a variety of mathematical areas during his long life. Many of these had co-authors, who are therefore said to have Erdös number 1. Mathematicians who have written a joint paper with someone with Erdös number 1 are said to have Erdos number 2, and so on.

Ideas about such informal networks lead naturally to the network of all networks, the Internet, and to ways to analyze its structure, shape, and "diameter." How, for example, are the Internet's nearly 1 billion web pages connected? What constitutes a good search strategy? How many links does the average web page contain? What is the distribution of document sizes? Are there many with, say, more than 1,000 links? And, perhaps most intriguingly, how many clicks on average does it take to get from one of two randomly selected documents to another?

A couple of years ago, Albert-Laszló Barabasi, a physics professor at Notre Dame, and two associates, Réka Albert and Hawoong Jeong, published results that strongly suggest that the web is growing and that its documents are linking in a rather collective way that accounts for, among other things, the unexpectedly large number of very popular documents. The increasing number of web pages and the "flocking effect" of many pages pointing to the same popular addresses, causing proportionally more pages to do the same thing, is what leads to a power law.

Barabasi, Albert, and Jeong showed that the probability that a document has k links is roughly proportional to $1/k^3$—or inversely proportional to the third power of k. (I've rounded off; the model actually predicts an exponent of 2.9.) This means, for example, that there are approximately one-eighth as many documents with twenty links as there are documents with ten links since $1/20^3$ is one-eighth of $1/10^3$. Thus the number of documents with k links declines quickly as k increases, but nowhere near as quickly as a normal bell-shaped distribution would predict. This is why the power law distribution has a fatter tail (more instances of very large values of k) than does the normal distribution.

The power laws (sometimes called scaling laws, sometimes Pareto laws) that characterize the web also seem to characterize many other complex systems that organize themselves into a state of skittish responsiveness. The physicist Per Bak, who has made an extensive study of them, described in his book

How Nature Works, claims that such $1/k^m$ laws (for various exponents m) are typical of many biological, geological, musical, and economic processes, and that they tend to arise in a wide variety of complex systems. Traffic jams, to cite a different domain and seemingly unrelated dynamic, also seem to obey a power law, with jams involving k cars occurring with a probability roughly proportional to $1/k^m$ for an appropriate m.

There is even a power law in linguistics. In English, for example, the word "the" appears most frequently and is said to have rank order 1; the words "of," "and," and "to" rank 2, 3, and 4, respectively. "Chrysanthemum" has a much higher rank order. Zipf's Law relates the frequency of a word to its rank order k and states that a word's frequency in a written text is proportional to $1/k^1$; that is, inversely proportional to the first power of k. (Again, I've rounded off; the power of k is close to, but not exactly 1.) Thus a relatively unusual word whose rank order is 10,000 will still appear with a frequency proportional to 1/10,000, rather than essentially not at all as would be the case if word frequencies were described by the tail of a normal distribution. The size of cities also follows a power law with k close to 1, the kth largest city having a population proportional to 1/k.

One of the most intriguing consequences of the Barabasi-Albert-Jeong model is that because of the power law distribution of links to and from documents on web sites (the nodes of the network), the diameter of the web is only nineteen clicks. By this they mean that you can travel from one arbitrarily selected web page to any other in approximately nineteen clicks, far fewer than had been conjectured. On the other hand, comparing nineteen with the much smaller number of links between arbitrarily selected people, we may wonder why the diameter is as big as it is. The answer is that the average web page contains only seven links, whereas the average person knows hundreds of people.

Even though the web is expected to grow by a power of 10 over the next few years, its diameter will likely grow by only a couple of clicks, from nineteen to twenty-one. The growth and preferential linking assumptions above indicate that the web's diameter D is governed by a logarithmic law; D is a bit more than 2 log(N), where N is the number of documents, presently about 1 billion.

If the Barabasi model is valid (and more work needs to be done), the web is not as unmanageable and untraversable as it often seems. Its documents are much more closely interconnected than they would be if the probability that a document has k links were described by a normal distribution.

What is the relevance of power laws, networks, and diameters to extreme price movements? Investors, companies, mutual funds, brokerages, analysts, and media outlets are connected via a large, vaguely defined network, whose nodes exert influence on the nodes to which they're connected. This network is likely to be more tightly connected and to contain more very popular (and hence very influential) nodes than people realize. Most of the time this makes no difference and price movements, resulting from the sum of a myriad of investors' *independent* zigs and zags, are governed by the normal distribution.

But when the volume of trades is very high, the trades are strongly influenced by relatively few popular nodes—mutual funds, for example, or analysts or media outlets—becoming aligned in their sentiments, and this alignment can create extreme price movements. (WCOM often led the Nasdaq in volume during its slide.) That there exist a few very popular, very connected nodes is, I reiterate, a consequence of the fact that a power law and not the normal distribution governs their frequency. A contagious alignment of this handful of very popular, very connected, very influential nodes will occur more frequently than people expect, as will, therefore, extreme price movements.

Other examples suggest that the exponent m in market power laws, $1/k^m$, may be something other than 3, but the point stands. The trading network is sometimes more herd-like and volatile in its behavior than standard pictures of it acknowledge. The crash of 1929, the decline of 1987, and the recent dot-com meltdown should perhaps not be seen as inexplicable aberrations (or as "just deserts") but as natural consequences of network dynamics.

Clearly much work remains to be done to understand why power laws are so pervasive. What is needed, I think, is something like the central limit theorem in statistics, which explains why the normal curve arises in so many different contexts. Power laws provide an explanation, albeit not an airtight one, for the frequency of bubbles and crashes and the so-called volatility clustering that seem to characterize real markets. They also reinforce the impression that the market is a different sort of beast than that usually studied by social scientists or, perhaps, that social scientists have been studying these beasts in the wrong way.

I should note that my interest in networks and connectivity is not unrelated to my initial interest in WorldCom, which owned not only MCI, but, as I've mentioned twice already, UUNet, "the backbone of the Internet." Obsessions fade slowly.

Economic Disparities and Media Disproportions

WorldCom may have been based in Mississippi, but Bernie Ebbers, who affected an unpretentious, down-home style, wielded political and economic influence foreign to the average Mississippian and the average WorldCom employee. For this he may serve as a synecdoche for the following.

More than a mathematical pun suggests that power laws may have relevance to economic, media, and political power as well as to the stock market. Along various social dimensions, the dynamics underlying power laws might allow for the development of more centers of concentration than we might otherwise expect. This might lead to larger, more powerful economic, media, and political elites and consequent great disparities. Whether or not this is the case, and whether or not great disparities are necessary for complex societies to function, such disparities certainly reign in modern America. Relatively few people, for example, own a hugely disproportionate share of the wealth, and relatively few people attract a hugely disproportionate share of media attention.

The United Nations issued a report a couple of years ago saying that the net worth of the three richest families in the world—the Gates family, the sultan of Brunei, and the Walton family—was greater than the combined gross domestic product of the forty-three poorest nations on Earth. The U.N. statement is misleading in an apples-and-oranges sort of way, but despite the periodic additions, subtractions, and reshufflings of the Forbes 400 and the fortunes of underdeveloped countries, some appropriately modified conclusion no doubt still holds.

(On the other hand, the distribution of wealth in *some* of the poorest nations—where almost everybody is poverty-stricken—is no doubt more uniform than it is here, indicating that relative equality is no solution to the problem of poverty. I suspect that significant, but not outrageous, disparities of wealth are probably more conducive to wealth creation than is relative uniformity, provided the society meets some minimal conditions: It's based on law, offers some educational and other opportunities, and allows for a modicum of private property.)

The dynamic whereby the rich get richer is nowhere more apparent than in the pharmaceutical industry, in which companies understandably spend far, far more money researching lifestyle drugs for the affluent than life-saving drugs for the hundreds of millions of the world's poor people. Instead of trying to come up with treatments for malaria, diarrhea, tuberculosis, and acute lower-respiratory diseases, resources go into treatments for wrinkles, impotence, baldness, and obesity.

Surveys indicate that the ratio of the remuneration of a U.S. firm's CEO to that of the average employee of the firm is at an all-time high of around 500, whether the CEO has improved the fortunes of the company or not, and whether he or she is under indictment or not. (If we assume 250 workdays per year, arithmetic tells us the CEO needs only half a day to make what the employee takes all year to earn.) Professor Edward Wolff of New York University has estimated that the richest 1 percent of Americans own half of all stocks, bonds, and other assets. And Cornell University's Robert Frank has described the spread of the winner-take-all model of compensation from the sports and entertainment worlds to many other domains of American life.

Nero-like arrogance often accompanies such exorbitant compensation. High-tech WorldCom faced a host of problems before its 2002 collapse. Did Bernie Ebbers utilize the company's horde of top-flight technical people (at least the ones who hadn't quit or been fired) to devise a clever strategy to extricate the company from its troubles? No, he cut out free coffee for employees to save money. As Tyco spiraled downward, its CEO, Dennis Kozlowski, spent millions of company dollars on personal items, including a $6,000 shower curtain, a $15,000 umbrella stand, and a $7 million Manhattan apartment.

(Even successful CEOs are not always gentlemen. Oracle's Larry Ellison, a fierce foe of Bill Gates, a couple of years ago admitted to spying on Microsoft. Amusingly, Oracle's sophisticated snoops didn't employ state-of-the-art electronics, but tried to buy the garbage of a pro-Microsoft group in order to examine its contents for clues about Microsoft's public relations plans. I'm talking real cookies here, not the type that Internet sites leave on your computer; scribbled memos and addresses on torn envelopes, not emails and Internet routing numbers; germs and bacteria, not computer viruses.)

What should we make of such stories? Communism, happily, has been discredited, but unregulated and minimally regulated free markets (as evidenced by the behavior of some accountants, analysts, CEOs, and, yes, greedy, deluded, and short-sighted investors) have some obvious drawbacks. Some of the reforms proposed by Congress in 2002 promise to be helpful in this regard, but I wish here only to express disquiet at such enormous and growing economic disparities.

The same steep hierarchy and disproportion that characterize our economic condition affect our media as well. The famous get ever more famous, celebrities become ever more celebrated. (Pick your favorite ten examples here.) Magazines and television increasingly run features asking who's hot and who's not. Even the search engine Google has a version in which surfers can check the topics and people attracting the most hits the previous week. The up-and-down movements of celebrity seem to constitute a kind of market in which almost all the "traders" are technical traders trying to guess what everyone else thinks, rather than value traders looking for worth.

The pattern holds in the political realm as well. In general, on the front page and in the first section of a newspaper, the number one newsmaker is undoubtedly the president of the United States. Other big newsmakers are presidential candidates, members of Congress, and other federal officials.

Twenty years ago, Herbert Gans wrote in *Deciding What's News* that 80 percent of the domestic news stories on television network news concerned these four classes of people; most of the remaining 20 percent covered the other 280 million of us. Fewer than 10 percent of all stories were about abstractions, objects, or systems. Things haven't changed much since then (except on the cable networks where disaster stories, show trials, and terrorist obsessions dominate). Newspapers generally have broader coverage, although studies have found that up to 50 percent of the sources for national stories on the front pages of the *New York Times* and the *Washington Post* were officials of the U.S. government. The Internet has still broader scope, although there, too, one notes strong and unmistakable signs of increasing hierarchy and concentration.

And what about foreign coverage? The frequency of reporting on overseas newsmakers demonstrates the same biases. We hear from heads of state, from leaders of opposition parties or forces, and occasionally from others. The masses of ordinary people are seldom a presence at all. The journalistic rule of thumb that one American equals 10 Englishmen equals 1,000 Chileans equals 10,000 Rwandans varies with time and circumstance, but it does contain an undeniable truth. Americans, like everybody else, care much less about some parts of the world than others. Even the terrorist attack in Bali didn't rate much coverage here, and many regions have no correspondents at all, rendering them effectively invisible.

Such disparities may be a natural consequence of complex societies. This doesn't mean that they need be as extreme as they are or that they're always to be welcomed. It may be that the stock market's recent volatility surges are a leading indicator for even greater social disparities to come.

9 | From Paradox to Complexity

Groucho Marx vowed that he'd never join a club that would be willing to accept him as a member. Epimenides the Cretan exclaimed (almost) inconsistently, "All Cretans are liars." The prosecutor booms, "You must answer Yes or No. Will your next word be 'No'?" The talk show guest laments that her brother is an only child. The author of an investment book suggests that we follow the tens of thousands of his readers who have gone against the crowd.

Warped perhaps by my study of mathematical logic and its emphasis on paradoxes and self-reference, I'm naturally interested in the paradoxical and self-referential aspects of the market, particularly of the Efficient Market Hypothesis. Can it be proved? Can it be disproved? These questions beg a deeper question. The Efficient Market Hypothesis is, I think, neither necessarily true nor necessarily false.

The Paradoxical Efficient Market Hypothesis

If a large majority of investors believe in the hypothesis, they would all assume that new information about a stock would quickly be reflected in its price. Specifically, they would affirm that since news almost immediately moves the price up or down, and since news can't be predicted, neither can changes in stock prices. Thus investors who subscribe to the Efficient Market Hypothesis would further believe that looking for trends and analyzing companies' fundamentals is a waste of time. Believing this, they won't pay much attention to new developments. But if relatively few investors are looking for an edge, the market will not respond quickly to new information. In this way an overwhelming belief in the hypothesis ensures its falsity.

To continue with this cerebral somersault, recall now a rule of logic: Sentences of the form "H implies I" are equivalent to those of the form "not I implies not H." For example, the sentence "heavy rain implies that the ground will be wet" is logically equivalent to "dry ground implies the absence of heavy rain." Using this equivalence, we can restate the claim that overwhelming belief in the Efficient Market Hypothesis leads to (or implies) its falsity. Alternatively phrased, the claim is that if the Efficient Market Hypothesis is true, then it's not the case that most investors believe it to be true. That is, if it's true, most investors believe it to be false (assuming almost all investors have an opinion and each either believes it or disbelieves it).

Consider now the inelegantly named Sluggish Market Hypothesis, the belief that the market is quite slow in responding to new information. If the vast majority of investors believe the Sluggish Market Hypothesis, then they all would believe that looking for trends and analyzing companies is well worth their time and, by so exercising themselves, they would bring about an efficient market. Thus, if most investors believe the Sluggish Market Hypothesis is true, they will by their actions make the Efficient Market Hypothesis true. We conclude that if the Efficient Market Hypothesis is false, then it's not the case that most investors believe the

Sluggish Market Hypothesis to be true. That is, if the Efficient Market Hypothesis is false, then most investors believe it (the EMH) to be true. (You may want to read over the last few sentences in a quiet corner.)

In summary, if the Efficient Market Hypothesis is true, most investors won't believe it, and if it's false, most investors will believe it. Alternatively stated, the Efficient Market Hypothesis is true if and only if a majority believes it to be false. (Note that the same holds for the Sluggish Market Hypothesis.) These are strange hypotheses indeed!

Of course, I've made some big assumptions that may not hold. One is that if an investor believes in one of the two hypotheses, then he disbelieves in the other, and almost all believe in one or the other. I've also assumed that it's clear what "large majority" means, and I've ignored the fact that it sometimes requires very few investors to move the market. (The whole argument could be relativized to the set of knowledgeable traders only.)

Another gap in the argument is that any suspected deviations from the Efficient Market Hypothesis can always be attributed to mistakes in asset pricing models, and thus the hypothesis can't be conclusively rejected for this reason either. Maybe some stocks or kinds of stock are riskier than our pricing models allow for and that's why their returns are higher. Nevertheless, I think the point remains: The truth or falsity of the Efficient Market Hypothesis is not immutable but depends critically on the beliefs of investors. Furthermore, as the percentage of investors who believe in the hypothesis itself varies, the truth of the hypothesis varies inversely with it.

On the whole, most investors, professionals on Wall Street, and amateurs everywhere, disbelieve in it, so for this reason I think it holds, but only approximately and only most of the time.

The Prisoner's Dilemma and the Market

So you don't believe in the Efficient Market Hypothesis. Still, it's not enough that you discover simple and effective investing rules. Others must not find out what you're doing, either by inference or by reading your boastful profile in a business magazine. The reason for secrecy, of course, is that without it, simple investing rules lead to more and more complicated ones, which eventually lead to zero excess returns and a reliance on chance.

This inexorable march toward increased complexity arises from the actions of your co-investors, who, if they notice (or infer, or are told) that you are performing successfully on the basis of some simple technical trading rule, will try to do the same. To take account of their response, you must complicate your rule and likely decrease your excess returns. Your more complicated rule will, of course, also inspire others to try to follow it, leading to further complications and a further decline in excess returns. Soon enough your rule assumes a near-random complexity, your excess returns are reduced essentially to zero, and you're back to relying on chance.

Of course, your behavior will be the same if you learn of someone else's successful performance. In fact, a situation arises that is clarified by the classic "prisoner's dilemma," a useful puzzle originally framed in terms of two people in prison.

Suspected of committing a major crime, the two are apprehended in the course of committing some minor offense. They're then interrogated separately, and each is given the choice of confessing to the major crime and thereby implicating his partner or remaining silent. If they both remain silent, they'll each get one year in prison. If one confesses and the other doesn't, the one who confesses will be rewarded by being set free, while the other one will get a five-year term. If they both confess, they can both expect to spend three years in prison. The cooperative option (cooperative with the other prisoner, that is) is to remain silent, while the non-cooperative option is to confess. Given the payoffs and human psychology, the most likely outcome is for both to confess; the best outcome for the pair *as a pair* is for both to remain silent; the best outcome for each prisoner *as an individual* is to confess and have one's partner remain silent.

The charm of the dilemma has nothing to do with any interest one might have in prisoners' rights. (In fact, it has about as much relevance to criminal justice as the four-color-map theorem has to geography.) Rather, it provides the logical skeleton for many situations we face in everyday life. Whether we're negotiators in business, spouses in a marriage, or nations in a dispute, our choices can often be phrased in terms of the prisoner's dilemma. If both (all) parties pursue their own interests exclusively and do not cooperate, the outcome is worse for both (all) of them; yet in any given situation, any given party is better off not cooperating. Adam Smith's invisible hand ensuring that individual pursuits bring about group well-being is, at least in these situations (and some others), quite arthritic.

The dilemma has the following multi-person market version: Investors who notice some exploitable stock market anomaly may either act on it, thereby diminishing its effectiveness (the non-cooperative option) or ignore it, thereby saving themselves the trouble of keeping up with developments (the cooperative option). If some ignore it and others act on it, the latter will receive the biggest payoffs, the former the smallest. As in the standard prisoner's dilemma, the logical response for any player is to take the non-cooperative option and act on any anomaly likely to give one an edge. This response leads to the "arms race" of ever more complex technical trading strategies. People search for special knowledge, the result eventually becomes common knowledge, and the dynamic between the two generates the market.

This searching for an edge brings us to the social value of stock analysts and investment professionals. Although the recipients of an abundance of bad publicity in recent years, they provide a most important service: By their actions, they help turn special knowledge into common knowledge and in the process help make the market relatively efficient. Absent a draconian rewiring of human psychology and an accompanying draconian rewiring of our economic system, this accomplishment is an impressive and vital one. If it means being "noncooperative" with other investors, then so be it. Cooperation is, of course, generally desirable, but cooperative decisionmaking among investors seems to smack of totalitarianism.

Pushing the Complexity Horizon

The complexity of trading rules admits of degrees. Most of the rules to which people subscribe are simple, involving support levels, P/E ratios, or hemlines and Super Bowls, for example. Others, however, are quite convoluted and conditional. Because of the variety of possible rules, I want to take an oblique and abstract approach here. The hope is that this approach will yield insights that a more pedestrian approach misses. Its key ingredient is the formal definition of (a type of) complexity. An intuitive understanding of this notion tells us that someone who remembers his eight-digit password by means of an elaborate, long-winded saga of friends' addresses, children's ages, and special anniversaries is doing something silly. Mnemonic rules make sense only when they're shorter than what is to be remembered.

Let's back up a bit and consider how we might describe the following sequences to an acquaintance who couldn't see them. We may imagine the 1s to represent upticks in the price of a stock and the 0s downticks or perhaps up-and-down days.

1. 0 1 0 1 0 1 0 1 0 1 0 1 0 1 0 1 0 1 0 1 0 1 0 1 0 1 0 1 0 . . .
2. 0 1 0 1 1 0 1 0 1 0 1 0 1 1 0 1 0 1 0 1 0 1 0 1 1 . . .
3. 1 0 0 0 1 0 1 1 0 1 1 0 1 1 0 0 0 1 0 1 0 1 1 0 0 . . .

The first sequence is the simplest, an alternation of 0s and 1s. The second sequence has some regularity to it, a single 0 alternating sometimes with a 1, sometimes with two 1s, while the third sequence doesn't seem to manifest any pattern at all. Observe that the precise meaning of " . . . " in the first sequence is clear; it is less so in the second sequence, and not at all clear in the third. Despite this, let's assume that these sequences are each a trillion bits long (a bit is a 0 or a 1) and continue on "in the same way."

Motivated by examples like this, the American computer scientist Gregory Chaitin and the Russian mathematician A. N. Kolmogorov defined the complexity of a sequence of 0s and 1s to be the length of the shortest computer program that will generate (that is, print out) the sequence in question.

A program that prints out the first sequence above can consist simply of the following recipe: print a 0, then a 1, and repeat a half trillion times. Such a program is quite short, especially compared to the long sequence it generates. The complexity of this first trillion-bit sequence may be only a few hundred bits, depending to some extent on the computer language used to write the program.

A program that generates the second sequence would be a translation of the following: Print a 0 followed by either a single 1 or two 1s, the pattern of the intervening 1s being one, two, one, one, one, two, one, one, and so on. Any program that prints out this trillion-bit sequence would have to be quite long so as to fully specify the "and so on" pattern of the intervening 1s. Nevertheless, because of the regular alternation of 0s and either one or two 1s, the shortest such program will be considerably shorter than the trillion-bit sequence it generates. Thus the complexity of this second sequence might be only, say, a quarter trillion bits.

With the third sequence (the commonest type) the situation is different. This sequence, let us assume, remains so disorderly throughout its trillion-bit length that no program we might use to generate it would be any shorter than the sequence itself. It never repeats, never exhibits a pattern. All any program can do in this case is dumbly list the bits in the sequence: print 1, then 0, then 0, then 0, then 1, then 0, then 1, ... There is no way the . . . can be compressed or the program shortened. Such a program will be as long as the sequence it's supposed to print out, and thus the third sequence has a complexity of approximately a trillion.

A sequence like the third one, which requires a program as long as itself to be generated, is said to be random. Random sequences manifest no regularity or order, and the programs that print them out can do nothing more than direct that they be copied: print 1 0 0 0 1 0 1 1 0 1 1 . . . . These programs cannot be abbreviated; the complexity of the sequences they generate is equal to the length of these sequences. By contrast, ordered, regular sequences like the first can be generated by very short programs and have complexity much less than their length.

Returning to stocks, different market theorists will have different ideas about the likely pattern of 0s and 1s (downs and upticks) that can be expected. Strict random walk theorists are likely to believe that sequences like the third characterize price movements and that the market's movements are therefore beyond the "complexity horizon" of human forecasters (more complex than we, or our brains, are, were we expressed as sequences of 0s and 1s). Technical and fundamental analysts might be more inclined to believe that sequences like the second characterize the market and that there are pockets of order amidst the noise. It's hard to imagine anyone believing that price movements follow sequences as regular as the first except, possibly, those who send away "only $99.95 for a complete set of tapes that explain this revolutionary system."

I reiterate that this approach to stock price movements is rather stark, but it does nevertheless "locate" the debate. People who believe there is some pattern to the market, whether exploitable or not, will believe that its movements are characterized by sequences of complexity somewhere between those of type two and type three above.

A rough paraphrase of Kurt Godel's famous incompleteness theorem of mathematical logic, due to the aforementioned Gregory Chaitin, provides an interesting sidelight on this issue. It states that if the market were random, we might not be able to prove it. The reason: encoded as a sequence of 0s and 1s, a random market would, it seems plausible to assume, have complexity greater than that of our own were we also so encoded; it would be beyond our complexity horizon. From the definition of complexity it follows that a sequence can't generate another sequence of greater complexity than itself. Thus if a person were to predict the random market's exact gyrations, the market would have to be less complex than the person, contrary to assumption. Even if the market isn't random, there remains the possibility that its regularities are so complex as to be beyond our complexity horizons.

In any case, there is no reason why the complexity of price movements as well as the complexity of investor/computer blends cannot change over time. The more inefficient the market is, the smaller the complexity of its price movements, and the more likely it is that tools from technical and fundamental analysis will prove useful. Conversely, the more efficient the market is, the greater the complexity of price movements, and the closer the approach to a completely random sequence of price changes.

Outperforming the market requires that one remain on the cusp of our collective complexity horizon. It requires faster machines, better data, improved models, and the smarter use of mathematical tools, from conventional statistics to neural nets (computerized learning networks, the connections between the various nodes of which are strengthened or weakened over a period of training). If this is possible for anyone or any group to achieve, it's not likely to remain so for long.

Game Theory and Supernatural Investor/Psychologists

But what if, contrary to fact, there were an entity possessing sufficient complexity and speed that it was able with reasonably high probability to predict the market and the behavior of individuals within it? The mere existence of such an entity leads to Newcombe's paradox, a puzzle that calls into question basic principles of game theory.

My particular variation of Newcombe's paradox involves the World Class Options Market Maker (WCOMM), which (who?) claims to have the power to predict with some accuracy which of two alternatives a person will choose. Imagine further that WCOMM sets up a long booth on Wall Street to demonstrate its abilities.

WCOMM explains that it tests people by employing two portfolios. Portfolio A contains a $1,000 treasury bill, whereas portfolio B (consisting of either calls or puts on WCOM stock) is either worth nothing or $1,000,000. For *each* person in line at the demonstration, WCOMM has reserved a portfolio of each type at the booth and offers each person the following choice: He or she can choose to take portfolio B *alone* or choose to take *both* portfolios A and B. However, and this is crucial, WCOMM also states that it has used its unfathomable powers to analyze the psychology, investment history, and trading style of everyone in line as well as general market conditions, and if it believes that a person will take both portfolios, it has ensured that portfolio B will be worthless. On the other hand, if WCOMM believes that a person will trust its wisdom and take only portfolio B, it has ensured that portfolio B will be worth $1,000,000. After making these announcements, WCOMM leaves in a swirl of digits and stock symbols, and the demonstration proceeds.

Investors on Wall Street see for themselves that when a person in the long line chooses to take both portfolios, most of the time (say with probability 90 percent) portfolio B is worthless and the person gets only the $1,000 treasury bill in portfolio A. They also note that when a person chooses to take the contents of portfolio B alone, most of the time it's worth $1,000,000.

After watching the portfolios placed before the people in line ahead of me and seeing their choices and the consequences, I'm finally presented with the two portfolios prepared for me by WCOMM. Despite the evidence I've seen, I see no reason not to take both portfolios. WCOMM is gone, perhaps to the financial district of London or Frankfurt or Tokyo, to make similar offers to other investors, and portfolio B is either worth $1,000,000 or not, so why not take both portfolios and possibly get $1,001,000. Alas, WCOMM read correctly the skeptical smirk on my face and after opening my portfolios, I walk away with only $1,000. My portfolio B contains call options on WCOM with a strike price of 20, when the stock itself is selling at $1.13.

The paradox, due to the physicist William Newcombe (not the Newcomb of Benford's Law, but the same mocking four letters WCOM) and made well-known by the philosopher Robert Nozick, raises other issues. As mentioned, it makes problematical which of two game-theoretic principles one should use in making decisions, principles that shouldn't conflict.

The "dominance" principle tells us to take both portfolios since, whether portfolio B contains options worth $1,000,000 or not, the value of two portfolios is at least as great as the value of one. (If portfolio B is worthless, $1,000 is greater than $0, and if portfolio B is worth $1,000,000, $1,001,000 is greater than $1,000,000.)

On the other hand, the "maximization of expected value" principle tells us to take only portfolio B since the expected value of doing so is greater. (Since WCOMM is right about 90 percent of the time, the expected value of taking only portfolio B is (.90 × $1,000,000) + (.10 × $0), or $900,000, whereas the expected value of taking both is (.10 × $1,001,000) + (.90 × $1,000), or $101,000.) The paradox is that both principles seem reasonable, yet they counsel different choices.

This raises other general philosophical matters as well, but it reminds me of my resistance to following the WCOM-fleeing crowd, most of whose B portfolios contained puts on the stock worth $1,000,000.

One conclusion that seems to follow from the above is that such supernatural investor/psychologists are an impossibility. For better or worse, we're on our own.

Absurd Emails and the WorldCom Denouement

A natural reaction to the vagaries of chance is an attempt at control, which brings me to emails regarding WorldCom that, Herzog-like, I sent to various influential people. I had grown tired of carrying on one-sided arguments with CNBC's always perky Maria Bartiromo and always apoplectic James Cramer as they delivered the relentlessly bad news about WorldCom. So in fall 2001, five or six months before its final swoon, I contacted a number of online business commentators critical of WorldCom's past performance and future prospects. Having spent too much time in the immoderate atmosphere of WorldCom chatrooms, I excoriated them, though mildly, for their shortsightedness and exhorted them to look at the company differently.

Finally, out of frustration with the continued decline of WCOM stock, I emailed Bernie Ebbers, then the CEO, in early February 2002 suggesting that the company was not effectively stating its case and quixotically offering to help by writing copy. I said I'd invested heavily in WorldCom, as did family and friends at my suggestion, that I could be a persuasive wordsmith when I believed in something, and WorldCom, I believed, was well positioned but dreadfully undervalued. UUNet, the "backbone" of much of the Internet, was, I fatuously informed the CEO of the company, a gem in and of itself.

I knew, even as I was writing them, that sending these electronic epistles was absurd, but it gave me the temporary illusion of doing something about this recalcitrant stock other than dumping it. Investing in it had originally seemed like a no-brainer. The realization that doing so had indeed been a no-brainer was glacially slow in arriving. During the 2001–2002 academic year, I took the train once a week from Philadelphia to New York to teach a course on "numbers in the news" at the Columbia School of Journalism. Spending the two and a half hours of the commute out of contact with WCOM's volatile movements was torturous, and upon emerging from the subway, I'd run to my office computer to check what had happened. Not exactly the behavior of a sage long-term investor; my conduct even then suggested to me a rather dim-witted addict.

Recalling the two or three times I almost got out of the stock is dispiriting as well. The last time was in April 2002. Amazingly, I was even then still somewhat in thrall to the idea of averaging down, and when the price dipped below $5, I bought more WCOM shares. Around the middle of the month, however, I did firmly and definitively resolve to sell. By Friday, April 19, WCOM had risen to over $7, which would have allowed me to recoup at least a small portion of my losses, but I didn't have time to sell that morning. I had to drive to northern New Jersey to give a long-promised lecture at a college there. When it was over, I wondered whether to return home to sell my shares or simply use the college's computer to log onto my Schwab account to do so. I decided to go home, but there was so much traffic on the cursed New Jersey turnpike that afternoon that I didn't arrive until 4:05, after the market had closed. I had to wait until Monday.

Investors are often nervous about holding volatile stocks over the weekend, and I was no exception. My anxiety was well-founded. Later that evening there was news about impending cuts in WorldCom's bond ratings and another announcement from the SEC regarding its comprehensive investigation of the company. The stock lost more than a third of its value by Monday, when I did finally sell the stock at a huge loss. A few months later the stock completely collapsed to $.09 upon revelations of massive accounting fraud.

Why had I violated the most basic of investing fundamentals: Don't succumb to hype and vaporous enthusiasm; even if you do, don't put too many eggs in one basket (especially with the uncritical sunny-side up); even if you do this, don't forget to insure against sudden drops (say with puts, not calls); and even if you do this too, don't buy on margin. After selling my shares, I felt as if I were gradually and groggily coming out of a self-induced trance. I'd long known about one of the earliest "stock" hysterias on record, the seventeenth century tulip bulb craze in Holland. After its collapse, people also spoke of waking up and realizing that they were stuck with nearly worthless bulbs and truly worthless options to buy more of them. I smiled ruefully at my previous smug dismissal of people like the tulip bulb "investors." I was as vulnerable to transient delirium as the dimmest bulb-buying bulbs.

I've followed the ongoing drama of the WorldCom story—the fraud investigations, various prosecutions, new managers, promised reforms, and court settlements—and, oddly perhaps, the publicity surrounding the scandals and their aftermath has distanced me from my experience and lessened its intensity. My losses have become less a small personal story and more (a part of) a big news story, less a result of my mistakes and more a consequence of the company's behavior. This shifting of responsibility is neither welcome nor warranted. For reasons of fact and of temperament, I continue to think of myself as having been temporarily infatuated rather than deeply victimized. Remnants of my fixation persist, and I still sometimes wonder what might have happened if WorldCom's deal with Sprint hadn't been foiled, if Ebbers hadn't borrowed $400 million (or more), if Enron hadn't imploded, if this or that or the other event hadn't occurred before I sold my shares. My recklessness might then have been seen as daring. Post hoc stories always seem right, whatever the pre-existing probabilities.

One fact remains incontrovertible: Narratives and numbers coexist uneasily on Wall Street. Markets, like people, are largely rational beasts occasionally provoked and disturbed by their underlying animal spirits. The mathematics discussed in this book is often helpful in understanding (albeit not beating) the market, but I'd like to end with a psychological caveat. The basis for the application of the mathematical tools discussed herein is the sometimes shifty and always shifting attitudes of investors. Since these psychological states are to a large extent imponderable, anything that depends on them is less exact than it appears.

The situation reminds me a bit of the apocryphal story of the way cows were weighed in the Old West. First the cowboys would find a long, thick plank and place the middle of it on a large, high rock. Then they'd attach the cow to one end of the plank with ropes and tie a large boulder to the other end. They'd carefully measure the distance from the cow to the rock and from the boulder to the rock. If the plank didn't balance, they'd try another big boulder and measure again. They'd keep this up until a boulder exactly balanced the cattle. After solving the resulting equation that expresses the cow's weight in terms of the distances and the weight of the boulder, there would be only one thing left for them to do: They would have to guess the weight of the boulder. Once again the mathematics may be exact, but the judgments, guesses, and estimates supporting its applications are anything but.

More apropos of the self-referential nature of the market would be a version in which the cowboys had to guess the weight of the cow whose weight varied depending on their collective guesses, hopes, and fears. Bringing us full circle to Keynes's beauty contest, albeit in a rather forced, more bovine mode, I conclude that despite rancid beasts like WorldCom, I'm still rather fond of the pageant that is the market. I just wish I had a better (and secret) method for weighing the cows.

Bibliography

The common ground and intersection of mathematics, psychology, and the market is a peculiar interdisciplinary niche (even without the admixture of memoir). There are within it many mathematical tomes and theories ostensibly relevant to the stock market. Most are not. There are numerous stock-picking techniques and strategies that appear to be very mathematical. Not many are. There are a good number of psychological accounts of trading behavior, and a much smaller number of mathematical approaches to psychology, but much remains to be discovered. A few pointers to this inchoate, nebulously defined, yet fascinating area follow:

Bak, Per, *How Nature Works*, New York, Springer-Verlag, 1996.

Barabasi, Albert-Laszlo, *Linked: The New Science of Networks*, New York, Basic Books, 2002.

Dodd, David L., and Benjamin Graham, *Security Analysis*, New York, McGraw-Hill, 1934.

Fama, Eugene F., "Efficient Capital Markets, II," *Journal of Finance*, December 1991.

Gilovich, Thomas, *How We Know What Isn't So*, New York, Simon and Schuster, 1991.

Hart, Sergiu, and Yair Tauman, "Market Crashes Without Exogenous Shocks," The Hebrew University of Jerusalem, Center for Rationality DP–124, December 1996 (forthcoming in *Journal of Business*).

Kritzman, Mark P., *Puzzles of Finance,* New York, John Wiley, 2000.

Lefevre, Edwin, *Reminiscences of a Stock Operator,* New York, John Wiley, 1994 (orig. 1923).

Lo, Andrew, and Craig MacKinlay, *A Non-Random Walk Down Wall Street*, Princeton, Princeton University Press, 1999.

Malkiel, Burton, *A Random Walk Down Wall Street*, New York, W. W. Norton, 1999 (orig. 1973).

Mandelbrot, Benoit, "A Multifractal Walk Down Wall Street," *Scientific American*, February 1999.

Paulos, John Allen, *Once Upon a Number*, New York, Basic Books, 1998.

Ross, Sheldon, *Probability*, New York, Macmillan, 1976.

Ross, Sheldon, *Mathematical Finance,* Cambridge, Cambridge University Press, 1999.

Siegel, Jeremy J., *Stocks for the Long Run*, New York, McGraw-Hill, 1998.

Shiller, Robert J., *Irrational Exuberance*, Princeton, Princeton University Press, 2000.

Taleb, Nassim Nicholas, *Fooled by Randomness,* New York, Texere, 2001.

Thaler, Richard, *The Winner's Curse*, Princeton, Princeton University Press, 1992.

Tversky, Amos, Daniel Kahneman, and Paul Slovic, *Judgment Under Uncertainty: Heuristics and Biases,* Cambridge, Cambridge University Press, 1982.

Index

D0056258

DRIFT HOUSE

The First Voyage

DRIFT HOUSE

The First Voyage

DALE PECK

Published by Bloomsbury Publishing, New York, London, and Berlin
Distributed to the trade by Holtzbrinck Publishers

Library of Congress Cataloging-in-Publication Data
Peck, Dale.
Drift House : the first voyage / Dale Peck.—1st U.S. ed.
p. cm.
Summary: Sent to stay with their uncle in a ship-like home called Drift House, twelve-year-old Susan and her two younger stepbrothers embark on an unexpected adventure involving duplicitous mermaids, pirates, and an attempt to stop time forever.
ISBN-10. 1-58234-969-X
ISBN-13: 978-1-58234-969-5
[1. Time—Fiction. 2. Uncles—Fiction. 3. Mermaids—Fiction. 4. Pirates—Fiction.
5. Stepfamilies—Fiction.] I. Title.
PZ7.P3338Dri 2005 [Fic]—dc22 2005047067

First U.S. Edition 2005
Printed in the U.S.A.
1 3 5 7 9 10 8 6 4 2

Bloomsbury Publishing, Children's Books, U.S.A.
175 Fifth Avenue, New York, NY 10010

All papers used by Bloomsbury Publishing are natural, recyclable products
made from wood grown in well-managed forests. The manufacturing processes
conform to the environmental regulations of the country of origin.

For
Noah and Rebecca Wertheimer,
and their parents.

And for
Nick Debs, who was generous
enough to share his dream with me.

"For all things proceed out of the same spirit,
which is differently named love, justice, temperance,
in its different applications, just as the ocean
receives different names on the several shores which it washes."

—Emerson

PART ONE

A Journey to the North

ONE

An Unexpected Announcement

After the towers came down, Mr. and Mrs. Oakenfeld thought it best that their three children go and stay with their uncle in Canada. The Oakenfeld family lived high on the Upper East Side, so Susan, Charles, and Murray understood very little of what was going on downtown. Their parents wouldn't even let them watch television, so it's understandable that the children were mostly concerned—at least at first—with how the move would affect school. Susan, in particular, has just joined the eighth grade debating club, and she was quite annoyed. When she was nine she had decided she would be a lawyer like Mr. Oakenfeld: she had been waiting to start debate for three whole years. Whereas Charles, in fifth grade, was secretly relieved. He was taking special classes at a magnet high school for science, and two days a week had to ride the West Side train all the way up to 205th Street in the Bronx, and he found the older boys were more than a little intimidating. At five, Murray was only in kindergarten and didn't care too much about all that. Of course, he still didn't want to leave his mother and father.

"But Uncle Farley has just moved into a gorgeous old house on the Bay of Eternity," Mrs. Oakenfeld told Murray. "He tells me there are pelicans and puffins, and tide pools with starfish, and the most beautiful sunrises you've ever seen. And the house has rooms and rooms and rooms—an enormous attic, and some kind of tower-gallery-type thingy for showing off his art collection, and every imaginable exotic plant in the solarium."

"Yay, Solar Mum!" Murray yelled, and ran off to pack his suitcase.

As the eldest—and a future lawyer to boot—Susan thought it was her job to be a bit more skeptical.

"The Bay of Eternity? Really, Mum, that's *quite* the queer* name for a place."

"Don't say 'queer,'" Charles said. "It's affected."

*Because Susan and Charles have a penchant for ten-dollar-words, we've provided a glossary to help decode some of their more affected arguments. The first word you'll probably want to look up is *penchant.*

I suppose I should explain right off that Susan was Charles and Murray's half sister. Before Mrs. Oakenfeld had come to America and married Mr. Oakenfeld, she had lived in England, only there she had been Mrs. Wheelwright. Lieutenant Wheelwright had died when Susan was just a baby, and she called Mr. Oakenfeld Daddy just as Charles and Murray did, and indeed she thought of him as her father. But some part of her clung to England, for though she had lived in America since she was two years old she insisted on calling Mrs. Oakenfeld Mum instead of Mom. In fact all the children called Mrs. Oakenfeld Mum, but only Susan said it with an English accent—a habit for which Charles teased her constantly.

"I say, Mum," Susan said again, ignoring her younger brother (who, as usual, had his nose buried in some boring-looking science magazine). "The Bay of Eternity *is* a queer name for a place. It sounds positively Vic*tor*ian."

"Don't say 'Vic*tor*ian.' " Charles drew out the second syllable somewhat longer than his sister had, then turned a page in his magazine—which, in fact, wasn't a magazine at all, but a catalog that offered replacement parts for old radios. The previous summer Charles had developed a passion for what he called antiquated technology, and the bedroom he shared with Murray was overrun with half-assembled (or, more accurately, half-disassembled) radios and telephones and one ancient television whose console was nearly as big as his dresser, even though the screen was smaller than the one on his father's laptop. "It's affected," Charles added, and flipped another page.

"It does sound a bit qu-quaint," Mrs. Oakenfeld said (though Susan was sure she'd been going to say "queer"). "But I think perhaps you shouldn't mention it to Uncle Farley. He might not see it that way."

Susan stuck out the side of her cheek with her tongue, as she often did when she was considering her options. She had pale thin cheeks and very short dark hair, so the lump her tongue made was quite pronounced.

"Does Uncle Farley have a telly?" she said finally.

"I believe they have television in Canada," Mrs. Oakenfeld said.

"I *meant*, does he have *cable*?"

Now Mrs. Oakenfeld understood. In addition to reading copious amounts of English literature—which is where she'd picked up somewhat affected words such as "queer" and "Vic*tor*ian"—her eldest child habitually tuned in to BBC World to keep her accent in top polish. She was especially fond of the news, and had been frustrated by not being allowed to watch it in recent days.

"I'll make you a deal," Mrs. Oakenfeld said. "If you go up to Uncle Farley's without any fuss, I'll see that you're provided with all the quality British programming you can watch."

A siren sounded in the street below the Oakenfelds' apartment, and the conversation lulled. Susan's cheek bulged as though she were sucking on a golf ball. Part of her was contemplating a move to a house she'd never seen inhabited by an uncle she'd never met, and part of her was contemplating just what, exactly, the siren in the street might indicate. It was a boggling proposition, and more than a little frightening, and only the strained expression on Mrs. Oakenfeld's face kept Susan from giving her mother the third degree. Although Susan claimed to have been waiting to start debating for three years, it was generally agreed in the Oakenfeld household that she had been very *actively* waiting, frequently annoying her parents and Charles with her cross-examinations.

"I suppose it will have to do," she said eventually. And she shook her mother's hand to seal the deal.

Charles turned a page noisily, but didn't say anything. As the well-behaved middle child, he was all too used to no one asking what *he* wanted.

TWO

Uncle Farley

"HE WILL BE AN OLD man," Susan announced, when at last the car turned left and Mrs. Oakenfeld said that this was the road her brother lived on. "He will have no hair except what grows out of his ears and nostrils, and he will wear spectacles and a brown cardigan with toast crumbs and sauce stains on it, even to bed. And he will be a Luddite."

Murray looked horrified. He was just beginning to understand there was no "Solar Mum" awaiting them at Uncle Farley's, and now he threw his arms around his mother's neck and said, "What's *that?*"

"A Luddite," said Charles, not bothering to look up from his book, "is someone who dislikes technology."

"Oh," Murray said, relieved. And a moment later: "What's technology?"

"Computers and video games and things like that," Mrs. Oakenfeld said, combing Murray's long black bangs out of his eyes. "Telephones and microwave ovens and cell phones." She reached over and patted Charles on the head, who suffered her attention like a disgruntled cat. "The kinds of things your brother likes."

Murray frowned.

"I think I would like to stay with you, Mummy."

"Don't be silly," Mrs. Oakenfeld said. "You know better than to judge someone before you've met him. You will have grand adventures at Uncle Farley's house. Which, I might add," she said, looking at Susan, "has *all* the modern conveniences."

"Mum," Susan said, switching tacks. "How come we've never visited Uncle Farley before?"

"Well, your uncle only moved here last year. And the Bay of Eternity *is* rather out of the way."

"I'll say," Charles said. "We've been driving for *days*."

"Normally we would take a plane, but . . ." Mrs. Oakenfeld shook her head. "As I said, Farley only moved to Drift House last year. Before that he was researching something in Rome, or was it Samarqand? I can never remember, he gets around so much with his studies. And your uncle *does* like his solitude. His taking you on is really an enormous favor to your father and me, which is why we want you to be on your best behavior while you're here."

"How long *will* we be here?" Susan said.

Mrs. Oakenfeld stroked Murray's hair. "Until it's safe to come back," she said, very softly.

Under different circumstances, Susan might have persisted. But when they'd driven out of the city three days ago, she had gotten a better picture of the sirens and the smoke and the soldiers with drawn guns. Down at the bottom of Manhattan, where once there had been a pair of silver towers, there was now a single mountain of ash and sparks, and though Susan couldn't quite grasp the full significance of that mountain, it still struck a pit of fear and sadness in her stomach. She had been on her best behavior since she'd seen it, and done what she could to keep her brothers under control—a difficult task, with Charles making fun of every other word out of her mouth. Her mother had rented a limousine for the drive, and at first it seemed glamorous, with all the children pretending to be this or that movie star or musician, waving to their fans in passing cars, or covering their faces to avoid the paparazzi. But after three days, even a limousine feels cramped, and everyone's nerves were at the breaking point. Susan had to bite her tongue to keep from interrogating her mother further, and instead watched as Mrs. Oakenfeld smiled brightly—if falsely, to Susan's perceptive eye—and rapped lightly on the glass.

"Driver, turn here. Through the gates."

"There are *gates?*" Charles said, looking up from his book as the car passed through an enormous archway of gray stone.

"Yes," Mrs. Oakenfeld said, "but Farley tells me they're so rusty he can't even close them. They're left over from colonial days, when the lord of the manor needed to make a formidable impression."

"Lord of the *manor?*" Susan said.

Mrs. Oakenfeld nodded at her daughter. "That looks like the old crest there," she said, pointing at a shield set into one of the open iron gates.

Though the paint had faded, Susan could still make out a blue pennant that bore the faint inscription DRIFT HOUSE in flecked gold letters, and below that the shape of a single-masted sailing ship set between a pair of crossed swords.

What looked like the wings and head of a dragon was emblazoned on the sail, and there was another dragon's head mounted on the front of the boat. Susan wasn't sure because the car went by so fast, but she thought the ship was the kind the Vikings sailed. This was worth noting, she thought, because Drift House was located in the province of Quebec, which, as she'd learned in history, had been settled by the French, not the Norwegians. Susan had gotten an A in history.

Which reminded her:

"Mum? What are we going to do about school?"

Mrs. Oakenfeld smiled wanly at her daughter. "Farley says there's a school in the village just a little ways up the road."

"*Village*?" Susan gasped. Horror at the prospect overcame her restraint. "How—how *provincial*." But she spoke quietly, not because she didn't want to upset her mother, but because her attention had been claimed by the dark thick trees that had suddenly closed in on both sides of the winding driveway.

Like most people raised in a city, Susan was familiar only with the kinds of trees that grew in straight lines in parks or along sidewalks. These trees were much bigger than those overgrown shrubs, taller and fatter and much, much darker: gigantic fir and spruce and larch, their needles clotting the air, their spines dark with pitch and moss. Here and there an oak or maple's broad limbs lightened the view, the leaves of the latter already tinted gold, and when Susan's eyes had adjusted to the light she saw that red creeper and grapevines laden with purple berries hung off nearly every tree like a spray of necklaces tumbling out of a jewelry box. All at once Susan realized that everything she'd ever heard about shadowy forests being scary or haunted was wrong. Under different circumstances, she felt she would have asked her mother to stop the car so she could get out and swing on a vine.

Mrs. Oakenfeld, meanwhile, was trying to reassure Susan about school. "Don't worry," she said, "Farley has agreed to tutor you himself. He speaks ancient Greek and Latin, and several other dead languages, and he was always winning prizes at school in science and math and history. You'll get quite the classical education."

Susan tore her eyes away from the window. Really, she didn't mind behaving. But she couldn't stand it when her mother talked to her like a child.

"No one actually *speaks* Latin or Greek," she pointed out. "That's why they're called *dead languages*."

"Perhaps it was French then, or Chinese. At any rate your uncle is terribly, terribly bright. You could spend a year with him and not learn the tenth part of what he knows."

Susan couldn't help herself. "Are we going to spend a *year* here?"

But before Mrs. Oakenfeld could answer, Charles cried out: "Look!"

The driveway had wound its way out of the forest at the top of a high hill, and down below an expanse of shining green lawn lay before them. The lawn was flat as a sheet on a freshly made bed, and stretched all the way to the crystal blue water of the bay, which was equally flat. Between the green grass and blue water lay a thin pale stripe of sand, and Susan had time to think that the sandy beach resembled a bar of light shining between two curtains that hadn't quite been drawn together, and then the car descended the hill and the view was blocked by the biggest house she had ever seen.

"Why, it's enormous!" Susan said.

"It's gargantuan!" Charles added. For, though he claimed to dislike Susan's affected manner of speaking, he knew more than a few big words himself.

Murray, who was much shorter than his brother and sister (even though he was sitting on Mrs. Oakenfeld's lap) saw the house last.

"Shipwreck!" he yelled, and clapped his hands.

In fact the house *did* look like a ship washed up on land. The main body of the building was three stories tall, with walls sided in white clapboard that hadn't been painted in so long the wood had turned dark and feathery gray, like the belly of a goose. Two short promenades on either side of the house led, on the left, to a squat octagonal tower, and, on the right, to a three-sided glass pyramid whose leading edge seemed to slice into the dark green grass like a prow. The roof of the tower, the promenades, and the main body of the house were all flat, and run all the way around with balustrades such as those found on the deck of an old-fashioned galleon. Only the roof of the pyramid was pointed (of course), and crowning its apex was a tall thin weathervane.

"That must be the gallery," Mrs. Oakenfeld said, pointing Murray toward the tower, "and that's the Solar Mum." Winking, she pointed to the pyramid.

"The word is *solarium*—"

Susan was cut off by a shout from Murray.

"And that's the cannon!" he exclaimed, pointing to what indeed looked like the heavy barrel of a cannon poking from the balustrade of the main roof. In fact there was a pair of them—one at each corner of the deck, aimed as if to protect the house from a two-pronged attack.

"Well, ah, yes," Mrs. Oakenfeld said, seeming a little taken aback by the sight of cannons on the house where she was sending her children to live. "The house *is* very old. But Farley says he's installed his art collection," she went on in a brighter voice. "Painting, sculpture, illuminated manuscripts, all sorts of funny little curios he's collected from here and there. I told you," she patted Murray on the head, "it's quite the grand place."

"Grand perhaps," Susan said, "but crooked."

"I beg your pardon," Mrs. Oakenfeld said. "Whatever do you mean, Susan?"

By now the car had stopped in front of the long straight walk, and Susan hopped onto the freshly raked gravel without waiting for the driver to let her out.

"Just look. The front door is over there, but this path leads sort of, I dunno, to a corner. It's all"—there was no other word for it—"crooked."

"Well, there must be a door *some*where," Charles said, climbing out behind his sister. He held his book closed with his finger between the pages so he wouldn't lose his place and hurried down the path after her.

Like the driveway, the path to the house was gravel, freshly raked, and straight as an arrow. It dead-ended at a pair of boxwoods trimmed to look like totem poles: a pyramid on top of a rectangle on top of an octagon, as if depicting the house set on end. The topiary looked like something that would flank the entryway to a country manor, but there was no door in sight: only the airtight seam where the two sides of the house met in a corner. High above, the snout of the cannon was just visible over the roof, and something that looked like green moss grew from the siding. Lichen, Charles thought, though he didn't say it aloud.

"This *is* rather odd," Mrs. Oakenfeld said, coming up behind them, Murray trailing after.

"And look," Susan said. "There are hedges and things—"

"Spirea, dear. And burning bush."

"Spirea and burning bush"—Susan didn't like the sound of that one—"all lined up at right angles like you'd see growing around the edges of a house. But the house itself is—"

"Crooked," Mrs. Oakenfeld said, peering at the bright red leaves of a hapless bush quashed under the side of the house. "Yes, it does indeed appear to . . . slant."

"There must be a door *some*where," Charles repeated. "Whoever heard of a front path leading to the *corner* of a house?"

There was clearly no door at the end of the path, but it was like Charles to ignore the evidence when the conclusion it pointed to was less than logical—as, for instance, the fact that the plants growing from the house's siding looked more like barnacles than lichen. Meanwhile, Susan had begun looking around the neat grounds: lines of hedge—spirea, as her mother had pointed out, and bright red burning bush—and flowerbeds, and paired rows of Japanese pagoda trees planted in lines that ran parallel to each other. Then she turned back to the corner of the house, all askew in relation to the garden's geometry. To her left the low tower of the gallery bulged across the lawn almost all the way to the driveway, while to the right the glass pyramid canted off toward the bay. It was almost as if someone had picked the house up off its foundation and put it back down carelessly, an impression that was reinforced by all the ivy that one would have expected to see growing up the walls fallen in tattered lines on top of the poor hydrangeas, netting them to the ground. The Oakenfelds picked their way through the tangled vines carefully so as not to trip over their cool green tentacles; Mrs. Oakenfeld put Murray down because she said he was too big to carry through such an obstacle course, but then she had to pick him up again because he was too small to make his own way, and when they finally made it to the front door everyone was winded except Murray, who wanted to go back through the "hop-stickle" course again.

"Why don't you ring, er, knock the, um, knocker instead?" Mrs. Oakenfeld suggested breathlessly.

Murray set to banging the knocker with a passion. It was big and brass and shaped like a ship's wheel, and he used both hands to smash it against the plate over and over until the door opened with a sudden whoosh, pulling the knocker out of Murray's hands, and nearly pulling Murray out of his mother's.

"I say!" said a tall bearded man in a blue shirt. "There's no need to bang the house down—Annie!" And all at once the man threw his arms around Mrs. Oakenfeld, and Murray too, since he was still in his mother's arms.

"Help!" Murray yelled. "You're squeezing me to death! Help, help!" But he clearly enjoyed being in the middle of such a warm embrace.

When at last Uncle Farley released Mrs. Oakenfeld and Murray and stepped back a pace, it was apparent he was nothing like the portrait Susan had painted in the car. For one thing, he was younger than his sister, and his hair, just like Mrs. Oakenfeld's, was thick and wavy and pale brown. He did wear spectacles (although perhaps we should not be so affected as Susan, and call them glasses), thin silver wires that nearly disappeared into his lightly furred round cheeks, but his blue eyes were so sharp they seemed to pierce right through the lenses. He was broad shouldered and a little rotund—what would have once been called stout—and all in all such a pleasant alternative to the withered ogre Susan had conjured that the children forgot about the crookedness of the house and instead followed him into the front hall. The room was just warm enough to make them realize there'd been a bit of fall chill in the air outside. It was getting on in September, remember, and it was Canada. The season was further advanced than it had been in New York City.

Drift House's entrance hall was so large that the living room of the Oakenfelds' Manhattan apartment could have fit into it with feet to spare. The hall stretched all the way from the front of the house to the back, and was paneled in pale burled beechwood, with three long Oriental rugs covering the floor in swirls of faded red and violet and green, and a pair of crystal chandeliers hanging overhead. Enormous paintings—the kind the Oakenfelds had only seen in museums—hung on the walls, and impressive pieces of silver and porcelain stood on highly polished side tables. There was even a white marble bust, blank eyed and serene, regarding the newcomers from an ebony plinth. Even Charles, who was normally only impressed by tangles of multicolored wires and tubes, mouthed an awed, "Wow!"

In fact, Drift House wasn't quite as grand as the children's initial reaction made it out to be, but growing up in New York City had given them a limited perspective on private dwellings, just as it had on trees. The house was large but hardly a mansion, with two good-sized parlors opening off each side of the central hallway on the first floor. On the second and third floors, the hallway was narrower, and ran the length rather than the width of the house. There were an awful lot of bedrooms up there, especially when you considered that Uncle Farley lived alone, but if you stopped to think of the place housing a provincial governor and his family and servants and guests and prisoners of war (Uncle Farley added this last with a twinkle in his eye, and Susan was pretty sure it was just for Murray's sake) then it actually began to seem a little cramped, as Uncle Farley explained while leading the Oakenfelds from room to room. He peppered his tour with things such as "It's been so long since I've seen you, Annie," and "Susan is the spitting image of you at that age," and "What's the quiet one reading?"—this last of Charles, who blushed and found that he was too shy to speak. Not that he could have gotten a word in edgewise, what with Murray's constant "What's that?" and "What's that?" and "What in the world is *that*?"

Uncle Farley led Murray by the hand as though giving him a private viewing that the rest of the Oakenfelds just happened to be tagging along on. He had to stoop to do so, but he never complained, and every once in a while he would swing Murray by the arm like the censers the alter boys swung at church at Christmas and Easter. It was only after they'd finished walking through all three floors of the house that it occurred to Susan to think it a bit odd that the first thing Uncle Farley should do is give them a grand tour—rather than, say, introduce himself properly to his niece and nephews, or offer them something to eat. But even though her stomach was rumbling, she sensed that if she lived at Drift House then she too would want to show it off. The only thing that really alarmed her was the fact that she hadn't seen a television *anywhere.*

"My goodness," Uncle Farley said to Murray, "was that your stomach I just heard, rumbling to shake the house down?"

Murray giggled and shook his head.

"Uncle Farley," he said, "can I see the Solar Mum?"

"What's this? Has someone's mummy been telling him stories?"

"Murray, darling," Mrs. Oakenfeld said, "there'll be plenty of time for exploring. Right now we should see about lunch, and then unpacking, and then I'm afraid Mummy has to get back to New York."

Lunch was served in the music room, which held a long gilded box mounted on legs, in front of which sat a red plush bench. The box looked to Susan like an ornate coffin, and she couldn't help exclaiming:

"What a queer-looking piano!"

As soon as the words were out of her mouth she braced herself for one of Charles's comments. But it was worse than she could have expected.

"It isn't a 'queer-looking piano,' " he said (Susan could hear the inverted commas in his voice, even if no one else could). "It's a harpsichord."

He seemed about to launch into a lecture on the technical differences between the two instruments when Uncle Farley said, "Who's for sandwiches?"

"Me!" Murray said. "I'm starving!"

Susan, red faced at being caught out by her younger brother (it was she who had taken piano lessons, after all: Charles studied violin), nervously set the metronome on the harpsichord to ticking before sitting down to what seemed like dozens of egg- and tuna-salad sandwiches stacked in neat triangles on a pair of covered silver platters, and ice-cold lemonade that Uncle Farley served in the kind of fancy crystal glasses the Oakenfelds only used when they had company. The meal would have been perfectly delicious had the metronome not been ticktocking relentlessly in the background, as if counting off the seconds until Mrs. Oakenfeld had to leave.

After they'd eaten, the Oakenfelds' driver helped Uncle Farley carry the children's suitcases up to the second floor. There was a bit of a scene when Charles said that if Susan was going to get a bedroom of her own then he should get a room of his own too, because he was practically as old as she was now, but Murray said he wanted to sleep with Charles as he did in their apartment in New York. An awkward moment passed before Charles assented to share a room with his younger brother, somewhat less than graciously.

"I hate being the middle child!" he said when the grownups were out of earshot.

Still smarting from the piano/harpsichord mistake, Susan said tartly, "Maybe you should run away and join another family then."

"No no no!" Murray said, sounding genuinely horrified. "I want Charles to stay!"

Charles, who was about to say he *would* run away if he felt like it, relented for Murray's sake. "It's all right, old boy, I'll stick around for a while at least."

"Don't say 'old boy,'" Susan taunted. "It's affected." But when Charles caught his breath she immediately regretted her words. "Old boy" was what their American father called Murray, simultaneously teasing his youngest son and his British wife, and Susan realized that Charles was just trying to relieve some of the stress of the tense situation: their mother *was* about to leave them, after all, with a man they had only just met.

All of a sudden Susan had a moment of panic. How long *would* their mother leave them? And what were they going to do—in the middle of nowhere, with no school, no friends, no discernible television even? She hurried down to the music room, where her mother and uncle were sitting next to each other on the couch, sipping tea from delicate white cups. Susan noted that they weren't laughing and chatting as you might expect a brother and sister to do after a long separation, but facing each other stiffly, with nervous expressions on their faces. As politely as she could, she asked to speak to her mother alone.

Mrs. Oakenfeld and Uncle Farley exchanged a glance, and then Uncle Farley said, "Of course, of course. I was just going to clear lunch away." And he gathered the dishes on a tray and walked out of the room.

Smiling, Mrs. Oakenfeld took her daughter's hand and pulled her to the couch.

"Yes, Susan?"

Susan didn't mince her words. "I don't want to stay here. I want to go home with you."

Mrs. Oakenfeld squeezed Susan's hand warmly. "Do you think I want to be separated from you and your brothers? I dislike it as much as you do. But I am afraid these are extraordinary times, and as such call for extraordinary measures. Your father and I simply don't think it's safe for you, right now, in New York."

Susan hated it when Mrs. Oakenfeld spoke to her like a child. But now that her mother was speaking to her like an adult she found she liked it even less—especially when those adult words made Susan think of that giant mountain of ash engulfing the whole city, and everyone in it.

"Then stay here," she said. "Why is it safe for you and not us? Please, Mum. Call Dad and have him come here. There's plenty of room. Please," Susan said again. "Don't leave us!"

Susan's voice had risen steadily while she spoke, and her

last words hung sharply in the air. Mrs. Oakenfeld waited till their echo had faded from the room before she spoke.

"Your father and I have responsibilities," she said, clearly laboring to keep her voice steady. "I wish it were otherwise, but this is the best we can come up with, at the moment. I know you don't understand now, but you will one day, when you're older."

"But what could be more important than being with us?"

Susan could feel her mother's hand tremble as it held hers. "Susan, please," Mrs. Oakenfeld said. "Please don't make this any harder. Right now I need you to be grown up. Not just for me. For your brothers. They're nervous too, even if they don't understand the seriousness of the situation as well as you do. They're going to take their cue from their older sister, so I need you to set a good example. If—if you break down, then it will just be that much harder for everyone."

Susan blushed hotly.

"I'm scared," she said.

"There's nothing for you to be scared of," Mrs. Oakenfeld said, smiling at her daughter. "You're safe as houses here."

"Not for us," Susan said. "For you and Dad."

Mrs. Oakenfeld didn't—couldn't—speak for a long time. Then, swallowing audibly, she withdrew her hand from Susan's and stroked her daughter's hair.

"Since your father—your real father—died, I have tried to shelter you as much as possible. Perhaps too much. I forget how wise you are sometimes. Wise and strong. But I promise you Dad and I will take care of ourselves. I hope you'll take equally good care of Charles and Murray."

Susan didn't like the situation. But she understood that her mother, like her, had no choice. And, sitting up straight, she said firmly:

"I promise."

Then suddenly it was time for Mrs. Oakenfeld to leave. Charles told Murray to keep a stiff upper lip or else Mummy might cry, but in the end she cried anyway, and so did Murray, although only a little. Mrs. Oakenfeld hugged Murray and Charles and Susan individually, and then she hugged Uncle Farley, and then she hugged all three of the children at the same time. There were kisses all around, and then Mrs. Oakenfeld was in the long black car. It slunk up the hill and disappeared into the trees.

Throughout his sister's leave-taking Uncle Farley had stood very quiet and patient. He watched the car until it was no longer visible, and only then did he turn back to face the three Oakenfeld children. They were standing very close to him, and even Susan had to crane her head far back to see her uncle's bearded face.

"Well now, children," Uncle Farley said. "I think it's high time I showed you the dungeon!"

THREE

The Talking Parrot

UNCLE FARLEY STOOD AS TALL and thick as one of the trees in the forest that had swallowed Mrs. Oakenfeld's car. The sun was at his back and his face was lost in shadow, and when his deep voice said, "I think it's time I showed you the dungeon!" Susan, the procedural sibling, prepared to grill her uncle on the particulars of Drift House's alleged dungeon, whereas Charles, the logical one, thought that Drift House couldn't *possibly* have a dungeon—could it? Only Murray seemed to understand that their uncle was joking, and he laughed out loud and jumped up and down between his siblings.

"Hooray! Dungeon!" He proceeded to make noises that

Susan thought were supposed to sound like the rattling of chains.

"Uncle Farley!" she said, still shaking. "You gave us—you gave Murray a very nasty turn!"

Charles, holding Murray's other hand, was going to tell Susan not to say "very nasty turn," but he was afraid that if he opened his mouth everyone would hear his teeth chattering.

"Oh, I think Murray will survive," Uncle Farley said now. "Won't you, old boy?" He tousled Murray's hair and Murray giggled and nodded his head.

"Uncle Farley," he said, asking the question both Susan and Charles were too shy to ask. "Is there really a dungeon?"

"No dungeon, I'm afraid. Just the wine and fruit cellar. If you're very good I'll show it you later."

"Yay, fruit seller!" Murray whooped. "I want to buy some fruit!"

"Come on then. We'll tour the grounds, and by the time we get back I imagine Miss Applethwaite will have whipped up some apple pie and ice cream for a snack."

Charles, hearing apple pie mentioned, perked up. "Who's Miss Apple-thw-thw-thw—"

"Yes, it's a tricky one, isn't it? Thwaite," Uncle Farley enunciated. "Apple-thwaite. She helps with the shopping and cooking and such. Your uncle is a man of many talents, but none of them extend to the culinary arts, I'm afraid."

Uncle Farley spent the next two hours leading the

children around the estate. The great lawn was even larger than it had appeared from the top of the hill, and every square inch of it was covered with the thickest greenest grass the children had ever seen. It was so long it fell over in little tufts and hillocks instead of being mowed short like a playing field, and it was damp too, so that within a few minutes everyone's shoes were wet and smudged with green. Uncle Farley took them all the way down to the bay, where a whitish-greenish rowboat was laid keel up like an endive fallen from a giant salad. The boat was very small, and looked even smaller when Susan measured it against the vastness of the Bay of Eternity. Up close the water didn't seem as flat as it had before, but dimpled like a hastily frosted cake, and out in the distance a dark line of clouds blended into the gray water. The breeze coming in off the bay felt stronger than it had when they first arrived, and quite cool as well. Susan wondered if it was going to rain.

She glanced at the dinghy again. "Do you take that little boat out there, Uncle Farley?"

"Oh, I end up on the water every once in a while," Uncle Farley said mysteriously. "Every now and again."

"Uncle Farley," Susan said, "the Bay of Eternity is a rather qu-qu—a rather peculiar name for a place, isn't it?"

Uncle Farley pursed his lips for a moment. "Not really. Not when you get to know it better." Just then a stiff, damp breeze misted the faces of everyone on the beach as it gusted in off the water. "Smells like rain," he said, confirming Susan's forecast. "We'd better turn back."

As they headed toward the house, Susan noticed for the first time that the hill they'd driven over was actually part of a line of cliffs that stretched in a crescent all the way around the house. With the sun well on its way to setting, the palisades were dark and seemed to cut the grounds off from the continent behind them, and gave the pale dwelling a lonely, even melancholy air. As they got closer to the building Susan saw what appeared to be a yew bush half squashed under the sharp corner of the solarium, which again gave her the impression of a ship's prow ripping right through the garden. Yes, the house definitely seemed to be crooked in relation to the grounds.

Speaking of things that were out of alignment: the breeze was blowing heavily at her back, but Susan noticed that the weathercock atop the solarium stood out in bold silhouette, even though the breeze should have turned it 90 degrees in the other direction. Perhaps it's merely ornamental, she thought, or rusted in place. Looking at it head-on, she realized that instead of the usual single arrow indicating the direction of the wind, there were three, crossed within the rooster's body, and she was just about to ask her uncle about this unusual design when the heavens opened up and big fat drops of rain—big fat *cold* drops of rain—began pelting the lawn.

"We'd better run for it!" Uncle Farley said, practically yelling to be heard over the din of the storm, which sounded like a hundred tubs being filled all at once. He caught Murray up in one hand and grabbed Charles with the other. Poor Charles: the rain immediately caused his glasses to fog over, and he stumbled along blindly in his uncle's wake.

It was only a short run to the house, but by the time they tumbled into the music room they were soaked. When they were safely inside Uncle Farley pulled the French doors closed and latched them. Although it was only five o'clock, it was nearly dark outside, and threads of mist coursed over the lawn and out across the bay.

"You'll get used to that, I'm afraid," Uncle Farley said. "It rains rather a lot up here."

"It's very rainy in England too," Susan said, as if they'd just come from London rather than New York. "Uncle Farley," she added, trying to make her voice sound as if her next comment were merely an afterthought, "I couldn't help but notice that the house is crook—"

"My goodness but I'm wet through," Uncle Farley spoke over her. He was standing by the metronome, which had stopped ticking, with a slightly confused expression on his face, and when he saw Susan staring up at him expectantly he shook his head back and forth like a dog, spraying Murray, who was still in his arms, with a fresh shower of drops, for which Murray rewarded him with a squeal. "What do you say we dry off and change clothes and have some of that soup?"

"I thought you said pie," Charles said in a rather dejected

voice. His shirt was too wet to clean his glasses, which always made him feel helpless and small, and he had been looking forward to something sweet.

Uncle Farley produced a handkerchief from the pocket of his trousers, miraculously dry, and handed it to Charles.

"Oh, if I know Miss Applewhite, she'll have prepared for every contingency—for, er, anything that might happen," he added, in response to Murray's blank stare. "We'll have soup now to warm up, and later on we'll see about pie. You'll find towels in the armoire in the upstairs hall. Hop to it now—we don't want you catching cold on your first day."

"Yay, hop-stickle course!" Murray shouted, and, skipping, ran for the stairs.

Susan thought Murray was in unusually high spirits given the fact that they'd been left to live with a virtual stranger. But just then something else was pressing more on her mind. When they were safely on the second floor, she pulled Charles aside.

"Charles," she said, "did you notice how Uncle Farley changed the subject when I tried to ask him why the house is crooked?"

Charles hadn't noticed, actually, being more concerned with the pie, but he figured he'd better say that he had.

"And I could have sworn Uncle Farley said the cook's name was Miss Applethw-thw-thw—"

"Applethwaite," Charles said. "He did."

"But just now he called her Miss Apple*white*."

"Did he?" Charles said. He honestly couldn't remember. "Are you sure you're not thinking of apple pie? Because he did mention apple pie."

Susan ignored this slander. "There's something fishy going on here," she said, and stuck her tongue in her cheek. Charles could practically see the gears turning in her brain, but it was obvious Susan couldn't quite make out what it was. "Never mind," she said, shaking her head. "I'll get to the bottom of it eventually." She cocked her head then. "Murray's being very quiet. We'd better make sure he's not fallen through a broken floorboard or something."

She stamped her foot on the floor, which resounded more solidly than her suspicions might have indicated. As if on cue, Murray's voice sang out from the room he was to share with Charles.

"New bathrobes! Hooray!"

New bathrobes? Susan thought. Really, was there anything her youngest brother *wouldn't* get excited about?

But as she slipped into her own thick fluffy blue robe, Susan had to admit it was luxuriously soft and warm. A small but toasty blaze had been built in her room's fireplace, and the robe had been hung in front of it and so took the chill right off her skin. Susan noted that her clothes had been unpacked as well, and folded neatly into dresser drawers or hung in the closet, both of which smelled warmly of cedar. Even her doll Victor Win-Win had been propped on the pillows of her bed. The sight of Victor Win-Win so exposed caused Susan to blush, because she hadn't wanted anyone to know she'd brought along her childhood companion—she'd had him ever since she lived in England—and, after apologizing to him for the indignity, she tucked him under the blanket.

After a brief sibling meeting, the children decided to go ahead and put on their pajamas since it had gotten so dark out. By the time Susan had dried off and changed clothes the boys had already headed downstairs. She took her time following, peeking at this and that object in the second-floor hallway. The first thing that caught her eye was a series of maps lining one wall. The maps depicted all the continents, but they were old—one was dated 1701, another 1642!—so the names and even some of the dimensions were different from what Susan (who'd gotten an A+ in the geography unit of her history class) was used to. On one little table she found a case containing a half dozen scrimshaw pipes, the bowls of which had been carved to resemble the figureheads of old ships: there were funny angels who had wings where their arms were supposed to be, and mermaids with long swirls of hair discreetly covering their chests, and even one dragon that reminded her of the figurehead on the ship in Drift House's coat of arms—no, not a dragon she saw when she picked it up, but some kind of bird.

Flecks of red and green paint still clinging to the pipe made her think the bird was a parrot, although it was a very fierce-looking parrot, with its beak wide open and its tongue sticking out as if in a battle cry.

The stairway itself was lined with portraits of stodgy old men in high-collared suits or military uniforms with swords and gold-braid epaulets, but as Susan's head cleared the landing she noticed a painting on the wall opposite the staircase that was somewhat different from the rest. It was larger than life, for one thing—the man in it must have been at least ten feet tall. For another, he was dressed not as a dignitary but as a pirate, with long black curls and a purple velvet frock coat and—aha!—a red and green parrot perched on his shoulder. He even had a pegleg: his right, which was propped up on a big chest that was probably supposed to be full of gold coins or something. Susan noticed that the chest had Drift House's coat of arms painted on it: it *was* a Viking ship, just as she'd suspected, and there were the two parrots' heads, one on the front of the ship and another on the big square sail. *Fierce*-looking birds, she noted again. But for some reason Susan doubted the peglegged man really was a pirate. His face was too, well, too *nice* really. Beneath his drooping mustache, his smile was broad and open and friendly, and a little mischievous as well.

She walked into the gaily colored drawing room just in time to hear Uncle Farley say, "Mr. Zenubian was by while we were outside. He stocked us up on firewood"—Uncle Farley pointed to a stack of split birch logs—"and got the fires going to keep us warm."

The fireplace was certainly cheery, but Susan found herself wondering if the old house lacked electricity. How would she be able to watch *BBC News*? How would Charles be able to log on to the Internet? But then she looked up and saw that the chandelier had lightbulbs in it, not candles, and the two torchères flanking the fireplace were both plugged into wall sockets. Phew, she thought. Aloud, she said,

"Mr. Zenubian?"

Uncle Farley looked up at her and smiled—a wide friendly smile that reminded her of the portrait in the hall. He had also changed into pajamas and robe, and set a table with four places very close to the firescreen. He was dressed so exactly like the children that Susan couldn't help but think that her uncle Farley looked a little like an overgrown child himself—albeit, with his beard, a somewhat furry one.

"Yes," he said now. "He and Miss Applethwaite practically run the place between them. I feel quite superfluous most times. Er, extra," he added, with a nod at Murray.

At this "Miss Applethwaite" Susan glanced at Charles, but her younger brother was too busy staring at four covered bowls on the table that, Susan now noticed, exuded a heavenly odor. The roaring fire offered its own smoky tang, and she felt her mouth water. The fire's crackle was a pleasant antidote to the patter of rain, but Susan couldn't help but feel that the sounds were a bit lonely as well. They were the only noises she could hear.

"Uncle Farley," Susan said, taking a seat at the table opposite her uncle. "Who is that man in the painting? The one dressed as a pirate?"

"His name's Pierre Marin," Charles said between bites of soup. "It says so right on the frame."

"I saw *that,*" said Susan (who hadn't actually). "I meant, who *was* he?"

"Monsieur Marin built Drift House toward the end of the seventeenth century. He was a bit of an eccentric, as the pirate costume probably indicates. Bit of an explorer, bit of an inventor, something of a philosopher as well. A jack-of-all-trades really."

"And what about you, Uncle Farley? What do *you* do?"

"You really should eat your soup before it gets cold, Susan. Consommé Celestine. It's quite tasty."

In fact the aroma was practically making Susan's nostrils twitch, but she was determined not to be put off. She took the lid off her tureen, revealing a bowl of pearly gold-tinged broth, but didn't pick up her spoon.

Uncle Farley squinted at her now. "Is there something in your cheek, Susan?" He stuck his own tongue in his cheek and pointed at it with his finger.

Susan blushed. "M-Mum always says a good conversationalist makes interested inquiries about her dinner companion."

"Oh, that's fine for strangers, but we're family. Ask me anything you like."

"Well, Daddy is a lawyer and Mum is an editor." She paused, swallowing a mouthful of saliva caused by the intoxicating vapor bathing her face. "But what are you?"

"Ah, what *am* I?" Uncle Farley's whiskers glistened where a bit of soup had spilled on them. "That's rather a different question, isn't it?"

"Susan said you were a Lud, a Ludi, a—" Murray couldn't remember how to say the word.

"A Luddite," Susan finished for him, feeling her blush creep up her cheeks.

At this, Uncle Farley's eyes glistened almost as much as his whiskers. He took another spoonful of soup and made a big show of savoring it; and if the rapid tinkle of Charles and Murray's spoons was any indication, the consommé tasted just as good as it smelled. "Susan isn't so very far off," Uncle Farley said, addressing Murray but still looking at his sister. "I am what used to be called a gentleman scholar, but is now more likely to be termed a dilettante. Rather like Pierre Marin, I make it my business to know a little bit about everything I can, and quite a bit about a few subjects that are near and dear to me, which do tend to be rather old-fashioned. My main area of interest is time."

"Time!" Susan said, for it was not a subject she had ever

given much thought to. "You mean you're a historian?"

"Not exactly," Uncle Farley said.

"Duh," Charles said. "He said *time*, not *history*. He's obviously a physicist." He smirked at Susan between rapid bites of soup, which caused him to dribble broth on his chin.

"Well, no, Charles, that's not quite right either," Uncle Farley said, handing his nephew a napkin and raising his eyebrows in amusement when Susan stuck her tongue out at her brother. "History and physics are indeed two ways of looking at time, but I prefer to take a third perspective. A more personal perspective."

By now Susan had to sit on her hands to avoid picking up her spoon. She wasn't even sure at this point why she wasn't eating, but she could be a very determined child when she set her mind to it.

"But how can you think of time personally?" she said. "I mean, it's not like you can control it."

"Control time? No, indeed, that is impossible. But people do think quite differently about time at various moments in their lives. Have you ever heard someone say that time seemed sped up to them, or slowed down? Or that they were experiencing déjà vu? I am curious about what produces these sensations, both inside the individual, and also outside."

The conversation had taken such an odd turn that even Charles had stopped eating. Only Murray's spoon continued to move rhythmically between his mouth and his bowl, accompanied by satisfied slurping noises.

"Are—are you a psychologist?" Charles asked now. "A psychologist of time?"

"I'm afraid there's no real term for what I do. But if you need to call me something, I suppose you could refer to me as a temperologist."

Susan, fearing she was losing control of the interrogation, suddenly changed the subject. "Uncle Farley," she said as forcefully as she could, "why is the house crooked?"

Before her uncle could answer a flash of lightning lit up the windows, followed closely by a peal of thunder loud enough to rattle the spoons in their tureens. The children all jumped in their chairs, and Susan noticed that Uncle Farley did too. In fact his face wore such a serious look of consternation that Susan was reminded of the nervous way he and Mrs. Oakenfeld had regarded each other earlier in the day, and the promise she'd made to her mother to be good, and to set an example for her younger brothers. And so, pulling her hands from her seat, she picked up her spoon and tasted her soup.

It was delicious.

Uncle Farley stared out the window at the dark sky for a long moment, then shook himself as if from a dream. "Miss Applewhite's really outdone herself this time, hasn't she?"

Susan looked up from her soup.

"Miss Applethwaite?" She glanced at Charles, but he was intent on his soup.

"Yes," Uncle Farley said. "The housekeeper. I thought I told you about her."

"Oh yes. Miss Apple*white*"—she practically yelled at Charles, who had tipped his bowl to get the last few drops of broth—"and Mister Zanzibar."

"Zenubian," Uncle Farley corrected her. He looked bewilderedly between the two older children, and Charles, looking up for the first time, returned his glance with an equally confused expression. Uncle Farley cleared his throat. "Yes, Miss Apple—Miss, um, Applewhite. And Mr. Zenubian. A resourceful pair indeed. However, they're neither of them mind readers, so you should be sure and tell me if you need anything so I can pass it on. Firewood, clean towels, and so forth. Now, Charles, I believe you requested apple pie?"

"Yes, please!"

As Uncle Farley lifted the lid off a pedestaled plate, Susan tried to catch Charles's eye to see if he had caught Uncle Farley's most blatant Applethwaite/Applewhite discrepancy. But Charles's gaze was transfixed by the steaming confection that emerged from beneath the lifted lid. It was surely the biggest apple pie Susan had ever seen, emitting wafts of sugary spicy steam from the mounded lattice of its crust, and even she had difficulty looking away from the generous portions her uncle began dishing out.

"One scoop of ice cream for you, Charles? Or two?"

"Two, please!"

"Susan, what about you? With ice cream, or just pie?"

The pie smelled so good Susan's mouth watered, and she decided she would wait to talk to Charles later.

"A la mode, please," she said. She had gotten an A in French as well—although, as we shall see later, her knowledge was a little more limited than she realized.

She was just savoring the last mouthful of dessert—the hot slightly tart filling of the pie swirling on her tongue along with the ice-cold sweetness of the ice cream—when Uncle Farley croaked, "Whaddaya say, Susie? Another slice to stuff in yer gob?"

"*What?!*" Susan exclaimed, turning toward Uncle Farley. But even as she did so, she realized the words had come from the opposite direction. She turned over her right shoulder to see who had spoken so crudely.

At first she didn't see the speaker because she was distracted by the walls. The drawing room, she suddenly saw—how had she missed it until now?—had been painted in a magnificent fresco that reached around all four sides of the room. As Susan turned her head from one wall to the next, she realized the fresco had been cleverly crafted to make it seem as if the viewer stood on the beach of an island in the middle of the sea. Two of the walls depicted a stretch of very white sand receding into the wide empty blue ocean, while the two others showed the sand giving way to a lush green plain from which sprouted strange, individual hills, each as separate from the other as cupcakes on a serving platter, although not nearly so uniform in size.

As Susan turned her head this way and that, she realized the 360-degree illusion was perfect: she felt as if she'd just washed up on shore in the little dinghy that Uncle Farley had shown them earlier. The more she stared at the fresco, the more real it seemed. She could practically smell the damp air (probably just the rain, she thought), hear the waves lap over the sand (she told herself it must be the water running down the storm gutters), feel the little boat wobble beneath her feet as she stepped out of it—and then, just as she was regaining her land legs, as it were, she saw the mermaid. She was a beautiful girl with long copper-colored hair and an iridescent tail sunning herself on a rock, and Susan almost gasped when she saw her, for it seemed that the painted girl's eyes looked right at hers with a knowing—Susan almost thought *impatient*—expression. Though the mermaid was only a few inches tall, Susan could see every one of the scales of her tail, could almost hear the plucked strings of the harp set off against the *whoosh* of the tide. Then, farther out in the water, a school of flying fish caught her eye, their finlike wings translucent in the bright afternoon sun, the little drops of water they shed as they flew sparkling in the air. In the farthest distance, where the water blended seamlessly into the cloudless sky, she could just make out a gray spume rising from the dark round body of a breaching whale. The whale was no bigger than her thumbnail, but so cleverly rendered that Susan could practically hear the hiss of escaping vapor and feel the mist wetting her cheeks.

"Take a picture, it'll last longer. Whatsamatter, Susie? Never had your oeils tromped before?"

Susan started at the voice, which she'd forgotten about as she'd been caught up in the scenes on the wall. Now a scratchy—and distinctly birdlike—laugh followed this latest pronouncement, and then a figure Susan had taken as part of the fresco's foreground ruffled its neck feathers and scratched the side of its head with a long, curved, and extremely sharp-looking talon. It was a green and red parrot, rather like the one in Pierre Marin's portrait—but much bigger.

"I—I know all about trompe l'oeil and perspective and, and chiaroscuro," Susan said, somewhat unsure of herself—she'd never been addressed by a parrot before. "I learned about it in my art class last year." (As you can probably guess by now, she got an A in art too.)

"Chiaro*scu*ro!" Uncle Farley's voice sang out. "Well done, Susan!" She heard the sound of his chair scraping away from the table but didn't turn from the eyes of the parrot before her, which seemed very yellow and large set in the bright red plumage of his head. "I see President Wilson has deigned to take note of us. Greetings, Mr. President. I gather you're feeling a bit *talkative* this evening."

For a moment the two of them seemed to stare each other down, and then the parrot—President Wilson, as we must now call him—spread his great green wings and beat them rapidly three or four times. "Rak!" he said, his voice more parrotlike than it had sounded a moment ago. "Polly want a cracker. Rak!" He laughed again, a low scratchy chuckle.

"What about some pie?" This was Murray's voice. He'd come around the table after Uncle Farley, his plate in one hand, spoon in the other. "Is it okay, Uncle Farley? I only have a little left, and I'm full."

Susan looked at her little brother. After all his exclamations and jumping about this afternoon, he seemed suddenly subdued. Not afraid,, but curious. He stared at President Wilson with an intense expression, almost as if he were trying to remember something.

Uncle Farley looked back and forth between Murray and President Wilson, and ended up shaking his head sternly at the latter.

"We don't usually feed President Wilson human food, as it tends to upset his stomach. But tonight it might be just the thing to *quiet him down*." This last was said directly to President Wilson, it seemed, and Uncle Farley was still looking at the parrot when he said, "I'd hate to have to put him *in his cage* and cover him up for the evening."

President Wilson beat his wings again, but didn't say anything. Murray's hair swayed in the breeze stirred up by the flapping, and he squinted a little but stood firm. When Uncle Farley lifted up Murray so he could feed President Wilson, Susan could see just how big the bird was. From the top of his bright red head to the tips of his long green and black tail feathers, he was longer than Murray, and when the bird reached to accept a spoonful of pie, Susan saw that his three-toed yellow foot—each toe capped by a sharp and slightly dirty talon—was nearly as big as Murray's hand.

But President Wilson didn't take the fruit directly. Instead he took the spoon in one of his feet and dexterously brought it to his beak. "Rak!" he screamed after snapping up the contents. "Mmmm mmmm good!"

"Susan," Charles called just then. "Come look!"

While Murray had been feeding the parrot, Charles had made his way to the frescoed wall beyond them. He was peering through his glasses at a tiny rendering of some kind of floating vessel.

"Susan! Quick!"

Susan was just about to go over to Charles when Uncle Farley's voice stopped her.

"My goodness, what a mess we've made. We'd better clear this up or Miss Applewhite—"

"Don't you mean Apple*thwaite*?" Susan said. She stared up at her uncle with the level, skeptical, cross-examination gaze she'd been practicing for her first year of debate—she'd learned it by watching the BBC's correspondents interrogate unreliable sources of information on the *World News* and practiced it on her parents countless times.

A wide strained smile showed through the fine brown hairs of Uncle Farley's beard as he met Susan's eyes, and he looked back and forth between her and Charles nervously. He tried sticking his tongue in his cheek, and Susan colored (and reminded herself *never* to stick her tongue in her cheek during an actual debate), but she refused to drop her gaze. On the perch beside Uncle Farley, President Wilson scratched his head and laughed again, but very quietly. At the sound, Uncle Farley's smile grew even wider and more desperate.

Susan glanced over at Charles now, who cleaned his glasses on the sleeve of his robe before inspecting the wall again, then looked back at Uncle Farley. Despite the beard and the still-wet hair and the old-fashioned pajamas, there was something about the set of his eyes and mouth that reminded her of her mother taking her leave this afternoon, and Susan was reminded yet again of the responsibility that comes with being the eldest child.

"Come on, Charles," she said reluctantly. "There'll be time to study the fresco tomorrow. We'd better help Uncle Farley clear the table."

Uncle Farley breathed a sigh of relief, audible over the heavy sound of rain still coming down outside and the lighter crackle of the fire inside the room. Within a few moments, he and the two older children had stacked the dishes onto a large wooden tray. The plates and crystal and silver clinked against each other when he lifted the tray and turned briskly toward the door. When all the children had followed him out of the room he set the tray down on a table in the hallway, then went back in the room and shut the light off and closed the double doors. The doors were glass, though, and in the light spilling in from the hallway chandelier Susan could just make out President Wilson on his perch, blending in again with the faint shapes of the fresco behind him. She had the distinctest sense that he was still watching her with his bright yellow eyes.

The sound of a lock turning brought her attention back to the hallway. Uncle Farley was pulling a big skeleton key from the door, which he dropped in the pocket of his robe.

"For President Wilson's sake," he said, not making eye contact with Susan or Charles. "He's very old, and needs his rest."

"He's really quite articulate, isn't he?" Susan said as he picked up the tray again. "Even for a parrot."

Uncle Farley walked past the portrait of Pierre Marin, with its own parrot perched on the pirate's shoulder, then turned into the dining room. "President Wilson is older than all of us put together. He's had more than ample time to learn his fair share of phrases."

"He seemed almost to be talking *to* me," Susan persisted, following her uncle into the dark room, where a long table smelled pleasantly of orange-scented oil.

"Quite an impressive trick, isn't it?" Uncle Farley set the tray on the dining room's marble-topped sideboard and reached for an ornately carved dowel rod mounted on the wall beside it. He tugged the rod and two panels slid open, one up, the other down, like a great mouth opening in the side of the room. When they reached their widest division the aperture was nearly five feet high and three feet wide. Uncle Farley hadn't turned on the light in the dining room, and the plush carpeted space beyond the opening receded into shadow.

"Yay!" Murray exclaimed. "Secret passage!"

"Nothing so exotic," Uncle Farley said. "I'm afraid it's just the dumbwaiter. It communicates to all the floors of the house, from the kitchen below us to the servants' quarters in the attic."

"Dumb waiter?" Murray said suspiciously, as if it might be related to the fruit seller and Solar Mum who had so far failed to turn up. He peered deep into the dark space, as if the unintelligent servant it housed might be hiding in the shadows.

"It's like a little elevator," Charles said to Murray. "But not for people. For dishes, and heavy things you don't want to carry up several flights of stairs. The carpet," he went on, because Murray was running his fingers over the padding that lined the chamber (it was all he could reach standing on his tiptoes), "is to muffle the rattling of dishes."

"Bravo, Charles," Uncle Farley said. "I say, you three *are* a bright lot. I can see I'm going to be kept on my toes for the next several month—" He broke off suddenly as he looked down at the three upturned faces, and Susan fancied that only she caught the sharp edge to his voice.

"Murray, come away from there," she said, for her little brother had dragged a chair up to the dumbwaiter and stuck his head inside it. "It is not a toy." She drew her little brother to her side and looked up at her uncle's face. In the dim rain-spattered light his bearded face seemed very high up, shrouded in worry and mystery that seemed more than a mere byproduct of the poorly lit room. Yet at the same time, his pajamas and bathrobe and nervously blinking eyes made him seem every bit as young and insecure as Charles, and even as she patted Murray's head she wondered if she were trying to calm her little brother, or her uncle. "Do you think we'll be here long, Uncle Farley?"

Uncle Farley drew the dumbwaiter's doors closed heavily, the top and bottom panels thumping together loudly in the dark room.

"I'm sure you'll be home before you know it. But in the meantime Drift House is really the best place for you. I'm sure we'll have any number of grand adventures before you leave. The time will just float by."

"It's fly by, Uncle Farley."

"Er, what's that, Susan?"

"You said float by. But the expression is fly by. 'Time will fly by.' " Given the fact that her uncle had pronounced himself a temperologist, the slip seemed particularly odd to her. But for a man who didn't seem to know the name of his own maid, perhaps it was just par for the course.

Then, as if in answer, she heard the faint sound of President Wilson flapping his wings on the other side of the wall.

"Rak!" the parrot's muted voice echoed faintly through the house. "Time's floating by. Rak!"

Uncle Farley smiled down at Susan, but it was a weak smile, and it was obvious he was still bothered by something. "It would seem President Wilson disagrees with you."

Susan sighed heavily. She had had enough debating for one evening.

"Uncle Farley," she said. "Do you think it might be possible to watch an hour of television before bedtime?"

"What, television?" Uncle Farley said, as if she'd asked for a cup of fungus tea, or something equally unsavory. "Rots the brain, television does. Why, I wouldn't even have one in my house!"

And, at this horrific pronouncement, another flash of lightning illuminated the sky beyond the curtains. A long low rumble of thunder vibrated the house from top to bottom, as if it could shake the whole crooked structure out to sea.

FOUR

Flood!

"SUSAN!"

The voice pushed into her ears like a tickling finger, and Susan pulled the pillow over her head.

"Susan, wake up!"

Susan turned on her side away from the voice. "Is it time for school already?"

"Susan, wake up! There's been a flood!"

Susan bolted straight up in bed. One time Murray had left the bathtub running for so long that the water had spilled over the edge and run down the hall into her room: the carpet had smelled of damp for weeks. But the floor seemed dry today—not to mention uncarpeted—and she looked around the strange room in confusion until her eyes came to rest on Murray. He sat on the bed with her, a look of wild excitement pinking his cheeks below his rumpled hair. Even worse, he had Victor Win-Win in his hands, and was making him dance like a puppet.

"Murray," she said, still half asleep. "Mum said you're not to come in my room in the morning just because you're watching telly." Delicately but firmly—she didn't want to rip her doll, after all—she took Victor Win-Win from her brother's hands and slipped him under the pillow.

"Americans say 'television.'" Charles's voice came from somewhere off to her right. And then, in a more serious tone: "I think you better have a look at this."

As Susan turned toward Charles, she took in the strange room again. The wallpaper was covered in a delicate pattern of pink and blue and yellow flowers—no, not flowers, she saw now, but fish, amorphous but oddly beautiful jellyfish and sea horses and anemones. There was a big maple chest of drawers, a writing desk with an old-fashioned inkwell and a stack of paper at the ready, and then Charles, standing in front of the big multipaned windows bathed in cold gray liquid light. He had an odd expression on his face, half awed, half alarmed, and when Susan saw him she hopped out of bed. Her toes curled up when they came into contact with the chilly floorboards and she minced over to him, Murray trailing behind.

"It's, like, *biblical*," Charles said, and he pushed his glasses up the bridge of his nose.

Susan pulled back the curtain. Immediately she had to stifle a gasp. As far as she could see—which wasn't very far, because of the thick fog—the ground had been swallowed by a great sheet of water dappled a million times over by the rain falling heavily from the sky.

"Flood!" Murray yelled redundantly from her side. He clapped his hands excitedly. "Flood, flood!"

What with the fog and the rippling sheet of water running down the window, Susan couldn't see even a trace of land. Not a bush, not even a tree. It was as if Drift House, true to its name, had washed out to sea. Indeed, as Susan stared at the rolling swells she could almost feel the floor shaking slightly beneath her chilly feet.

"Has anyone been in to see Uncle Farley?"

Charles shook his head, but his gaze remained fixed on the view through the glass. "Will you look at it? It's endless!"

"Nonsense," Susan said, struggling to keep her voice steady. "It's just the fog. Why, you can make out the hills we came in over—just there." In fact Susan could see no such thing, but she was hoping her words might convince herself. "Now come along, Murray. Let's you and I go wake Uncle Farley and see what he has to say about all this."

Of the three children, only Murray seemed unconcerned

by the plain of water beyond the window. Indeed, he actually seemed excited.

"Can we go out in the boat later?" he said to his sister as they walked out of the room.

"I'm afraid we already *are* in a boat," Susan said—but she muttered it under her breath so Murray didn't hear.

Holding her little brother by the hand, she ventured out into the long hallway. It was even cooler out there than it had been in her bedroom, but the old rugs that lined the floor were wonderfully textured, just scratchy enough that she wanted to rub her soles back and forth on them, yet not so hard that her feet couldn't sink in a bit and so warm her toes. As Susan walked toward the closed doors at the other end of the hallway, past the pipes and the framed maps and the stairs, she imagined that she could still feel the house rocking a bit like a boat, though when she tried to concentrate on the sensation it seemed to evaporate. She tried to remember which room was her uncle's, then had the answer revealed to her by a sudden blast of music from behind one of the doors—a jangly disco-y sort of sound that seemed distinctly out of keeping with what she'd imagine her uncle's taste to be—especially at that hour of the morning. She knocked on the door as politely as she could. What time was it? she wondered. She hadn't seen a clock anywhere.

"Uncle Farley?"

After a long pause the music switched off. Then a voice that sounded half relieved, half nervous, said: "Come in."

Susan pushed the door open halfway.

"Uncle Farley? We wanted to tell you that, well, we just thought: you should know that, well, it seems there's been a bit of a—"

"Flood!" Murray yelled from beside her.

Susan nodded her head. "Well, yes. A flood."

Uncle Farley proffered one of his nervous smiles that Susan was getting used to.

"So I've seen."

He was sitting up in bed, having obviously been awake for a while. In fact, judging from the dark circles under his eyes, Susan had the impression that he hadn't slept at all. There were several books scattered around the bed, big leather-bound volumes with faded gilt lettering, as well as loose sheets of paper that Susan thought—she couldn't quite tell from across the room—looked a bit like the antique maps hanging in the hallway. Uncle Farley gathered all this up now and piled it on the bedside table, where Susan noticed an old-fashioned radio—the ones that are made out of wood and shaped like a Gothic arch—that she presumed had been producing the disco sounds she'd heard a moment ago. The pointed top of the radio just poked above the stack of books and papers Uncle Farley piled on the table.

"Well," Uncle Farley said, pushing his glasses up his nose

exactly as Charles did. "It's really nothing to be concerned about. Quite common actually. One of the hazards of living at sea level. It just means we'll have to occupy ourselves with indoor activities today. Don't worry, there's still plenty you haven't seen. The gallery, for one thing, and the solarium. Although it's not very sunny today, I'm afraid."

"Then there's nothing we need to do? No . . . bailing?"

For the first time since yesterday afternoon Uncle Farley's laugh seemed completely unforced. "Oh my goodness no. Drift House was built by a shipmaker—that is, Pierre Marin, whose picture you saw last night, hired a shipmaker to build the house for him. It's as watertight as any oceangoing vessel."

Just then the house gave a sort of lurch. There was no other word for it—the floor beneath Susan's feet seemed to shift several inches to the right. There was an ominous creaking noise as well, as of watertight seams coming ever-so-slightly apart.

Murray yelped and dashed for the bed, Susan not far behind. A moment after they'd landed there was a third plop, and Susan pulled her head from a tangle of sheets and blankets and saw Charles trying nonchalantly to push his glasses up his nose.

Beside Uncle Farley, the antique radio had begun playing again, a quieter kind of music that Susan recognized from the old movies her parents liked to watch after they'd put the children to bed, Technicolor films whose reds and greens always seemed oversaturated to her eyes, like the plumage of the parrot downstairs. Then something small but heavy fell to the floor behind her, and there was a distinct rolling noise as it made its way across the floorboards.

Uncle Farley looked at the three children on his bed. He seemed at a loss for words.

"Uncle Farley," Murray broke the silence. "What was *that*?"

"Oh, nothing, nothing. Just the house settling, you know, in the rain."

With a little click the object that was rolling across the floor came to a stop. Susan could see it just there, a brass cylinder about six inches long.

"There we are, see?" Uncle Farley said. He made an attempt to smooth Murray's hair, but his hand froze as the object began rolling the other way. As it came closer Susan was able to see that it was a little collapsible telescope—a spyglass.

"Hey!" Charles said suddenly, hopping off the bed and walking to the bedside table. "Is that a Philco Tombstone from the thirties? Cool!"

At the word "tombstone" Susan looked up anxiously, but then she realized Charles was referring to the radio on the bedside table, whose arched shape did in fact resemble a gravestone. Uncle Farley looked similarly confused.

"Charles is a bit of an electronics buff," Susan explained, even as she climbed off the other side of the bed to retrieve the rolling spyglass. "For his science project last year he built a model of the original telephone based on someone—"

"Meucci," Charles said, pushing aside books and papers to have a closer look at Uncle Farley's radio.

"Meucci's design. Charles says this Meucci was the real inventor of the telephone, and Alexander Graham Bell only came after. Uncle Farley," Susan added in the most casual voice she could muster, even as she pulled open the spyglass and stared at the magnified image of her uncle's rather startled-looking left eye. "Does it feel like the house is *swaying* to you?"

"What, what?" Uncle Farley's head whipped back and forth between Charles and Susan. "Of course not. Settling, I tell you. It's just a bit of settling."

"Weird," Charles said from the table. "This radio isn't genuine at all."

"Er, how can you tell?" Uncle Farley said.

"The Philcos were made of veneer—a model this size would've only weighed three or four pounds. But this feels like it's been carved from solid wood. It weighs a ton. And the logo's wrong too—look, it's a bird or something. A rooster, maybe."

Susan peered at the tiny bird-shaped icon with three arrow-tipped lines trisecting it. It looked familiar for some reason. Maybe it was a rooster, she thought. Or maybe, she realized, it was a *parrot*.

Just then, when Charles's hands were nowhere near the knobs, the music stopped playing. Charles stepped back from the radio slightly, as if he might be accused of breaking it.

"Er, yes," Uncle Farley said, not quite meeting his nephew's eye. "I've been meaning to fix that. Who knows, perhaps you'll be able to assist me."

"Cool!" Charles said.

"This seems very nautical," Susan said now, climbing back on the bed with the spyglass in her hand.

"Er, naughty, nautical, what?" Uncle Farley said, glancing away from Charles to look at the object in Susan's hands. "Oh, yes, it's a bit of a theme around here. The sea, and all that. Charles, what *are* you doing?"

Charles was attempting to slide Uncle Farley's bedside table away from the wall.

"Charles!" Susan said sharply. "Come away from there. That's quite rude."

"Who wants breakfast?" Uncle Farley cut in. "I'm sure Murray does."

"Breakfast!" Murray sang out. Or yelled really, Susan thought, guessing that her little brother was feeling excluded by all the talk of radios and telephones and spyglasses.

"Let's just see what Miss um, Miss Applethwaite"—Uncle Farley glanced nervously at Susan—"has cooked up for us, shall we? She has an uncanny knack for knowing what people want." He threw back the covers and stepped from bed. "Now, let me guess. Murray, I bet you'd like pancakes."

"Pancakes!" Murray yelled, but then he shook his head. "No, waffles!"

"Waffles for Murray. And Susan?"

"Muesli, please. With fresh fruit if there is any."

"I'm sure Miss Applethwaite can find something. Charles?"

Charles, who thought the word "muesli" was affected (though he liked the taste of it just fine), said, "Cereal, please," but Susan was too busy fiddling with her spyglass to notice the look he shot her.

As Uncle Farley walked to the far side of the room the house lurched again, and he staggered a step to the right and had to catch himself on the side of a wingback chair.

"Whee!" Murray yelled, as if he were on a ride at an amusement park.

"Really, Uncle Farley, this house does a *lot* of settling." Susan lifted the spyglass to her eye and stared at the back of her uncle's head as if she could read the thoughts that flowed beneath the mussed locks of fair brown hair.

Instead of answering, Uncle Farley walked to a dowel on the wall, just like the one in the dining room. Susan tried to imagine the house's layout in her head, and realized that, yes, Uncle Farley's bedroom was right on top of the dining room. Uncle Farley pulled on the rod and the dumbwaiter opened to reveal a huge tray of covered dishes. Immediately, a bacony smell flooded the room.

"Bacon!" Murray said, jumping from the bed.

Uncle Farley carried the tray to a small table near the windows where, Susan noticed, trails of water were still running thickly down the glass. But then her attention was drawn to the dishes Uncle Farley was setting out: waffles for Murray and muesli and sliced fruit for her and Charles—kiwi and fresh peaches, her favorite, and blueberries, his. There was also bacon and eggs for Uncle Farley, and syrup and butter and marmalade, and orange juice, and tea. It was, uncannily, *exactly* what they'd asked for, and Susan was about to quiz Uncle Farley on this when she saw one additional item.

"Crumpets!" she exclaimed. She dropped the spyglass on Uncle Farley's bed and practically ran to the table.

"Don't say 'crumpets,'" Charles couldn't help himself from saying as he followed her. "They're just English muffins."

"I'm afraid Miss Applethwaite would take Susan's side on this one, Charles. These are genuine crumpets, based on a family recipe from Derbyshire."

"Uncle Farley," Susan said, her cheek filled for once not with her tongue but with a mouthful of hot crumpet and melting butter. "Is Miss Applethwaite working in the kitchen right now?"

Uncle Farley didn't look up from the thick pat of butter he was spreading on his own steaming crumpet. "Oh yes. It's her little domain down there. And I should tell you she *hates* to be disturbed."

"But won't the basement be full of water?" Susan said. "I mean, with the flood and all?"

Uncle Farley put his knife down a little heavily. "Oh, well, well, no. These old houses, you know. Quite watertight."

"But didn't you say the house was settling just now? That doesn't seem so watertight to me."

As if in direct support of Susan's words, the house "settled" again, a quick shift several inches to the left. A creaking shudder shook the entire building, accompanied by a vibration that clinked the silver against the china and made the windows rattle in their frames. For the next several moments there was just the sound of Murray's fork as he brought piece after piece of syrup-soaked waffle to his mouth. Even Charles was looking at Uncle Farley expectantly, his jaw working slowly as he tried to chew a crunchy mouthful of muesli as quietly as possible.

"Well, yes, you see. I mean, the house." Uncle Farley smiled wanly. "It settles, you know, as a unit. All in one piece, as it were."

Susan had once seen a newscaster say "Hmph" when an interviewee gave an equally dubious answer, and it took all her manners to hold back a similar exclamation. Instead she turned back to her meal in silence, listening as Uncle Farley suggested a few activities the children could do to pass the time during the rain—among them, hide and go seek, at which

Murray clapped his hands and shouted "Yay!" At one point while Uncle Farley spoke, Charles nudged Susan lightly in the arm and nodded toward her glass of juice. The orange liquid could be seen swaying slightly back and forth, climbing half an inch up one side of the glass and then sliding down and climbing up the other. As Susan watched the swaying liquid she began to feel the sensation more distinctly than she had earlier. The house was definitely rocking from side to side.

But since Uncle Farley didn't seem to want to acknowledge it, she reached for her butter knife instead. There would be time for more investigating later. Right now: crumpets!

FIVE

The Voice in the Radio

AFTER BREAKFAST, THOUGH, AS THEY headed back to their rooms to bathe and change into their clothes, Charles pulled Susan aside.

"So you suspect it too!" he whispered.

"Of course," Susan whispered back, not sure what Charles was referring to, but not wanting to be caught out. But then, when Charles wasn't forthcoming, she said, "Suspect what?"

"That we've washed out to sea!"

Susan had felt something was amiss, but this idea struck her as preposterous, and she said so.

"Don't be preposterous, Charles!"

"Don't say 'preposterous,'" Charles said, pushing his glasses up his nose. "Because it's not at all. Now listen. Last night, when you and Uncle Farley were busy with that freaky parrot thing—"

"President Wilson."

Charles shuddered. "Yes, whatever. While you were playing with him, I checked out a scene on that funny wall painting."

"Fresco," Susan said.

"Stop interrupting!" Charles said. "So listen. I checked out a scene of the fresco, and guess what I saw?"

Susan didn't say anything, lest Charles accuse her of interrupting again.

"I saw us!" Charles said excitedly. "And we were on this house—"

Susan couldn't help herself. "*On* it?"

"Yes, *on* it. On the roof of the tower part. It's flat, you know. Except it wasn't a tower really, it was the quarterdeck on the back of a galleon—er, the poop deck," Charles said, stammering and blushing slightly. "The main house was like the covered cabin amidships, and the solarium, well, you know, it's all pointy, like the prow of a ship."

Susan hated it when Charles resorted to jargon—especially when she wasn't familiar with it—but all she said was, "How could you make that out? I looked at that picture myself. It was minuscule."

"But that's just it," Charles said. "When I glanced at the picture, it seemed like a little sketch. But when I really stared at it I could make out the tiniest details—I could even see your tongue in the side of your cheek."

Susan pulled her tongue from her cheek. But even as she did so she remembered her own similar experience looking at the fresco, how she could make out the scales on the mermaid's tail, and the drops of mist spewing from the whale's blowhole.

"The first clue," Charles went on, "was that lichen I saw on the house when we arrived. Only now I don't think it was lichen at all—it was barnacles. And then there was the fact that the house was all out of line with the garden and the path, as if it had washed up crooked. And then this morning Uncle Farley told us it had been built by a shipmaker. And finally—"

For his last bit of evidence Charles simply stood in the middle of the hallway. The house was undeniably rocking back and forth: his pajamas could be seen moving on his body as if there were a draft.

"It's even on the longitudinal axis," Charles said. "Exactly like a boat."

"Well," Susan said doubtfully. "It *is* called Drift House."

"And Susan! When I looked behind Uncle Farley's bedside table? I saw that his weird Tombstone wasn't plugged in!"

Susan shuddered at the name of the radio. "Are you sure you didn't unplug it when you pulled the table out?"

Confusion clouded Charles's eyes for a moment. "I hardly moved the table—"

But Susan put up her hand.

"This is prepos—this is ridiculous, Charles. Radios that play when they're not plugged in, and pictures of us on the drawing room wall, and—and houses drifting out to sea! There's something funny going on here, but let's not get carried away. I'm going to take a bath," she continued when Charles opened his mouth to protest. "And you are too. We can have a look at the fresco afterward. And Charles," she added, "see to it that Murray bathes as well, will you?" Before Charles could protest that it wasn't his job to "see to" Murray, Susan pivoted on her heel and marched to her room.

Somewhat sulkily, Charles made his way to the room he shared with Murray—the room he'd been forced to share! It was so like Susan! When *she* noticed Uncle Farley referring to the invisible housekeeper as both Miss Applethwaite and Miss Applewhite, there was a mystery. But when *he*, using far more scientific methods, noticed several oddities that all pointed toward one conclusion, he was being ridiculous! Just because she was the oldest didn't mean she was the smartest, Charles thought. *He* was the one who had been invited to take special classes with high schoolers, not her. And making him help Murray take a bath! Charles would have stamped his foot at the indignity of it, if he hadn't been afraid of losing his balance on the tilting floorboards.

But after the children had bathed and changed their clothes (actually, Charles hadn't bathed, although he had run a bath for Murray in the gigantic bathtub, which came equipped with a pretty awesome assortment of floating toys, including a propeller-driven submarine), they reconvened in the downstairs hallway, where they discovered that the doors to the drawing room were still locked.

"Crap!" Charles said half under his breath, startling Susan so much that she didn't think to admonish him. He rattled the handle angrily, only to jump back when a loud "Rak!" emanated from the shadowy chamber: the curtains had been drawn over the windows, and the glass doors themselves were covered by gauze curtains, rendering the room beyond all but invisible.

In another moment, Charles was running up and down the hallway, pulling open drawers, lifting up vases, opening the lids of little boxes. "There must be a key somewhere."

"Charles!" Susan hissed, glancing up the stairs to see if Uncle Farley was coming. "Stop that at once! You're being terribly rude!"

But Charles was half submerged beneath the rug and didn't respond. When, finally, he emerged, his hair and clothes were covered with lint, but something shiny glinted in his hand.

Susan's yelp betrayed her own excitement.

"You found it!"

"Just a quarter, actually," Charles said. He squinted at it. "Canadian even. Huh? What's that, Murray?"

For Murray was tapping him on the shoulder, his hand extended solemnly. There, protruding over each side of his tiny palm like a seesaw on a fulcrum, was a large, slightly rusty skeleton key.

"You genius!" Charles said. "Where did you find it?"

In response, Murray merely turned and pointed silently at a big brass umbrella stand at the opposite end of the hallway. After his boundless energy of the past twenty-four hours it was odd to see him so subdued, and Susan was about to ask if anything was wrong when she had the distinct impression of eyes staring at the back of her head. She whipped around, only to see the portrait of Pierre Marin looking down at her, his smile, she imagined, the tiniest bit less open today.

Just then Uncle Farley's voice came from the upper landing.

"Children? Are you down here?" The stairs creaked as he began to descend.

Charles looked around wildly a moment, then snatched the key from Murray's hand and dropped it down the front of his shirt. But his shirt wasn't tucked in (Charles had resolved to be as sloppy as possible today, in defiance of Susan's bossiness earlier) and the key clattered loudly onto the floorboards—which were bare, because the rug was all bunched up from his investigations. While Charles hastened to straighten the rug, Susan jumped for the key, her foot landing on it just as Uncle Farley's face came into view. He was carrying a large canvas bag in one hand. Susan could just make out the top of the Tombstone radio, as well as several big books and rolled sheets of paper.

"What's going on down here? Did someone drop something?"

In answer, Charles held out his hand with the quarter in it. Uncle Farley's eyes flitted between the two older children. Then he said, "Charles, you're absolutely covered in lint. Are you sure you bathed this morning?"

"Oh yes, Uncle Farley!" Charles answered enthusiastically, as if his uncle had asked him if he'd enjoyed last night's pie. "I just, I, um, I dropped one of my shoes under the bed and had to crawl under to get it. I must have picked up the lint there."

"Lint under the bed? How unusual. I'll have to speak to Miss, um, to someone about that. Well now, listen. I have a few things to attend to in the gallery, so I'm afraid I'm going to have to ask you to entertain yourselves this morning. Should you need me, I'll be just through the library there." Uncle Farley pointed to the doors opposite the drawing room.

"Uncle Farley," Charles said, peering into the bag the older man held. "Are you going to work on your radio? I've never seen the inside of an old Tombstone."

If her foot hadn't been concealing the key Murray had found, Susan would have kicked Charles in the shin. Now was hardly the time for him to engage Uncle Farley in a discussion of old radios!

"Oh, ah, well, yes," Uncle Farley stammered. Susan had noticed that Uncle Farley got as jumpy when the subject of that radio came up as he did when she put him on the spot about the housekeeper's name. But still, this wasn't the time. Though she'd only been standing on the key for a few moments, it felt like an eternity. She was afraid her leg was going to fall asleep.

"Well, as you surmised earlier," Uncle Farley continued, gesturing with the bag in his hand, "this isn't a genuine Tombstone at all. Just a fancy reproduction—and a faulty one at that, always turning itself on and off." He looked down at the top of the radio, as if afraid it might do just that at any moment.

Charles looked distinctly disappointed. Susan felt as if the key beneath her foot were vibrating—she was convinced Uncle Farley could see her leg shaking, as if she were trying to hold down a squirming dog.

"Perhaps we can go on the Internet later and order you a genuine old radio to work on," Uncle Farley continued, and Charles's face brightened immediately.

"Really? Cool!"

"Right then. I'm sorry to disappear on your first day, but this thing keeps waking me up in the middle of the night, and anyway, I'm sure you three can find ways to occupy yourselves in a big old house. Susan, could you come with me for a moment?"

By now it felt as if the key were twisting itself into the bottom of her foot, as if it were trying to carve a keyhole in flesh and bone. Susan did her best to quell the shaking in her leg.

"Where, Uncle Farley?"

"I just want to show you how the dumbwaiter works. Lunch will be ready whenever you want it. All you have to do is retrieve it."

Uncle Farley hesitated a moment, as if waiting for Susan to lead the way. Susan kept her foot pressed firmly on the key.

"After you!" she said as brightly as she could.

Uncle Farley looked at her strangely for a moment, then shook his head and turned toward the dining room. As soon as he had, Susan motioned to Charles to take over for her. She was afraid the key might stick to the rubber sole of her tennis shoe when she lifted her foot, but luckily it didn't. Murray, who had watched the whole scene play out with wide eyes, seemed about to say something when Susan pressed her finger to her lips. She caught him by the hand and led him into the dining room after Uncle Farley.

"—you have to do," Uncle Farley was saying as they came into the room, "is give this bellpull a little tug when you're ready to eat." He gestured to a sort of flat, tassel-ended cord Susan hadn't noticed last night. "Mind that you only pull once. You-know-who"—he grinned weakly—"gets a little stroppy when it goes off repeatedly."

"I'm sure we can figure it out, Uncle Farley," she said. "Murray," she went on, "you must stop putting your head in there!"

For that's exactly what Murray was doing. He had dragged a

chair over just as he had last night, and put his head right into the dumbwaiter. Even now he was calling out, "Hello? Hello-o-o?"

"Susan's right, Murray," Uncle Farley said, hoisting Murray from the chair. "I'd rather you didn't play around the dumbwaiter." He looked at his wrist exactly the way you would if you wore a watch—only Uncle Farley wasn't wearing a watch. When he realized what he'd done he looked up sheepishly. "Er, um, yes. I really must be off."

Back in the hallway, they found Charles still standing on the key and staring up at Pierre Marin's portrait anxiously. Susan fancied that he wasn't looking at the pirate, but at the big bird on his shoulder.

"Why Charles," Uncle Farley said, "you look positively bereft standing there like that."

"Just waiting for Susan and Murray." Charles smiled brightly, and if it had been lighter in the room—if the rain hadn't still been coming down as thick as a waterfall outside—Uncle Farley might have seen the nervous sweat standing on his nephew's brow.

Or perhaps he wouldn't have, because he seemed lost in his own thoughts as he headed through the library toward the gallery with his canvas bag containing the radio and the old books. For a moment Susan found herself wondering why Uncle Farley had all those old books if in fact the radio wasn't an antique, but then she was distracted by Charles, who leapt off the key, snatched it up, and raced to the drawing room doors.

He stopped dead in his tracks though, when a voice solemnly called out, "Charles!"

Charles looked up as if he expected to see Uncle Farley, or the mysterious Mr. Zenubian, or even Pierre Marin, stepped from his portrait on his one good leg. But it was only Murray, his eyes wide the way a little brother's will go when he realizes his older siblings are up to no good. He pointed at the key in Charles's hand, and in a voice that sounded almost unnaturally deep, intoned, "Bad!"

"M-Murray!" Charles stammered. "M-m-my word, old boy. You gave me a very n-n-nasty turn!"

Susan didn't know if she should make fun of Charles for saying "my word," "old boy," or "very nasty turn"—and she was too busy trying to swallow the lump in her throat to figure it out. Instead she turned to Murray.

"Murray darling," she said, kneeling down before him. "Didn't you say you wanted to play hide and go seek when Uncle Farley suggested it earlier?"

It took a moment for "Murray darling" to lower his arm and turn away from Charles. He turned toward Susan, blinking rapidly. "What?"

Perhaps if Susan had been less intent on breaking into the drawing room she would have noticed something odd about Murray's behavior, but all she said was, "We've got this whole big house to run around in. Can you think of a better place to hide away?"

"Yay!" Murray yelled, seemingly back to normal. "Hide and go seek!"

"As the eldest," Susan said, "I'll be it first. So why don't you and Charles go hide while I count to one hundred?"

"A hundred!" Murray shouted. In New York Susan only counted to twenty-fourteen—she said it wasn't fair that she should have to count higher than Murray could.

Susan closed her eyes. "One . . . two . . . three . . ."

She could hear whispering voices, scampering footsteps up and down the hall. Then, finally, one light tread—it had to be Murray's—tiptoeing up the stairs. She was all the way up to seventy-two when Charles tapped her on the shoulder.

"Sorry," he said, pointing just inside the library doors at a rather large leather-sided chest. "He wanted to hide in there. I had to persuade him to go upstairs."

"We shall have to make this up to him," Susan said. "Perhaps Miss Applethwaitey-white will put some extra cookies on his plate."

"We can worry about that later," Charles said. "Right now, let's get in that room!"

As Charles tried to stick the key in the keyhole, Susan noted the almost manic cast of his features. If Murray had been unduly subdued, then Charles, normally so thoughtful and cautious, was nearly frenzied. It was unlike him to get worked up about anything—usually he was content to bury his nose in a book, or tinker methodically on some electronics project or other. Well, she supposed, these *were* fantastic circumstances. If she wasn't mistaken, the rocking motion of the house had increased quite a bit. For the first time she found herself seriously considering the fact that Charles might be right—that they might have drifted out to sea!

"It's . . . a little . . . tight," Charles said, working the key Murray had given him into the keyhole. "I . . . hope it's . . . the right . . . one." But that was seeming less and less likely: Charles was using both hands to turn the key, and nothing was happening.

"Leverage!" he cried. "I need leverage!" He looked up and down the hall, then ran to the umbrella stand and pulled out one of the few metal-stemmed models. Most of the umbrellas seemed to be old wooden contraptions with tartan or striped awnings, but the one Charles selected was newer, with a metal stem and black nylon cloth—Susan had bought any number of them on the streets of New York when she'd gotten caught in the rain on the way home from school. Thinking of New York and school and walking home along the park gave her a pang of homesickness, but she was too curious about what Charles was going to do with the umbrella to give the feeling much thought. For a moment she thought he might actually be heading outside, but all he did was slip the pointed metal tip of the umbrella into the large hole at the end of the key, which was still sticking out of the keyhole. Susan saw that he was fashioning a kind of gigantic crank.

"Why Charles Oakenfeld!" Susan said. "You *are* clever."

Charles didn't acknowledge her. He was absorbed in his task, pulling on the long umbrella with both hands.

When, after a moment, the key still hadn't turned, Susan said, "I don't know, Charles. Do you think we should force it?"

A loud *CRACK!* was her only answer. The umbrella lurched sideways, the door popped open, and Charles fell to the floor.

"Rak!"

The commotion had obviously woken President Wilson.

Charles, standing a little shakily on the seesawing hallway floor, attempted to pull the key from the door, but it wouldn't budge.

"I'm afraid it's stuck."

Susan frowned. "Well, we're both in for it now. We might as well have a look at what we came for, before we're grounded for the rest of our lives."

"Do you think he *will* ground us, Susan?" Charles said, sounding more like the timid brother she was used to. "Oh, why did I do such a dumb thing! I—I just had to get in here."

"Uncle Farley seems a little out of sorts himself. I doubt he'll even notice. Now, let's see this scene you were talking about before."

Charles led Susan over to the wall, giving President Wilson on his perch a very wide berth. For his part the parrot hardly deigned to notice them, opening a single yellow eye and glaring at them a moment before closing it again. His feathers were all puffed up and his feet were splayed widely on his perch. For a parrot, he breathed awfully loudly. It almost sounded like snoring. Susan was eying the bird suspiciously, wondering if he was faking it, when Charles said, "Susan!"

She hurried over to him at the wall opposite the windows. "What is it?"

"Look!"

Susan looked. On the wall was the scene Charles had described just after breakfast: Drift House, sailing on the open sea like some great galleon. There, on the roof of the gallery tower—what had Charles called it? the poop deck?—were the three Oakenfeld children, their skin bright in the afternoon sunshine, their hair blowing in a cool ocean breeze, and all looking in different directions as though searching for something. Murray even had the spyglass that she'd seen in Uncle Farley's room earlier in the day, and as Susan stared at her little brother's face she could almost feel the breeze on her own cheeks.

"Well?" Charles cut into her thoughts.

"Well what? Isn't this what you said you saw? Last night?"

"But last night it was over there!" Charles pointed to the wall beyond President Wilson. It was the wall that contained the fireplace, and Susan noted that Mr. Zenubian had cleaned the coals from the grate at some point. How *did* Drift House's servants manage to do their work without ever being seen?

"You must be remembering wrongly, Charles."

"I'm not. I remember exactly, because I had to pass far too close for comfort to that, that *bird*."

President Wilson shifted a little on his perch and said something in his sleep. It sounded a bit like "parrot," or maybe "pirate," followed by a funny noise, half sigh, half gurgle.

"*And,*" Charles went on, "last night we were alone in the picture. But now look."

He pointed: the children were still alone on the deck of the gallery, but four thick leafy vines were attached to the solarium at the front of the house, one to each corner and the fourth to the apex, looped around the weathervane mounted there, and when Susan followed the vines into the water with her eye she could just make out, below the surface of the sea, the bodies of a small school of lissome fish. No, not fish at all—mermaids! From the tautness of the vines, Susan could tell the mermaids were pulling Drift House. Immediately, she looked around for the rock she'd seen last night. It was still there, but the mermaid that had been sunning on it was gone. Her lyre lay abandoned on the sand next to a conch and a pair of starfish.

"These fish weren't there last night," Charles said.

"They're not fish, Charles. They're mermaids."

Charles peered through his glasses. "Good gravy!"

"Charles," Susan said, "I think we may be missing the

forest for the—er, that is, we may be missing the sea for the fishes."

"What do you me—" Charles broke off when he saw what Susan was pointing at. Then he said the only thing he could think of: "Oh no."

For a few feet farther along the wall, on the side of the painted house that was opposite the mermaids—behind it—the sea spiraled down into a huge whirlpool, an enormous tunnel of roiling water that descended into blackness from which only great plumes of vapor erupted. The mist was so palpable that Charles had to take off his glasses and clean them on his shirt.

"Well," Susan said, "at least the mermaids seem to be pulling us *away* from the whirlpool. We'll be all right, I'm sure."

Charles turned to Susan. "D'you think that's what will—"

Before he could finish, the sound of music flooded the room. Susan and Charles looked around wildly, but before either of them could find the source of the sound President Wilson began flapping his wings on his perch.

"Rak! Rak!" he screeched. "Radio! Radio!"

Susan glanced through the open doors of the drawing room, through which she could see across the hallway to the library and the covered atrium that led to the gallery.

"Radio! Radio!" President Wilson shrieked at the top of his lungs. Small, parrot lungs, but no less powerful for that.

"President Wilson, sshh!" Susan said. "Charles, close the door."

As Charles ran to shut the door, Susan headed over to President Wilson.

"Radio," he croaked, albeit more softly.

"They won't stay closed!" Charles said from the door. "I must have broken the latch!"

"Nice birdy," Susan whispered. "Quiet birdy."

She could have sworn that President Wilson rolled his eyes. "RADIO! RADIO!" he screeched, even louder than before.

"Susan! Uncle Farley is going to hear for sure!"

Susan turned back to President Wilson. She abandoned her singsong tone. "All right, President Wilson. Be quiet or I'll put you in your cage."

In fact she had no idea where the cage was, but she was hoping President Wilson didn't know that. She and the parrot stared at each other for a long moment, and then the latter fluffed up his feathers, let out a little moan, and closed his eyes. Within a few moments a thin snore whistled out of his closed beak.

When he was quiet, Susan and Charles rushed over to the source of the music. It was a large radio tucked into a corner of the room. In appearance it was similar to the one in Uncle Farley's room—a Tombstone, as Charles had pointed out, with two knobs, a single dial, and one large speaker set into the face and covered by delicately carved woodwork. It even had the same logo as the smaller radio, a rooster- (or, Susan reminded herself, glancing at President Wilson, *parrot-*) shaped bird etched in profile on each of the knobs. Susan instinctively recognized the sound coming from it as big band–style music from the 1940s, though she couldn't have said why. It also occurred to her, if only vaguely, that each time the music had come on this morning it had seemed to be a bit older than what had played before, as if the radio was tuning not so much down the dial but down the ages—back in time, as it were.

When they were right beside it Susan saw also that this radio was a lot bigger than the one upstairs, which was tiny and rather harmless looking. In fact it was almost as big as—

"A coffin," Susan and Charles whispered at the same time.

"Well, not for a grownup maybe," Charles said. "But it would hold one of us just fine."

"A very cheering thought, Charles, thank you," Susan said. And then, after a moment in which there was nothing but the radio's soft, disquieting music, "Well? You're the radiologist."

Charles rolled his eyes. "Radiology is X-rays. A completely *different* kind of wave."

"What*ev*er," Susan said, and then, hearing the grating, American-sounding diphthong in her voice, said, "Bother that. Can you see how this thing works? I have a feeling that these radios have something to do with what's going on."

Charles had moved around to the back of the radio. "Just this," he said, and held up the unplugged plug of the radio's cord. "Not by electricity."

He began fiddling with the knobs on the radio, but, as with the one upstairs, nothing happened. The music still emanated from the single speaker, a smooth, almost watery sort of sound that, given the circumstances, was hardly soothing.

"Charles, wait."

"What?"

"Listen."

"What?" Charles paused a moment. "It's just this dorky old music."

"Sshh! And don't say 'dorky.' "

A moment later, Charles's face suddenly brightened. "I hear it too!"

Beneath the music, a voice could be heard speaking.

Susan cupped both her ears. "Something . . . something . . . wait! Drift House! . . . something . . . something . . . request location! Charles," she exclaimed, "Charles, someone's looking for us!"

For a moment Charles's face was as bright as his sister's, but then it darkened. "You know what that means, don't you, Susan?"

"What are you talking about, Charles?"

"If someone's looking for us then we really *must* be lost. We're lost at sea!"

For a moment there was nothing but the deceptively gentle rolling back and forth beneath their feet. If they'd really been on a boat the rocking would have been calming, but given the fact that they were in their uncle's house it was not so pleasant at all.

The hiss of music continued to spill from the radio, behind which both children could now hear plainly, "Echo Island to Drift House. Do you copy, over? This is Echo Island broadcasting to the transtemporal vessel Drift House. If you copy please confirm location, over."

Susan and Charles looked at each other.

Susan gulped. "Transtemporal?"

Charles pushed his glasses up the bridge of his nose. "Well, 'trans.' That means between, or across. And of course 'temporal' means—"

Susan remembered what Uncle Farley had called himself last night—a temperologist. "Time," she said.

"Across time," Charles whispered. "Holy cow!"

"Echo Island to Drift House. Please confirm location, over."

All at once both Susan and Charles began beating on the side of the radio.

"Hello! Hello! We're here, hello! Hello, we're here, please come in, hello!"

"This is Echo Island to Drift House. Do you read me, over?"

"Oh it's no good," Charles said. "They obviously can't hear

us." He smacked the side of the radio one more time. "There *must* be a way to figure this out."

"Well," Susan said, when it looked like Charles was going to strike the set yet again, "let's not break—hello, that's it!"

Charles looked up at her blankly.

"It's not that we don't know how to work it. It's broken! Just like the one in Uncle Farley's bedroom!"

"Of course," Charles said. "I thought he was just sneaking into the gallery to listen to it in private. But he really *was* going to fix it."

"Or try to anyway."

At the sound of Uncle Farley's voice, both Susan and Charles nearly jumped out of their skins. They turned to see him standing in the doorway, his arms splayed against the frame to keep his balance on the rocking floor. The worst part, as Susan would tell Charles later, was that he didn't even scold them.

"Children," he said sternly, "have you any idea where your little brother is?

SIX

The Golden Locket

THE RADIO HAD GONE SILENT again.

For once Susan and Charles were entirely in sync.

"Uncle Farley!" they both exclaimed.

Let us try to picture the scene a little more clearly: Charles (who, as we know, hadn't bathed for two days) was covered in lint from searching under the hallway carpet for a key. His hair stuck out from his face in a hundred directions, and he'd torn his pants when he fell while cranking open the door. Susan was somewhat less disheveled—but Susan was the older child, and the chagrin that showed on her face for having failed in her responsibilities more than made up for her facade of neatness. Her tongue filled up her cheek, and her cheek was covered with a bright red blush of shame.

"Children," Uncle Farley repeated as if they were preschoolers themselves. "Where is Murray?"

Susan looked at Charles, who was looking at her. She looked back at Uncle Farley.

"We, um," Susan stammered, "that is, we were playing hide and go seek."

"And you thought perhaps Murray had hidden inside that radio?"

If it was possible, Susan blushed even more hotly than she had been.

"Well, er, no, that is—"

Uncle Farley held up his hand, but as he did so the house gave a sort of twisting lurch and he had to grab on to the doorframe. Both the children fell on their seats and even President Wilson had to flap his wings to remain on his perch.

"Rak!"

"And you," Uncle Farley said, addressing the bird. "You could have at least *warned* me."

"Rak!"

"Oh, don't make excuses. That is a particularly *human* trait"—here Uncle Farley looked sharply at Susan—"and doesn't suit parrots at all."

"Uncle Farley," Susan said, her composure slowly returning,

"Charles and I, that is, well, we heard voices. A voice, that is. Coming from the radio."

At this, Uncle Farley visibly blanched.

"It's a radio? That's what they do?"

Susan thought he sounded unconvinced by his own words.

Now Charles spoke up from behind Susan. "The voice was calling for us, Uncle Farley."

"Calling?"

"It asked for Drift House. And, um, there's also the painting."

"Fresco," Susan said, almost in a whisper.

"The fresco. It, well, it seems to be *moving.*"

For a moment Uncle Farley just stood in the doorway, swaying slightly with the motion of the house. Then he sighed loudly.

"There isn't time to go into it now. I'll explain everything after we find your brother—which I suggest we do before he falls overboard."

At this last word, the two Oakenfeld children stood up unsteadily on the tilting floorboards and made their way to their uncle.

"Right. I'll take the third floor. Charles, you take the second, Susan, you look around down here. We'll meet in the hall in fifteen minutes."

Then the only sounds in the house were the dull thuds of footsteps on all three floors, rumbling noises as things were slid (Susan had a particularly rough time with the chest in the library that Murray had tried to hide in earlier) and the lonely echo of the two children and their uncle calling out into the quiet rooms. "Murray?" "Murray, where are you?" "Murray, darling, please come out. Miss Applethwaite has promised to make whatever you want for lunch if you come out." Other than that there was just the occasional creak that sounded up and down the house whenever it pitched particularly sharply. Susan thought the sound a bit ominous at first, but as she got used to it she realized it sounded like a recording of whales she'd heard once, and then the sound came to seem less ominous than lonely—as if the house, like the three searchers, was calling out for someone. Susan also noticed that the sun was growing steadily brighter outside, a sharp, thick, silvery light that pushed through the fog and curtains and warmed the big rooms of Drift House. The brighter illumination made peering into dark corners and closets easier, but it still didn't shed any light on the basic situation:

Murray was gone.

Some time later, the two children and their uncle rejoined one another in the first-floor hallway beneath the bright eyes of Pierre Marin and the parrot on his shoulder. As they came together, Susan noticed again the key sticking out of the drawing room door. As she looked at it now, it seemed gigantic—clearly much bigger than the key Uncle Farley had used last night. What had they been thinking? They *hadn't* been thinking, Susan thought. Neither she nor Charles nor Murray had behaved normally since they'd arrived at Drift House. Right then and there she resolved to be more careful in her dealings with this strange place, and to remember her responsibilities, and her promise to her mother. She only hoped it wasn't too late for Murray.

Uncle Farley was panting slightly from his run up and down two flights of stairs, and the fabric of his dark blue shirt had gone dark with sweat at the underarms. Though he was a big man, it was becoming clear he wasn't used to physical exercise, and he looked *quite* disgruntled as he glowered down at Susan and Charles. He seemed every bit as tall as the figure in the painting behind him.

"Well?" he said, as if one of the children might have found Murray and put him in her pocket (Susan was convinced Uncle Farley blamed her for Murray's loss). Somewhat petulantly (for she did, in fact, blame herself) she turned her pockets inside out. Charles, seeing her, did the same, and the Canadian quarter he'd found earlier in the day went clattering over the floor.

Seeing the two children looking so dejected, Uncle Farley's face softened a little.

"Oh, buck up, you two. He's bound to turn up somewhere. Charles, pick up that coin you dropped—you never know when these things will come in handy—and then let's search the solarium."

The three were just about to set off when a loud "Rak!" emanated from the dining room.

"Is that deuced bird on the loose?" Uncle Farley said. "I'm going to tether him to his perch if he continues to be such a nuisance."

He led the way into the dining room, where President Wilson sat on the back of a tall wooden chair—the very same chair, Susan noted absently, that Murray had pulled over to the dumbwaiter earlier in the day. The parrot, Susan saw, was swaying a little, in sync with the swaying house.

"President Wilson!" Uncle Farley exclaimed. "Just look at what you're doing to that chair!" Indeed, President Wilson's talons had scored the darkly varnished wood with deep pale scratches. "If Miss Applethwaite sees what you've done she'll pull out your tail feathers and turn them into a duster!"

President Wilson rolled his eyes.

"Oh *will* she?" he said in a voice that seemed far less parrotlike than the voice he'd been using. "Really, Farley, now is hardly the time for more of your inane charades." He sighed, a high-pitched raspy sound reminiscent of a thumbnail being drawn along the teeth of a comb. "As I've tried often to tell you," President Wilson continued, "your problem is that you always get caught up in the big picture and miss the little details. Rather like this one with her tongue in her cheek," and he nodded at Susan.

Susan was so stunned by the change in President Wilson's

demeanor that she forgot to pull her tongue from her cheek. The words rolled out of the parrot's mouth—er, beak—as smoothly and eloquently as if they'd come from a lawyer or a newscaster, with only a slight squeak in the upper register giving any sign that it wasn't a person speaking. Both she and Charles stared at him with eyes fit to pop from their heads.

"Ah, well, yes," Uncle Farley was saying, wringing his hands and glancing sidelong at Susan and Charles. "Here's a good birdy now—"

"Farley," the parrot interrupted him. "The time for pretending is long passed." President Wilson shook his head, looking like a teacher disappointed by a pupil who had once showed promise but had since proved incapable of real learning. "To think that a parrot of my distinction, education, and service should end up with a human such as you to look after! It is positively deplorable!"

"Now see here, President Wilson—"

"*Farley*," President Wilson spoke over him yet again. "Have you looked outside recently, or has your nose been buried behind too many bookcases and under too many beds?"

All three humans turned toward the window. By now they'd gotten used to the even roll of the floor beneath their feet, and so had almost forgotten the flood that had started the day. But now—

"Holy cow!" Charles said, sounding for all the world like one of President Wilson's "Raks!"

"Charles!" Susan said.

But Charles was running to the window.

"Susan! It's . . . I mean . . . we really have! We've washed out to sea!"

For the fog was gone. Not a cloud was visible in the pale blue sky. Bright hot light beamed down on the other side of the glass, but where twenty-four hours ago there had been a sea of lush green grass, there was now an endless expanse of gently rolling cerulean water.

Have you ever awakened on a cool morning with the blanket pulled all the way up to your chin? And looked down that length of wool or cotton (let's say the blanket is blue) at the bumps of your feet *all the way* down at the other end of your body? And felt almost as though those bumps weren't your feet, but just bubbles in the fabric? For one weird dizzying moment it feels like you've lost your body, like you're just a pair of eyes. Not even eyes: you're just *sight,* without a body to do the seeing. And then you wiggle your toes: the blanket moves, you feel it against your skin, all at once your feet are *yours* again, as well as the body that lies between them and your chin. The waking-up moment is over, but while it had lasted it had been a kind of magic, an unsettling but exciting reminder that you are a vital part of things, an integral piece of the fabric of the world. For Charles, looking out the window at the unexpected expanse of water was like that feeling: he wasn't frightened, just slightly disconnected from it. He had only to wiggle his toes and everything would fall into place.

Touching the window gingerly, as if opening it might sink the house, Charles raised the sash. Immediately the smell of water pushed into the room, and a faint lapping sound could be heard as waves washed against the side of the house. Charles leaned out the window slightly, as if there might be land in the periphery. But no matter how far he turned his head to one side or the other there was just blue and blue and more blue. He could feel his heart hammering in his chest, but it wasn't fear. No: it was excitement, exultation even, the thrill of pure joy. As Charles looked out of his uncle's dining room window at the sea that had sprung up around it, he felt as if he were the luckiest boy in the world.

"Rak!"

President Wilson called everyone's attention back to the room.

"If you're done sightseeing," he said, "I believe there is the matter of a lost child."

With a guilty pang, Charles remembered Murray. He looked at the deep water one last time—it couldn't be, he told himself, Murray wouldn't have . . . he didn't even know how to swim! Gulping, he turned to face Uncle Farley and President Wilson.

Uncle Farley was shaking his head. "Well," he said to the

parrot, "you scratching up the chairs isn't helping matters. Now come down—"

"Not the chairs," President Wilson interrupted Uncle Farley again. "The *chair*. Just *one chair*," he continued when Uncle Farley still stared at him dumbly. Even Charles, who found President Wilson more intimidating now that he'd revealed he really could talk, was starting to wonder if the parrot were not so bright after all.

President Wilson glared around the room. *"A chair,"* he said one more time, *"not where it's supposed to be."*

"President Wilson—" Uncle Farley began, but just then Charles exclaimed:

"The dumbwaiter!"

"Finally!" President Wilson sighed dramatically.

Uncle Farley's gaze flashed between Charles by the window and President Wilson on the chair, and then he said, "Oh, rubbish!" and stomped to the dumbwaiter. He yanked it open so violently that the upper and lower panels slammed into the stops with a loud bang and almost snapped shut again.

There, curled up asleep on the padded chamber, was Murray. His thumb was in his mouth, and the loud bang seemed not to have disturbed him at all.

At the sight of Murray looking so peaceful, everyone in the dining room got quiet. Then, gently, Uncle Farley lifted Murray from the dumbwaiter and held him in his arms.

As he passed through the dumbwaiter's portal into the

bright room, Murray's eyelids fluttered open. He blinked rapidly, looking at everyone with big solemn eyes, and then he threw his arms around Uncle Farley's neck and squeezed tightly. Somewhat startled, Uncle Farley hugged him back, then walked to the table and sat Murray on its edge.

"Murray," Susan said, "we were worried sick about you!"

"Hey," Charles said. "He's got something around his neck!"

Uncle Farley had walked back to the dumbwaiter, whose doors had fallen softly closed. He opened them and peered inside as if there might be another Oakenfeld child hidden inside; then, apparently satisfied that all was as it should be, he pulled his head out and said, "Er, what's that, Charles? What's Murray found?"

Charles pointed: Murray was wearing a thin gold chain with a heart-shaped pendant, also gold, hanging at the end of it. Murray looked down at it, then up at Uncle Farley, but he didn't say anything.

"It's a locket, Uncle Farley," Susan said now, striding over to the table. "It should open and—"

She was about to snap it open when Charles said, "I think maybe you better ask first."

Susan looked over at Charles, then back at Murray. Swallowing, she said, "Is it—is it okay, Murray?"

Murray looked at his sister with eyes still bleary from sleep. For a moment it seemed as if he were trying to remember who she was. Then, smiling slightly, and slightly sadly, he nodded.

"It's okay," he said in a voice nearly as raspy as President Wilson's.

Susan examined her little brother for a moment. He looked exactly the same as he had the last time she'd seen him, yet there was something different about him as well, and not just the locket around his neck. She hesitated so long that Murray smiled again and said, "It's okay, Susan. You can open the locket. Nothing will happen."

Susan hadn't thought anything would happen when she opened the locket. Now, somewhat nervously, she used her fingernails to pry it open. There, staring back at her with rather resolute expressions, were tiny black-and-white pictures of herself and Charles.

"Why, it's us!"

The images were so closely cropped it was hard to make out anything in the background. All she saw was sky. Maybe a hill behind Charles's head? It did look vaguely familiar, but she couldn't place it. She was just about to ask Murray where he'd gotten the locket—not to mention the pictures—when her brother threw his arms around her shoulders.

"I missed you! Oh, Susan, I missed you so much!"

Susan wrapped her arms around her little brother's warm body. "Murray," she said, trying to laugh away her nervousness. "I'm sorry we played that trick on you, but we were only apart for a little while. No more than an hour or two."

Murray sat back on the table.

"It felt like *forever*. It felt like *my whole life*."

"There, there, Murray," Uncle Farley said. "All's well that ends well." He slid the chair with President Wilson on it back to its place at the table. President Wilson half spread his wings and jumped lightly from the chair to the tabletop, sliding and wobbling a bit on the polished wood until his claws found a purchase. Uncle Farley made a bit of a face at this, but didn't say anything. Charles, who'd been walking over to the table to claim his own hug from Murray, looked at President Wilson and hung back a little.

"I trust we'll have no more hiding in the dumbwaiter," Uncle Farley said. He surveyed his audience. "Well," he said, "I suppose the president and I have some explaining to do."

"Lu-u-u-cy!" President Wilson exclaimed. "You got some 'splaining to do!"

All four humans looked at the parrot blankly, and under their collective gaze President Wilson scratched his red head with a claw.

"Well, I *am* a parrot, you know. Clowning just comes naturally to us."

"Er, right," Uncle Farley said. "Well, ah, yes," he said. "Perhaps a little food before we start? To help settle our stomachs."

And, turning to the dumbwaiter, Uncle Farley slid open the panels. There, where Murray had been not two minutes before, sat a tray of gleaming covered dishes. Susan and

Charles looked at each other with wide eyes; only Murray seemed unsurprised.

"Yay," he said, albeit not as enthusiastically as he'd done at breakfast. "I haven't eaten in *years*."

"Uncle Farley," Susan said, "how did Miss Applethwaite do that?"

"Mmmm, yes, Applethwaite. Or Applewhite." Uncle Farley chuckled at himself. "Not sure how I managed to mess *that* one up. Anyway, have a seat and it'll all come out. As President Wilson said, there's no use for secrecy anymore."

Soon enough the four humans were seated at the table with steaming omelettes before them, while President Wilson had an enormous dish of delicious-smelling honey-roasted peanuts all to himself. Charles, seated closest to President Wilson, tried to ignore the bits of nut that spattered onto his plate as the parrot crunched his meal in his powerful, sharp beak.

"The truth is," Uncle Farley said when everyone was settled, "we know very little about Drift House's history, and even less about its, ah, attributes. What it can do." He gestured at the window, and the vast body of water beyond it. "What *that* is."

"Let's not oversimplify," President Wilson said, spewing still more bits of nut onto Charles's plate. "We know very well what *that* is," he said, and he used a wing as Uncle Farley had used his arm to indicate the water outside the window. "It is the Sea of Time."

"The Sea of Time!" Susan and Charles said at the same moment.

"Well, yes," Uncle Farley said to President Wilson. "Yes, we know it's called the Sea of Time, but what we don't know is what that means. Do we, Mr. President?" he added, in a slightly barbed voice.

President Wilson suddenly seemed to find his dish of nuts particularly compelling.

"President Wilson!" Susan exclaimed. "Is that you? In the painting in the hall?"

It would seem impossible for a creature whose cheeks are feathered in bright red plumage to blush in embarrassment, but that is exactly what President Wilson appeared to do. His normally loud, clear voice muttered something unintelligible into his dish of peanuts.

"I'm sorry, President Wilson," Uncle Farley said. "We didn't quite catch that." He seemed clearly to be enjoying the parrot's discomfort.

"I said it's my great-grandfather Xerxes."

"Ah," Uncle Farley said with relish. "So you're *related* to him."

"Yes," President Wilson said. He snapped a nut in his beak.

"You could even say you're *descended* from him."

"Yes. You could say that."

"You could even say that the parrot on Pierre Marin's shoulder, the great parrot Xerxes, *passed something down*—"

"Oh, enough, Farley!" President Wilson said, flapping his wings in frustration. "I forgot, all right? Yes, I did not remember my lessons. *You* try spending the better part of a century acting like a foolish unthinking bird to four successive generations of imbecilic humans and see how much information *you* manage to retain. Under the circumstances," he said in a slightly self-pitying voice, "I think I did rather well."

Susan, who had followed this exchange between parrot and uncle with bewilderment, now asked President Wilson, "Were you supposed—that is, are you Drift House's historian?"

"It's hard to be a historian when you don't remember—"

"Ninety-seven years, Farley!" President Wilson cut him off again. "For ninety-seven years I had to endure 'Hello, hello, hello?'" President Wilson said this in exactly the voice you use when you're trying to get a parrot to repeat what you're saying. "'Hello, hello, hello?' 'Polly want a cracker?' 'Do the little dance for us, birdy, do the little dance.'" Consciously or unconsciously President Wilson's head bobbed up and down as he said this last, and he did a little side-to-side shuffle. Now he drew himself up short. "It was enough to drive even the most disciplined parrot to distraction!"

Uncle Farley seemed to relent a bit.

"Well, fortunately for President Wilson, I was able to uncover a few pieces of Drift House's history in the course of my own researches. Pierre Marin built it in 1675, and it seems to have been the culmination of a lifetime of study into the nature of time."

"Why, that's just like you!" Charles said.

Uncle Farley smiled, as if pleased Charles had remembered.

"Yes, we do share some interests in common. I had even come across his name on a few occasions, in a footnote to this or that seventeenth-century text, but never paid him much attention. Marin's investigations, you see, were, depending on your point of view, either much more practical than my own, or simply insane."

"Insane!" Susan said. "What do you mean?"

"Well, Marin was something of a transitional figure, caught between the Age of Exploration and the Age of Enlightenment. Such men believed no place, be it physical or mental, lay beyond the power of man to reach, and as such—"

"Oh for heaven's sake, Farley!" President Wilson cut him off. "Get to the point already. Pierre Marin discovered a way to travel through time."

Susan and Charles looked at each other.

"That's what the radio said!" Susan said then. "It called Drift House a trans, a transtempo—"

"Transtemporal vessel," Charles finished for her. "Is that it, Uncle Farley? Is Drift House some kind of time machine?"

"Well, yes. But not in the science fiction, H. G. Wells sense of the term. Drift House doesn't possess the power to manipulate time, or our passage through it. Rather, it allows its occupants to access—" and here Uncle Farley nodded at the vast body of water beyond the window.

"The Sea of Time," Charles whispered in an awed voice.

"Uncle Farley," Susan said, "are you saying that we can, can *sail* in Drift House to any point in time?" And, when her uncle nodded at her, she added, "But . . . but *how*?"

At this, Uncle Farley frowned at President Wilson, who, despite the rigidity of his beak, seemed somehow to grimace back at his human judge.

"Oh, don't look so satisfied, Farley. I'm sure I remember that part of my lessons correctly. In fact, I think *you* broke it."

"Broke what?" Charles said. And then, remembering: "You mean the radio."

Uncle Farley nodded. "According to President Wilson, either the large radio in the drawing room or the smaller set you saw in my room is all that's needed to pilot the house. Unfortunately, he doesn't remember how they do this."

"Not true! I told you: turn the dials."

"'Turn the dials,'" Uncle Farley mimicked the parrot, "hardly seems complex enough to negotiate the vicissitudes of time travel."

"The *what!*"

"The viciss—"

It can be quite illuminating to see other people exhibit behavior that you yourself are prone to, and Charles, seeing Uncle Farley and President Wilson bicker exactly as he and his older sister did, felt distinctly embarrassed for them. And, clearing his throat quietly, he said, "Perhaps I can help, Uncle Farley. I do know quite a bit about radios."

Uncle Farley smiled at Charles, even as he stood up and walked to the dumbwaiter. Charles couldn't imagine how he could still be hungry (he'd managed to polish off his omelette even as he spoke, though Charles and Susan had hardly touched their plates). But when Uncle Farley slid the panels open it wasn't a dish of food that appeared before him, but the Tombstone radio they'd seen him take into his study earlier.

Susan gasped.

"Did . . . did Miss Applethwaite get that for you?"

"Maybe," Uncle Farley shrugged. "If I'd ever seen her, I'd ask. But who or what Miss Applethwaite is is yet another of the questions President Wilson is unable to answer. All he remembers is her name." Uncle Farley laughed at himself. "I'm afraid I didn't do such a good job at remembering it myself. Apple*white*? What was I thinking?"

Before Susan could ask him more about this, Uncle Farley loosened a catch on the back of the radio and opened it up. This time it was Charles who gasped.

Susan peered into the radio but saw nothing gasp-worthy. "It's just an empty box."

"Don't you get it?" Charles said. "It's supposed to be a radio, but there's no tube, no transformer, no coil—no *radio*. It's just an empty box."

"But it still manages to produce noise," Uncle Farley said. "And I believe it has other attributes as—Susan? What are you doing?"

For Susan had jumped out of her chair.

"The drawing room!" she exclaimed, and ran into the hallway.

Uncle Farley and Charles hurried after her, Murray coming rather more slowly behind.

"Susan!" Uncle Farley called after his rushing niece as she vanished through the drawing room's broken door. "What on earth is going on?"

When Uncle Farley and Charles arrived in the drawing room they found Susan standing in the middle of the floor, apparently staring off into space. No, not space, but at the fresco. There was the house with the children standing on the roof of the tower, the long vines still attached to the solarium and leading to the murky shapes of the mermaids under the water. There was the enormous whirlpool spewing mist and shadow that the mermaids appeared to be rescuing the children from. And there was the strange island with its odd, individual hills dotted across its green surface.

"Murray," she said to her youngest brother as he ambled into the room, "could I look at your locket again, please?"

Docilely, Murray walked over to Susan, and, somewhat gingerly, she snapped the locket open. She peered at the tiny pictures, then looked again at the fresco. Just then President

Wilson flapped into the room, landing somewhat heavily on his perch.

"Too many . . . nuts," he panted.

"Susan," Uncle Farley said, ignoring the parrot. "Would you please explain yourself?"

"I'm not sure how," Susan said. She gestured at the locket, and then at the fresco—at the portion that depicted the island. "It's . . . well . . . they're the same."

"What do you mean, the same?" Uncle Farley said.

"This hill," Susan said. "The one you can see behind Charles's head? I'm pretty sure it's that hill right there."

Uncle Farley and Charles crowded close around Murray, peering first at the locket and then at the fresco. The two hills did in fact appear to be the same hill. As his sister and brother and uncle oohed and aahed around him, Murray endured their examinations like a well-behaved (if slightly bored) patient suffering a routine physical checkup.

"How—how extraordinary!" Uncle Farley said.

"Murray," Susan said then, "did you somehow . . . go there?" And she indicated the island painted on the wall of the room.

"More to the point," Charles added, "did *we*?"

"I—I really don't understand," Uncle Farley said. "This is baffling. It defies comprehension. It—" He cut himself off. "Murray," he said in a slightly calmer tone of voice, "where did you get this locket?"

But Murray simply looked up at Uncle Farley and Susan and Charles. He didn't appear to be withholding something from them. Rather, he seemed to be waiting.

"Murray?" Susan said. "Can you tell us what happened?" She heard her voice, the kind of voice that you use to address a five-year-old, which is what Murray was. But the look in his eyes seemed old beyond his years, and she felt much more like a child asking a grownup for information.

"Come on, Murray," Charles said encouragingly. "I'm sure we can help you figure it out. It sounds like you had a really cool adventure," he added, his voice sounding falsely bright.

But Murray only stood there. Waiting.

And then, just when the silence seemed unbearable, a slow knocking filled the house.

THUD—THUD—THUD.

At the sound, Susan's mind immediately filled with a frightening image: Pierre Marin, dressed in his pirate robes, stepping out of his picture and walking down the hall toward her. It was his pegleg that made the ominous knocking. When she glanced at Charles, she could tell from his wide-eyed expression that he was thinking exactly the same thing.

THUD—THUD—THUD.

This time she recognized the sound: it was the big brass knocker on the front door, the one Murray had banged fit to shake the house down a mere twenty-four hours earlier. Susan glanced over at Uncle Farley, who was looking at President Wilson, who in turn was looking at Murray.

Murray waited until everyone was looking at him before he spoke.

"The mermaids are here."

He didn't sound happy about it at all.

"But Murray," Susan said, "how do you know this?"

Murray only fingered his locket.

THUD—THUD—THUD.

For a moment no one moved. And then Susan said, "Well, I'm going to answer the door. Perhaps the, the mermaids can give us some answers."

She started toward the hall when Murray cleared his throat behind her.

"There's one thing I think you should know."

The tone of her little brother's words froze Susan in her tracks. She turned back to him. "What's that?" she said in the lightest voice she could manage.

"If you answer that door, you'll die."

PART TWO

Adrift on the Sea of Time

SEVEN

Diaphone

A CHANDELIER HUNG ABOVE THE drawing room table. It wasn't turned on: now that the rain had ended, the afternoon was warm and clear, and in the bright light the drawing room's vivid fresco seemed like nothing more than an ordinary wall painting. But the chandelier still managed to call attention to itself, albeit discreetly, by the soft tinkle of its crystals. They bounced against each other like wind chimes as the chandelier swayed back and forth in the gently rocking house, and Susan found herself staring at it as though it were a hypnotist's swinging watch. She let the gentle motion and the sweet musical sound of the tinkling crystals soothe away the harsh words her five-year-old brother had just uttered—until the lulling noise was obliterated by the rapping that sounded once again throughout the house.

THUD—THUD—THUD.

For a moment Susan felt very far away, as if she were much, much more grownup than she was, and looking down at her twelve-year-old self expectantly. At first all she saw was how small she looked framed by the big doorway, half in, half out of the drawing room. But then she saw that her tongue wasn't in the side of her cheek, and she brightened somewhat, and even managed to smile a little.

"Murray darling," she said in a voice much more steady than she felt on the tilting floorboards. "You're going to have to be brave now, and try to tell us what you meant by what you just said."

Poor Murray: he looked so small in the middle of that big room, with Uncle Farley's hulking form towering over him. He snapped the mysterious locket closed and pressed its two halves tightly together, as if he could seal up the mysteries they hinted at.

"Oh, I don't know, Susan, I don't know. It's just, I have the strangest . . . I mean, they *feel* like memories, but how can they be memories if they never happened?" Murray's cheeks were puffy and red with frustration, and with the effort of holding back his tears. "Susan," he wailed, "I'm only five years old!" But however much he *looked* like a five-year-old, it was very clear that he didn't feel like one. Or, for that matter, talk like one.

Now Uncle Farley leaned close to Murray. "I can see you're confused, Murray. Frankly, we all are. But what you just said is quite disturbing, and it would help tremendously if you could tell us a bit more. Later we'll try to figure out where these, as you say, memories come from. But right now I think it's important to find out what exactly you *do* know. Now, are the"—Uncle Farley glanced at the dark shapes swimming below the water in the fresco before he said the word—"are the mermaids going to harm your sister?"

Murray looked at Uncle Farley, and then he looked at Susan, and then he looked back at Uncle Farley.

"I don't *think* so," he said.

THUD—THUD—THUD.

Susan tried not to jump at the knocking. Whoever was doing that banging sounded strong enough to knock the door down if someone didn't open it pretty soon.

"You don't sound so sure of yourself," she said.

There was a look of intense concentration on Murray's face.

"I think they will ask you to do something. To help them." His eyes closed with his effort. "To rescue . . . someone." Suddenly his face relaxed, and he opened his eyes. "Yes," he said. "The mermaids won't hurt you. In fact we need them, to do that." He pointed at the scene on the fresco in which the mermaids were pulling Drift House away from the great whirlpool in the center of what they now knew was the Sea of Time. "But in exchange for saving us, they will ask you to save one of them, and that's how you . . ." He let his voice trail off.

Susan allowed her brother's words to sink in for a moment. How could he possibly know such things? But as she looked at his face she knew he was telling the truth. She turned to the fresco then, and as she looked at the enormous vortex that threatened Drift House and its precious human cargo, she imagined that she could feel the house's rocking increase, as if the current that carried it were speeding up, becoming more violent.

She looked at her uncle, and saw that he too was examining the fresco.

"I don't think we have any choice," she said to him. "We have to let them in, unless we want *that* to happen."

Uncle Farley tore his eyes away from the fresco with difficulty.

"It certainly does look that way."

Charles, who had watched this exchange with a look of dumbfounded horror on his face, suddenly burst into speech.

"This is crazy! Murray obviously fell asleep and had a bad dream. He can't possibly know any of these things. Come on, Uncle Farley," he said, grabbing his uncle's hand. "Let's try to fix the radio. You yourself said you thought it controlled the house somehow, and I know a lot about radios. I'm sure I can help you. We don't need m-m-mer—" His rational scientific brain wouldn't let him say the word out loud, and instead he pulled on his uncle's arm again.

But Uncle Farley refused to budge. "Charles," he said simply. "It's too late." And he nodded at the doorway behind his nephew.

Charles turned. Susan had gone to answer the door.

She hurried down the hall. The knocking hadn't come in a while, and as afraid as Murray's words had made her of meeting the mermaids, she was even more afraid that they'd gone. And of course a part of her was excited too. Mermaids! She, Susan Oakenfeld, was going to meet real live mermaids!

Just then, the seesawing motion of the floating house tripped her forward, and she practically fell against the door handle. She grasped it firmly, turned, and pulled the door wide—just in time to receive a splash of water full in the face. She was still wiping the water from her eyes when the loveliest, most musical voice she'd ever heard said, "So you *are* home!"

When Susan could focus again, she saw first the sea, the same sea that Charles had seen from the dining room windows. And, like Charles, Susan found all that water mysterious at first, a little frightening even. How could there be so *much* of it? But the water was so blue and dazzling, so sweet smelling, that it was impossible to be frightened for long.

"Why, it's fresh water!" she said out loud.

"Of course it's fresh water. You don't think we'd live in *brine*, do you?"

Susan started a little, and looked down. And there was a face as beautiful as the voice that issued from it—the face of a girl who looked to be about her own age, with a pointed chin and a small pouty mouth and endless streams of copper-colored hair that fanned out in the water around her.

The girl smiled when Susan's eyes met hers. "I was knocking for*ever*," she said. "I was beginning to think we'd called—that is, that the house had come out on its own."

There was a shadow in the water deep beneath the girl's head, something green and wide and rippling, like a flag blowing in a breeze.

"Excuse me," Susan said. "But are you the mermaid?"

The girl in the water frowned slightly. "Well, now, I don't know."

"What do you mean?" Susan said. "How can you not know if you're a mermaid or not?"

"Ah, well, that's a different question, isn't it? Am I *a* mermaid? Yes, certainly. But am I *the* mermaid? That I don't know. What mermaid did you have in mind?"

"W-well, I don't know," Susan said, now thoroughly confused. "Murray was just talking about mermaids, and there was one in the drawing room fresco last night, and, well, I don't know. I've never met a mermaid before."

"Never met a mermaid!" The girl in the water—the mermaid—seemed shocked by this. "Where on earth are you from?"

"I'm from New York," Susan said.

"Is that anywhere near Copenhagen?"

"Not exactly. Copenhagen is in Europe."

"Ah," the mermaid nodded her head. She seemed deep in thought for a moment, and then she said. "New York is not in Europe?"

"New York is in New York." Susan shook her head. "It's in America." The mermaid seemed about to speak again when Susan said, "Please? Are you the mermaid in the drawing room fresco?"

And all at once the water around the mermaid began to churn and froth, and the mermaid seemed to lift up slowly, as if she were being pulled by strings. Her upper body was bare and pale, clothed only in a halter made of clamshells and beaded string and some kind of white flower, her tummy smooth and white as a sheet of paper (no navel, Susan noted, though she wasn't sure what that meant) and then all the sudden there was the rest of her.

Her tail.

The scales started just below her hipbone and they were bigger than Susan would have expected—nearly as big as a guitar pick—but they quickly got smaller, smaller than Susan's fingernail and then as small as crystals of salt, and every bit as iridescent. Though the tail was as green as the leaves on the elm trees in Central Park, if Susan looked at one scale at a time she saw white and purple, yellow and blue, the colors changing as the scales undulated over the strange rippling muscles that made it move back and forth and held the mermaid suspended on the water like a dolphin dancing on its flukes. Susan thought it was the most beautiful sight she was ever likely to see in her life. But she had only been on the Sea of Time for a few hours, and it held many more beautiful things in store for her.

"I am Diaphone," the mermaid said in a voice that was at once absolutely clear and yet also sounded like bells in the distance, as if the mermaid were both talking and singing at the same time. And then all at once she sank back in the water—not quickly, like a person falling, but slowly, gracefully, like a candlewick being dipped carefully in wax.

Susan tried to speak and discovered she had been holding her breath. She had to let it out and suck in another, and then she said, "My name is Susan. Susan Oakenfeld," she added, feeling that "Susan" was inadequate next to a name like "Diaphone."

Without thinking about it, Susan had stuck out her hand, and Diaphone rose a little in the water to shake it. Her hand was cold but not clammy—not like a fish, but like a girl who's been swimming in cool water all morning. When she squeezed Susan's hand, Susan had the impression of incalculable strength, as if the mermaid could have lifted her like a doll.

"It is a great honor to meet you, Susan Oakenfeld. My sisters and I have been waiting for you for a long, long time."

EIGHT

The Proposition (part 1)

A SHORT WHILE LATER THE three Oakenfeld children stood on the roof of the gallery, where Susan and Charles were engaged in a heated argument over which side was port and which starboard.

"Now listen," Charles said. "If we think of the house as a ship, then the solarium is obviously the prow, and this is the poop deck."

"Poop deck," Murray said, and giggled quietly. He was standing apart from his siblings at the railing. He had Susan's spyglass and was looking out over the wide blue sea.

"So that would make that side"—Charles pointed to

Murray—"the right side. And the right side is port. I'm sure of it."

"And I'm sure it's the left," Susan said. "The right side is starboard." In fact Susan wasn't sure at all, but ever since Diaphone had told her the mermaids had been waiting for her she'd been feeling nervous, and when she was nervous she got irritable, and when she got irritable she was prone to disagree with people just to be contrary—especially Charles, who had to be right about *everything.*

"I'm telling you, the right side is port," Charles insisted, because Susan's peevishness was catching, and he too was feeling contrary. "*Port.*"

"Saying *port* doesn't make you any more right, Charles. Just because you *emphasize* the word doesn't mean you *know* what you're talking about."

"Poop deck," Murray said again.

"It's the *right* I'm telling you!"

"Eh, what's this?" Uncle Farley said, emerging from the glass doors that led to the second-floor hallway. "Port's the left side, Charles. Always the left."

In his left hand—his port hand, if you will—Uncle Farley carried President Wilson on his perch. President Wilson had said he was feeling too full to fly, and he had refused to sit on Uncle Farley's shoulder because he was not a *pet* parrot, but rather the only rational member of this household. He rode on his perch like a prince on a litter, feathers fluffed, eyes closed, and he didn't open them until Uncle Farley, huffing slightly, set the perch down on the deck.

"Silly humans," President Wilson said then. "Always bickering over the least consequential things." He stood up now, stretched one massive green and red wing and then the other, and shook his head rapidly. "Now then. Where is the fish-girl?"

"She said to meet her here," Susan said. "She said she would be right back, although it feels like it's been more than an hour."

"Well, I don't like it," Charles said. "We're standing here exactly as we were in the picture in the fresco." Although he didn't say it aloud, Charles was nervous about the fact that things were playing out as the fresco had predicted. Because if the fresco was right, then that might mean Murray was right, and if Murray was right then Susan might . . . Charles wouldn't even let himself think it.

"I think the drawing room walls might more accurately be termed a mural, Charles," Uncle Farley said now. "A fresco is traditionally painted on wet plaster—"

A sound from President Wilson cut Uncle Farley off. It was sort of like a human "hmph," but coming from a parrot's beak it sounded more like a piece of cellophane being wadded up. If you have ever sat next to someone opening a roll of candy in a movie theater, you know the sound exactly.

"Charles's fears," President Wilson said, "although not

surprising, are nevertheless unfounded. One thing I do remember from my childhood lessons is that the drawing room walls do not predict the future per se. No one can know the future unless he's been there and come back, which, obviously, none of us has done." He looked around at his audience as if daring anyone to contradict this last assertion, his eyes landing finally on Murray, who had put down his spyglass and was rubbing the locket around his neck. "Rather," President Wilson said, tearing his eyes away from the mysterious gleaming amulet with difficulty, "the walls show what is *most likely* to happen, based on the nature of everyone involved. The drawing room—"

"President Wilson," Susan interrupted, "are you saying the drawing room *knew* about the mermaids last night? Before we even washed out to sea?"

"I wouldn't say it *knew*," President Wilson said, as if a thinking room were the most natural thing in the world. "Rather, it took into account the fact that none of us knows how to control Drift House and then looked for someone else who could have summoned us onto the Sea of Time. Since the mermaids are the most organized and mobile of the sea's thinking inhabitants, logic would suggest that they brought us out here. Indeed, the situation is so clear it's a wonder I didn't reason it out earlier."

"A wonder indeed," Uncle Farley muttered, but before President Wilson could retort Susan exclaimed, "The mermaids brought us here! Of course!"

From the railing Murray said, "Poop deck" one more time, and then he turned to his siblings.

"Actually, Charles," Murray said, using the new grownup voice that made Susan feel very nervous (not to mention guilty), "I'd be less concerned with the fact that the drawing room walls predicted the future than with the fact that they were right."

Charles, who, like Susan, was intimidated by the new Murray, stammered a little when he said, "Wh-what are you talking about, Murray?"

Murray sighed. He seemed significantly less distraught than he had earlier, as if he'd grown used to this new version of himself. "To tell you the truth, I'm not one hundred percent sure. But I've been thinking about what happened to me in the dumbwaiter, and I'm pretty sure I somehow made my way into the future." As he said these last words he looked at President Wilson, who looked at him skeptically but held his tongue. "A pretty long way into the future, I think."

Susan, whose head was already reeling with the idea that the mermaids had called Drift House onto the Sea of Time, found it nearly impossible to wrap her head around this new idea. "Why do you think that?"

"It's hard to explain," Murray said, rubbing the locket around his neck. "It's just that ever since I woke up in Uncle Farley's arms I've had a constant feeling of déjà vu—of feeling like I've done all this before," he said in response to the confused look on Susan's face.

Susan remembered her uncle using the term yesterday. She had been too shy to ask him what it meant, but it was even weirder to learn the definition from her five-year-old brother.

"I suppose we'll just have to go with that," Uncle Farley said now. "Stranger things are happening around here. But I'd like to understand why it bothers you that the drawing room mural predicted the future accurately."

Murray looked at his sister sadly. "Well, as I said earlier," he began quietly, "in that future, Susan died."

"Oh blast!" Uncle Farley said. "This is all my fault. I should never have agreed to take you on."

"Oh, stop your whinging, Farley," President Wilson said. When the children had interrupted him earlier, he had promptly closed his eyes and pretended to be asleep. But at Murray's last words he sat up on his perch again, and now fixed his bright yellow eyes on the youngest Oakenfeld. "There's something on your mind, boy. Speak."

"It's hard," Murray said. "I can't remember anything specifically, but I have the strongest feeling that Susan's death can be prevented."

"Um, *how*?" Susan said, rather eager to know what needed to be done.

"Well," Murray said, "it's obvious you have to do what the

mermaids ask you to do. If not, we'll all die in that whirlpool thing. But I don't think you have to do it this time around the same way you did it the first time. You can"—Murray shook his head at the strangeness of it—"you can change the future."

"That is ridiculous!" President Wilson said. And then, realizing his tone had been somewhat stern, he said, "I'm sorry, child. Though I sense that you yourself believe what you say, the truth is that history, the past as well as the future, can*not* be changed."

"But they're always doing it in movies," Charles said, eager to grab on to something he could treat analytically. "People always go back in the past, and then they return to their own time and discover that everything's different—that their presence in the past somehow changed things. Or people suddenly show up from the future, with the goal of fixing some problem that hasn't even happened yet."

"Like *The Terminator*!" Susan said. (She was not allowed to watch violent movies like *The Terminator*, but had stayed up late at a slumber party at her friend Jeanne-Marie Gladwell's house and seen it there.)

"Right!" Charles said, nodding so rapidly he almost shook his glasses off. "Physicists call it the ripple effect!"

"The 'ripple effect' is just a theory, Charles Oakenfeld," President Wilson said in as droll a voice as possible, "and I venture to add that there are very few Hollywood directors who are actually *familiar* with time travel."

Charles, chagrined, dropped his eyes, but Murray insisted, "I think you're wrong, President Wilson. I think that's why I came back. I'd figured it out, you see. I figured out how to save Susan." His voice was calm, but his little fingers were rubbing his locket furiously. "All I have to do now is *remember* it."

There was an awkward silence, and then, just as Susan opened her mouth to speak, a voice came from behind and below her.

"Stand back, Susan Oakenfeld!"

Susan barely had time to step back from the railing before a gold and green form shot up from the sea below. It soared into the air above her head and then came down with amazing ease ending up perched on the railing as if it had been sitting there all day. It was, of course, Diaphone.

"I hope I didn't startle you," she said, though her voice—again Susan noted the beautiful musical undertone—indicated exactly the opposite. There was a proud, slightly mischievous gleam in her eye as she smoothed her copper-colored hair off her face. "There is no other way, really, for the fish tailed to negotiate your ungainly waterless world."

Charles, who hadn't seen the mermaid before, stared at her with a weird, almost bewitched smile on his face. So did Uncle

Farley, Susan noticed, though Murray and President Wilson both seemed unimpressed by her beautiful voice and face and form, not to mention the grand entrance she had just made. For her part, Diaphone barely glanced at anyone besides Susan, though the latter thought the mermaid's eye had lingered a little when it came to Murray.

"Explain yourself," President Wilson declared now. "How dare you summon us from our world without our permission?"

Diaphone just stared at President Wilson for a moment, then all at once she burst into peals of laughter, clapping her hands together in her amusement.

"A talking bird-thing!" she cried. "How funny! You there," she said, waving a hand at Uncle Farley. "Bring him to me, slave."

Uncle Farley's face turned bright red beneath his whiskers, and he sputtered, "I can assure you, miss, that I am by no means—"

But President Wilson cut him off. "Bring me near the fish-girl and I'll scale her with my own two feet!"

Diaphone laughed again, so exuberantly that she slapped the flukes of her tail against the deck with a wet smacking sound.

"I think we should find out what she wants," Susan said then.

"You are a sensible girl," Diaphone said, stifling her last laughs. "It must be torture for you, having to deal with so many obstinate males."

As soon as she spoke, Susan realized she *was* the only girl on board. She didn't know why, but the thought made her feel lonely somehow, and, despite Diaphone's haughty demeanor, she found herself warming to the mermaid.

"Please," she said. "Will you tell us why you brought us here? And what we have to do to get home?"

"That's what *I* just said," President Wilson grumbled under his breath.

If Diaphone heard President Wilson, she ignored him, addressing her words directly to Susan. "We need your help, Susan Oakenfeld."

"But what could *I* possibly do for *you*? I'm not even a good swimmer."

It pained Susan to admit this, but it was true: she had gotten a B in the swimming unit of physical education, and then only because she had written a paper for extra credit.

"Swim?" Diaphone said, suppressing another burst of laughter. "I have seen that thing you humans call swimming." She flapped her slender arms around her beautiful face as though she were trying to get somebody's attention across a crowded room. "Why, you can no more swim than I can walk on dry land."

"Is that what you need Susan for?" Charles said. "To do something on dry land?"

Diaphone looked at Charles as if he were another novelty like a talking bird. Then, addressing Susan, she said, "Your boy-slaves are very poorly trained. I could be of some assistance to you with that, if you'd like." She smacked her tail against the deck again, and Susan felt the powerful vibrations through the wood beneath her feet.

"Th-they're not my slaves," she said, more nervous than ever. "That is my brother Charles, my brother Murray, and that is my uncle Farley. And that is my, I mean, our—"

"My name is President Fernando Wilson," President Wilson said regally. "And I am my own bird, thank you very much."

"Fernando?" Charles said, but President Wilson only stared at him.

Diaphone was silent for a moment, and then she said, "It is long since the mermaids' Royal Court did away with slaves, for precisely such impertinence. We tried sewing their mouths shut, but that of course made it impossible for them to eat. I don't know whether to admire your fortitude, Susan Oakenfeld, for bearing such oafish creatures around you, or to attempt to educate you out of your ignorance.

"Nevertheless," she continued in a firmer voice, "your household is not my concern. The Royal Court of Her Most Aqueous Empress Queen Octavia, Undisputed Sovereign of the Sea of Time and All Its Tributaries and Headlands, and All the Vessels That Sail upon It, makes a claim upon your services. One of our sisters has been captured and detained against her will by that ragtag group of brigands who style themselves the Time Pirates. We know that Drift House and the Time Pirates have in erstwhile days been in league together, but that relationship has grown tenuous in recent centuries. Therefore we demand that you compel the pirates to release our sister Ula lu la lu. If they will not give her up voluntarily, Her Most Aqueous Empress Queen Octavia, Undisputed Sovereign of the Sea of Time and All Its Tributaries and Headlands, and All the Vessels That Sail upon It, compels you to liberate her by force. In exchange, the Royal Court of Queen Octavia will save Drift House from its impending destruction."

Diaphone's speech had rolled by so fast that it was all Susan could do to keep up. Indeed, only her final word made any real impression on Susan's brain.

"Destruction?"

Diaphone's little mouth curled into a tiny smile that sent chills up and down Susan's spine.

"It is obvious to any sea-living creature that Drift House is floating aimlessly, with neither rudder, oar, nor sail. Already it has strayed into the current of the Great Drain, and soon enough it will be pulled into the drain's vortex, where it and everything it contains will most surely be sucked into the void from which there is no return."

So there it was. It was what they'd figured out on their own, of course, but hearing it from Diaphone's lips made it sound much more final. An image of the whirlpool on the drawing room walls flashed in Susan's mind. The Great Drain. Even the memory of it seemed to spin her round and round as though she were already caught in the twisting tunnel of water, and for a moment she was so dizzy she thought she might fall down.

Shaking her head to clear it, Susan said, "I shall have to speak to my family."

Diaphone considered this for a moment. Then she said, "A leader who inspires her slaves to work for her rather than merely forcing them to do her bidding must surely get better labor out of them. I am impressed at your wisdom, Susan Oakenfeld, and more convinced than ever that you are the girl-child who can save our dear comrade."

Susan blushed confusedly at this compliment, even though all she'd been planning to do was ask for advice.

"But be warned," Diaphone continued. "If you decline to help us, then you and the possessions you seem to hold so dear will die before the second fall of darkness." Diaphone paused, and proffered another of her little smiles to Susan. "I shall return for your answer shortly, Susan Oakenfeld. Be wary that you do not tarry overlong in exhorting your slaves, lest you drift so close to the Great Drain that not even my sisters and I can pull you clear."

Then, with a soundless ripple of her powerful tail, Diaphone flipped into the air and disappeared over the railing.

A moment later there was a splash, and a few drops of water flew all the way up the deck.

As soon as Diaphone was gone, Susan felt a tap on her leg. She looked down to see Murray holding her spyglass out to her.

"I think you should have a look at this."

Susan took the spyglass warily, again made nervous by Murray's strange, grownup demeanor. Murray pointed off to starboard—or was it port? Already Susan had forgotten. He pointed off to the right.

Susan placed the spyglass to her eye. The first thing she saw was mist, and for a moment she thought she was seeing yet another image from the drawing room mural—the whale that had poised on the edge of the sea. But then she realized this was far larger than any geyser shot through a blowhole. As she panned the spyglass back and forth, she saw that the patch of mist was miles wide. It was only when she adjusted the focus that she could make out its edges. The surface of the Sea of Time seemed to just drop away, as if over a waterfall. The sight was so startling that Susan gasped and nearly dropped the spyglass overboard.

"What is it?" Charles demanded. "Let me see!"

Susan handed him the spyglass wordlessly. With her unaided eye, she could just make out the mist on the horizon. It looked like a tiny, harmless cloud.

"It's rather alarming, isn't it?"

Susan looked up at Uncle Farley and nodded. "What would *you* do, Uncle Farley?"

For a moment Uncle Farley stood there with an uncertain look on his face. But then he did something that only an adult can do: he smiled a big warm smile and spread his arms, and, after hesitating a moment, Susan threw herself against his soft warm belly.

"Asking one of your slaves for advice, Susan?" Uncle Farley said, and Susan could feel his laugh against her chest.

"I don't think that slaves business was funny at all," Charles said now, bringing the spyglass away from his eye.

"Quiet, slave!" Susan ordered, and she and Uncle Farley laughed again.

Charles stamped his foot against the deck, but it didn't have nearly the impact Diaphone's tail had made a few moments earlier.

"Be serious! We have a real situation here."

"I hate to agree with the boy," President Wilson said from his perch, "but he's right."

"No doubt he is right," Uncle Farley said, "but there's no use getting all bent out of shape about it. The Great Drain's still a ways away. What say we get the dumbwaiter to whip up some grilled cheese sandwiches and then we can discuss our options?"

"I don't see as we *have* any options," Charles grumbled,

"and we just ate lunch." But then, thinking he could probably get another slice of apple pie, he followed his family back inside.

A few minutes later they were seated around the table in Uncle Farley's bedroom. Uncle Farley had his grilled cheese, Charles his apple pie, and Murray shocked everyone by asking for coconut sherbet, though when Susan asked him where he'd ever heard of such a thing he only fiddled with the locket around his neck and said he didn't know. President Wilson asked for a light mix of sunflower and fennel seeds as a *digestif.*

"Now then," Uncle Farley said when everyone was served. "I should tell you that I have only managed to take the house out one other time since I moved in last year. It was at that point that I discovered, as the two of you suspected earlier, that the radios that control Drift House's course are not working properly."

"So our hypothesis was correct!" Susan said.

"Don't say 'hypothesis,'" Charles said, still a little peeved over the whole slaves business.

Uncle Farley tousled Charles's hair, an indignity that Charles suffered in silence. "Your *theory,*" Uncle Farley said, winking at Susan over Charles's head, "was indeed correct. I was quite unable to steer the house, either through time, or, more important, through the water. Nor could I make contact with the various voices that came over the airwaves occasionally. President Wilson and I were stuck out here for nearly a week."

"Was that all it was?" President Wilson said, cracking a seed shell in his beak. "It seemed like a century."

"But you must have figured out a way to control the house, Uncle Farley," Charles said. "Or else how did you get back?"

"A good question, Charles. The truth is, I have no idea. For several days we just drifted along aimlessly. The weather was perfect, as it is now, and of course the dumbwaiter kept us well fed. I saw it as an opportunity to catch up on some reading. But then we were caught up in a current at some point—of course it was the Great Drain, but since I hadn't heard of it at that point I had no idea what was happening. I could see the clouds of mist rising from it, but that was all. It was only when we were perilously close that I was able to see the extent of the vortex. It's really quite unlike anything I've ever seen before, or want to see again—a hole in the sea bigger than Central Park. The house was shivering and twisting about in the current, and I can tell you I thought all was lost, when all of a sudden we just stopped. It was almost as though we had hit a rock under the water. But then the house began to move backward, away from the drain. In less than a day we were back at the Bay of Eternity, where, rather inelegantly, I'm afraid, the house drifted back to its original spot—a little crookedly, as Susan so perceptively pointed out."

Susan, who normally would have delighted at receiving such a compliment, now merely blushed a little, and then said, "But what caused the house to reverse direction, Uncle Farley?"

"I'm afraid I've been unable to figure that out, Susan. My only theory is that there's some kind of fail-safe built in to save the house from catastrophe. I doubt that, however."

"Why?" Susan asked.

"Well, one of the things President Wilson does remember is that the house is supposed to return to the exact moment when it left. Otherwise, people might notice it was gone. Or, if you stayed away for many years, you might come back to find things incredibly different, but you yourself unchanged—it would raise people's suspicions, to say the least. But when President Wilson and I returned, we discovered that a week had passed in the real world, just as it had for us. So that suggests to me that the house's mechanisms weren't controlling our journey, but some other force."

"Uncle Farley!" Susan said excitedly. "What if *you* didn't take the house out? What I mean is, what if the mermaids summoned the house onto the Sea of Time, and then returned it to Canada afterward?"

Uncle Farley considered this for a moment. "But why would they do such a thing? It seems like a lot of effort just to toy with someone's life."

"Diaphone said the mermaids were waiting for me. What if they thought I'd arrived for some reason, and called the house out, only to discover I wasn't there after all?"

Again, Uncle Farley was silent for a moment, and then he laughed slightly. "Well, it's no more nonsensical than anything else that's happened." He shook his head. "I don't know, Susan, I really don't know."

"Yes, but what if there *is* a fail-safe?" Charles said now. "What if the mermaids are just bluffing to get us to help them?"

As always when he was at a loss for words, Uncle Farley looked as though he wished he had a fork in his hands, or a piece of buttered toast.

"I don't see as how we can just wait and find out, Charles," he said. "Our lives hang in the balance, I'm afraid."

"Speak for yourself, human," President Wilson said, and there was a touch of Diaphone's haughtiness in his voice. He flapped his wings a few times, as if to demonstrate that he could always fly away.

Susan changed tack a bit. "What do we know about these Time Pirates, Uncle Farley?"

"Precious little, I'm afraid. President Wilson seems to recall something about a battle that took place between Pierre Marin and someone called Captain Quoin, but the occasion or outcome of that battle is unknown."

"Oh, bother!" Susan said. "What's the point of having grownups around if they can't tell you what you need to know?"

Uncle Farley bowed his head. "I'm afraid I'm inclined to agree with you, Susan." He sighed. "President Wilson told me I should never have agreed to let you stay here until I'd learned more about Drift House. But the situation in New York just seemed so serious, and of course I never dreamed that the house could be summoned out by someone on the Sea of Time."

There was a moment of strained silence, punctuated only by the sound of Murray sipping the last of his sherbet. Then Susan said:

"I suppose there's nothing else for it then. I shall have to accept the mermaids' proposition, and attempt to save this Uly, Ulae—"

"Luly lally—" Charles said.

"Ala alu—" Uncle Farley tried.

"Ula lu la lu," President Wilson said then. "Ula lu la lu, oh where oh where are you?" When everyone looked at him in wonder, he averted his eyes slightly, and said, "I *am* a parrot, after all."

The sky had darkened considerably before Diaphone returned to Drift House. Uncle Farley had retired to the gallery again, to fiddle some more with the broken radio, and President Wilson was sleeping off his meal. The three Oakenfeld children gathered on the poop deck, passing the spyglass back and forth, looking either at the low cloud on the horizon, or else in one or another direction over the open sea. But no matter where they looked they saw nothing but water.

"Susan," Charles said at one point. "Have you noticed there's no sun?"

"What?" Susan said. "But it's setting, it's . . ." Her voice trailed off when she realized that, though it *was* getting darker, there was no direction from which the fading light came. Instead there was just a general dimming.

"Have you noticed we don't cast any shadows either?" Murray said.

The two older children whirled around like dogs chasing their tails. But it was true: they had no shadows.

"How queer," Susan said. Some part of her waited for Charles to admonish her as he would have done yesterday, but instead the familiar, beautiful voice came from the port—starboard?—the left side of the house.

"Stand back, Susan Oakenfeld!"

A moment later a bright form flashed through the air in a spume of water, and there was Diaphone, seated demurely on the railing. There was a little smile on her face. She obviously enjoyed the effect of these entrances on the children.

"Susan Oakenfeld," she said now. "Have you made your decision?"

Susan looked at Charles and Murray for a moment. Murray's face was stony and unreadable above his golden locket, while Charles nodded at her supportively, if nervously.

She checked to make sure that her tongue wasn't in her cheek, and then she said, "I have."

"Then as the appointed representative of Her Most Aqueous Empress Queen Octavia, Undisputed Sovereign of the

Sea of Time and All Its Tributaries and Headlands, and All the Vessels That Sail upon It, I compel you to tell me what your choice is."

"Please, Miss, Miss Diaphone," Susan began nervously. "What will you do if I help you save Ula, Ulay—"

"Ula lu la lu."

"What will happen if I save her?"

"It is as I said. We will tow your vessel away from the Great Drain."

Susan nodded, and then she said, "If you please, Miss. That is, well, that's not enough."

Diaphone's expression didn't change, but Charles hissed, "Susan!"

Susan ignored him.

"If I save your U-Ula, your sister, then I want you to take Drift House back to the Bay of Eternity, in Canada."

Diaphone's pout turned down at the corners slightly, becoming a genuine, if small, frown.

"You expect the Royal Court of Her Most Aqueous Empress Queen Octavia to act as tugboats for your pathetic vessel? Be wary, girl-child, that you do not cross the line into insolence."

Susan shook her head.

"I certainly didn't mean to be insolent," she said. "But it's no good, you see, if all you do is pull us clear of the Great Drain. We'll still be adrift, with no way of getting back to Canada."

"Susan," Charles hissed again. "What are you *doing*?"

But Diaphone's mouth had turned up again, and she was regarding Susan with the same amused pout she'd made upon her entrance.

"You are brave," she said at length, "for a girl-child, at any rate. Although your slaves are *most* unruly. The bargain shall be struck."

She curled up her tail then. Susan expected her to smack it back down on the deck, but she held it there until finally Susan realized she was supposed to respond somehow. Taking a guess, she walked across the deck and touched one of her feet to the wide flukes of Diaphone's tail. She balanced on one foot until Diaphone nodded and relaxed her tail, and then Susan stood on two feet again.

"The light grows dim. I shall take the terms of our agreement to Queen Octavia, and when it is bright again we shall be back to take your vessel to the Time Pirates."

Susan turned and looked at the cloud of mist on the horizon—it was hard to see in the fading light, but it seemed to be somewhat larger.

"Please, Miss Diaphone," she said, turning back to the mermaid. "Is there enough time for that?"

"Let us hope so," Diaphone said. "For your sake."

And then, with a flick of her tail, she was gone.

NINE

Strategizing

IT WAS A LONG NIGHT. In part it was long because none of the children knew when it was going to end. A place without a sun to go down or come up did not exactly inspire confidence that the light would ever return. Though Uncle Farley assured the children that the days on the Sea of Time had roughly the same proportion of light and darkness as they did in southern Canada during the summer—that is, relatively long days with eighteen or nineteen hours of light and five or six hours of darkness—the children found the pitch black unsettling. There was no moon, either. It was absolutely and completely—I mean *totally*—dark. Were it not for the soft lap of waves against the side of the house, the children could have believed they were floating in outer space.

Additionally, Charles noticed his watch wasn't working.

He and Susan were alone in the drawing room, Uncle Farley having taken Murray up to bed (he'd said he could go alone, but Uncle Farley vetoed that idea, taking a firm grip on Murray's hand). While they waited for him to come back, Susan and Charles stared at the walls of the drawing room to see if it had any fresh insights as to what lay in store. But the mural showed only the same tropical island scene it had shown yesterday: the empty green plain dotted with those funny, individual hills. Susan felt that the perspective was slightly different, however, in a way she couldn't quite put her finger on. In fact, she was less interested in the slightly altered mural than she was in Charles's stopped watch, but Charles stared hard at the walls as if he could puzzle out the change, his brow furrowed in concentration.

"Maybe the battery's gone dead?" Susan suggested now—referring to Charles's watch, of course, not the mural.

"It doesn't *have* a battery," Charles said, glancing at his sister with an annoyed look on his face. "It's the kind you *wind*, and I've *wound* it, and it's not working. There's something weird about this picture," he added. "I can't quite figure it out, but it's definitely different."

"Maybe you broke it," Susan said, still focused on the watch. "Maybe you wound it too far." She reached for Charles's watch and tried to take it off his wrist.

"I did *not* wind it too far," Charles said, thrusting his hand in his pocket to get his watch away from Susan. "*Something* is making it not work. It's almost as if it's shrinking."

"Your watch is shrinking?"

The children looked up to see their uncle walking into the room with yet another tray laden with food—popcorn, as their noses had already told them, the kind with caramel, the kind with cheese, and the plain kind, loaded with butter.

"My watch isn't shrinking!" Charles nearly shouted. "The island is!"

At this outburst, Susan let out an exasperated sigh and turned to look at the mural. Her sigh promptly turned into a gasp.

"It is!"

It was true. Where before the island had stretched off to the left and right farther than the eye could see, now its edges were visible, if only barely, and its strange solitary hills seemed to have diminished slightly.

Uncle Farley set the fragrant tray of popcorn on the table.

"I'm not sure what's going on with the mural," he said, "but I think I can clear up the mystery of Charles's watch. You see, as near as I can understand it, there are periods of light and darkness on the Sea of Time, but there aren't really days as such. For whatever reason, even though time does pass here, and in the real world, it doesn't affect us. Were we to stay for ten years, I'm pretty sure you wouldn't grow a day older or an inch taller."

"Are we going to be here for *ten years*?" Susan and Charles said at practically the same time.

"Way to go, Farley," President Wilson said from his perch, where he'd appeared to be snoozing.

"Oh, oh well," Uncle Farley stuttered, reflexively grabbing for some popcorn and stuffing it into his mouth. "I was only making a for-instance," he said. "I'm sure this nasty business will all be concluded tomorrow, and we'll be back on dry land before nightfall." He looked around nervously at the children. " 'Tomorrow,' " he said, making quotation marks with his fingers. He indicated the tray. "Popcorn?"

"It's definitely getting smaller," Susan said, pointing to the island. Its entire outline could be seen now. "You can practically see it shrinking."

"Or sinking," Uncle Farley said, reaching for more popcorn. "You children really should try this. I don't know how Miss Applethwaite, er, the dumbwaiter does it. It almost as if the popcorn is dusted with cheese—"

"It's not shrinking *or* sinking!" Charles cut off his uncle. "Don't you see? We're rising above it. It's like we're looking down on it from an airplane."

As soon as Charles spoke, Susan realized he was right. Why hadn't she thought of it before?

The island was becoming significantly smaller before their eyes. It was indeed as if the watchers were floating over it. "Not in an airplane," Susan hurried to say aloud. "More like a helicopter or something, a hot-air balloon. Uncle Farley," she added, "can Drift House fly as well as float?"

"Well, I, um, that is, I never thought to . . ." Uncle Farley's voice trailed off. He looked at President Wilson.

President Wilson looked back and forth between the humans and the mural.

"I can fly," he said finally. "Perhaps it's showing my perspective?"

Charles ignored him. "Look!" he exclaimed.

The picture had changed with stomach-churning rapidity. The island, which had receded until it was smaller than the table they sat at, suddenly disappeared, as did the water and the sky and everything else. Silently, but with a blurring motion that made you imagine a whooshing noise, all four walls of the room faded to a swirling deep blue so dark it was almost black.

"It—it's just gone off," Uncle Farley said.

Charles looked at his watch, then at his uncle.

"Maybe it's wound down?" he said, trying to keep the frightened edge out of his voice.

"The mural?" Susan said sarcastically. "Or your watch? Oh, never mind," she said before Charles could snap at her. "I'm going to bed."

Everyone turned and looked at her expectantly, but all she said was, "My brain is too full to take in one more thing. And it's going to be a long day tomorrow. I imagine we could all use some sleep."

The darkened walls made the room that much darker as well, but Susan could still see Uncle Farley's concerned expression. But all he said was, "There are candles in the hall. Be sure to take one to light your way upstairs."

Susan nodded, easing her way out of the eerily twilit room. It was only when she was safely in the hallway with a bright candle in her hand that she allowed herself to admit what had really driven her from the drawing room: that the dark swirling blue wasn't a malfunction or a powering down of the mural's function, but an entirely new scene—an underwater scene, as if from deep beneath the sea. As if she'd failed, and the house and all its occupants had sunk to a watery grave.

"*No*!" she said out loud, in a fiercely determined whisper. "I won't let that happen! I won't!"

There is something singular about walking down a long dark hallway with only a single flickering candle to light your way—it takes far more concentration than Susan would have guessed. She discovered she had to walk slowly or else the candlelight would tremble and scatter like a handful of rice kernels thrown in the air. She tried shielding the candle with her hand, but if she did that it defeated the purpose, since her palm contained all the light. But if she walked at a steady moderate pace then she found herself enclosed in a little globe of light. Susan tried to imagine the globe of light as her own personal force field.

In her bedroom she found a small fire burning in the grate—the night *was* a little cool, she realized. Fresh pajamas had been set out for her, and Victor Win-Win was propped on top of the pillows. When she lay down next to him she thought she would never be able to sleep, her head was so full of conflicting thoughts and emotions—the mermaids, and Murray's strange golden locket, and the weird scenes on the drawing room mural. It was all so overwhelming! But the cotton sheets were soft and warm, the pillow fluffy and embracing, and the down comforter rested atop her with the weightless warmth of a cloud. Victor Win-Win had the warm musty smell that dolls get only after years of close association with their humans, and as Susan thought of that she had a sense of herself, in a way she never had before, as a being who exists in time. Who accumulates history, and learns from the past, and makes plans for the future. It was a little bewildering, but it was also strangely comforting: being in time was just another way of saying you were alive, and Susan realized she liked being alive. It seemed an obvious thing, but somehow actually acknowledging it made her feel better, and, hugging Victor Win-Win once and then hiding him under the pillow again, she closed her eyes. The gentle rocking and lapping of the house was like being in a cradle. I'll just rest my eyes, Susan thought, and plan out what to do tomorrow—and then, when she opened her eyes, the room was filled with light, and there was Murray beside her bed, just as he had been yesterday.

Or not "just": because yesterday he had been an exuberant five-year-old, jumping up and down yelling, "Flood!" And this morning he stood still, a solemn expression on his face, and when Susan's eyes met his, he said,

"Good morning, Susan. Uncle Farley says do you want breakfast?"

"Oh, bother that man and his *food*. We're all going to get fat as Christmas turkeys!"

"Uncle Farley says that we won't gain any weight no matter how much we eat while we're here. He says our hair won't grow, or our fingernails, and we won't even have to use the bathroom."

"Oh, Murray," Susan said, still amazed at hearing such articulate sentences coming from her little brother's mouth. "Murray, are you *in* there?"

And Murray smiled, as sweet and genuine a smile as Susan had ever seen on his face. "I'm still here, Susan. I'm just a little wiser. If I seem subdued, it's only because I'm worried about *you*."

Susan smiled back at her brother. "I'll be fine, Murray, don't you worry about me." And she allowed her little brother to lead her by the hand to Uncle Farley's bedroom, for breakfast.

If the previous evening had seemed long, then this day

proved even longer. For the mermaids did not come with first light—didn't come by breakfast, or lunch, or tea (in the absence of working clocks, Charles pointed out, your stomach seemed to be about the only way of measuring time, although the truth is none of the children felt very hungry, and only ate because they were nervous, and because the meals the dumbwaiter produced were so enticing). They ate in the formal dining room rather than the cozier drawing room, because the walls of the latter were too dark and foreboding for comfort. In the absence of any magical signs, the children made their way frequently to the poop deck, where they looked nervously at the growing cloud of mist on the horizon. No spyglass was needed now. It had grown tall as a snow-covered mountain range, a great white cliff that soared hundreds of feet in the air. What was more, they could hear it too, a muted but distinct roar that competed with the gentler sound of lapping waves—which, for that matter, wasn't quite so gentle anymore. The water had acquired a roiling current, and as the day progressed the house pitched and rolled at an ever greater pace. The side-to-side motion made walking anywhere a bit of a fun-house experience, but not so fun if you knew what was causing it. By afternoon tea, all of Drift House's occupants were feeling a bit seasick, and even Uncle Farley seemed uninterested in the steaming scones the dumbwaiter had produced.

"Where *are* they?" Charles said. "This is getting perilous."

"Don't say 'perilous,'" Susan said, but gently. "You'll

frighten yourself." Indeed, of all Drift House's occupants, only Susan seemed calm. "Don't you see?" she said now. "They're trying to scare us by making us wait. They want us to realize how much we need them."

"But we *do* need them. That thing"—Charles pointed to the great cloud that was visible through the music room's windows—"is getting ready to swallow us."

"Yes, but *they* also need *us*," Susan said. "To rescue their Ula hula hoop."

"Ula lu la *lu*," President Wilson said blearily. He'd been sleeping all day—he'd started out on his perch, but after being pitched off it by the lurching house he'd settled down on an ottoman like a roosting chicken. "Can you humans remember *nothing*?"

"Hmph," Uncle Farley said. "If a certain *someone* in this room had a better memory, we might not *be* in this predicament."

"My goodness, Farley," President Wilson said in a completely different tone of voice. "Are those roasted hazelnuts? Perhaps you could—ah, thank you." And he buried his beak in the dish Uncle Farley set down before him.

"Don't you see, Charles," Susan said, "this is the substance of diplomacy. Both sides need something from the other, and thus both sides are protected. They'll be here in time. I have every confidence of it."

Charles, who could figure out how to rewire an electronic circuit just by examining it, now looked at Susan as if she were speaking Greek. "Whatever," he said. "I just hope they get here soon."

Charles's words proved prophetic: only a few moments later the familiar *THUD—THUD—THUD* resounded through the house. Everyone at the table jumped except Susan—even President Wilson flapped crazily off his ottoman, upending his dish of nuts and ending up in Uncle Farley's plate.

"Rak! We're under attack!" he croaked. Then, looking around sheepishly, he grabbed a piece of scone off Uncle Farley's plate and began nibbling at it nonchalantly.

Uncle Farley frowned at the parrot. "Really, President Wilson, I hope you don't plan on making this an habitual occurrence."

"They're here, they're here!" Charles shouted over him. He jumped from his chair and ran wobbily on the tilting floorboards to the front door. When he pulled it open, he saw Diaphone's familiar face floating in the churning water below him. "We thought you'd never come!"

A look of irritation flashed across Diaphone's face.

"Where is Susan Oakenfeld, boy-slave?"

"Now listen," Charles began. "I'm *not*—"

"I'm here," Susan said then, coming up behind Charles. She lay a hand on his shoulder. "Perhaps now is not the time," she said softly.

Charles stepped back reluctantly, muttering something about how he was tired of being called a slave.

"Susan Oakenfeld," Diaphone announced, "the Royal Guard of Her Most Aqueous Empress Queen Octavia, Undisputed Sovereign of the Sea of Time and All Its Tributaries and Headlands, and All the Vessels That Sail upon It, has come to tow your ailing craft to safety in accord with the bargain struck in yesterlight."

"And afterward?" Susan said. "After we save your friend? You'll tow us back to Canada?"

"Rescue Ula lu la lu, and the mermaids shall keep their end of the bargain."

"Very well then," Susan said. "What should we do?"

A mischievous smile flickered over Diaphone's face.

"Brace yourselves."

She turned away from the house then, and opened her mouth, and a sound such as Susan had never even dreamed of passed through the mermaid's lips. It was so beautiful it made Susan's knees weak. She wanted to sit down and lean against the wall and close her eyes.

A moment later three shapes leapt from the bubbling water with the quickness of trout snatching flies from the air. They were so fast Susan could hardly recognize them as more mermaids—but she was able to see that they trailed long green vines behind them like streamers. Then, remembering the picture in the drawing room, Susan called out:

"Everyone to the solarium!"

Uncle Farley, emerging from the music room with a

squawking President Wilson wobbling and flapping crazily on his perch, looked confused at the sight of the three Oakenfeld children dashing off toward the prow of the house.

"Eh, what's this? What's going on?"

"Come along now, Farley," President Wilson screeched. "Follow, follow!"

The house was pitching back and forth now, and the children stumbled through the drawing room and down the covered passageway and then pulled up short in the open arched door of the solarium, holding on to the doorframe to stay upright. They had not yet seen the solarium, so you will understand if for a moment they forgot what was going on outside the house. After a moment, Charles said:

"Holy cow!"

"Mmmm, ah, yes," Uncle Farley said, panting up behind him. "It's quite impressive, isn't it?"

"Impressive" hardly did it justice: there, in front of the children, was a grove of trees stretching all the way to the top of the glass pyramid that enclosed them—which was *a lot* higher than any of them would have thought, and so big they couldn't see the far wall from the doorway. The grove was composed of dozens of long straight trunks that seemed to get all tangled up at their bases and then again at their canopies. Some of the trunks were thin as jump rope while others were thicker than the Greek columns on the front of the Metropolitan

Museum, but all of them were woven together so tightly at their bases that they seemed like one tree. In fact:

"Why, it's a great banyan!"

"Well done, Susan," Uncle Farley said, breathing heavily as he leaned against one of the trunks (which, Susan knew, although she didn't say it out loud, are really a banyan's airborne roots).

The banyan dominated the solarium, but under its widespread leaves other plants grew, frondy things that looked like baby palms, and big bushy ferns, and sweet-smelling flowers and shrubs and a few tiny trees. Orchids and vines sprouted from and twined among the branches of the banyan itself, and a soft carpet of moss stretched underneath. Murray, who was barefoot—and who, Susan noted, seemed remarkably calm—said, "Squishy!"

Just then a thump sounded on the glass overhead. Through the banyan's thick leaves, Susan could see that one of the mermaids had lassoed the weathervane on the top of the pyramid—the long green rope could just be seen snaking down the glass to the waterline.

The children ran up to the leading corner of the solarium now, just in time to see a mermaid with streams of bright red hair coil one end of her rope around a thick metal cleat. When the mermaid saw them through the glass she flashed them a look of such complete disregard that the children all started back, and Susan felt a stab of fear in her stomach—that was hardly the look of someone with your best interest at heart. Susan felt like a kitten in a pet store window being stared at by a person walking a great big dog. Then:

"Susan!"

Susan turned and looked at Charles, then turned back to the window to see what he was pointing at. The red-haired mermaid had disappeared under the water with the free end of her rope, but Susan hardly noticed. All she saw was the immense white wall of mist that filled up her entire field of vision. She could now see that it wasn't merely a soft cloud of water vapor, but rather that it was as solid and thick and squidgy as taffy, and here and there shot up in the air in great bellows of exploding steam. They were so close to the drain she could even see the edge of the sea itself, dropping off as though over the end of the world.

Suddenly Diaphone's face burst from the waves. Her long copper-colored hair streamed wetly down her back, and Susan could just make out the mighty undulations of her tail as she beat against the tugging current. She made a motion to open the window, and Susan, seeing that the panel was latched and hinged, unlatched it and let it fall open.

"Grab hold of something!" Diaphone shouted over the roar of the Great Drain. "We're dangerously close! We're going to have to turn your vessel roughly!" Almost before she'd finished speaking, Diaphone dove back beneath the water.

Susan acted quickly. "Charles, Murray," she called over her shoulder. "Find something to hold on to!" She grabbed one of the exposed roots of the banyan tree. Charles and Murray followed her lead. All three children wrapped arms and legs around the tree's roots like firemen sliding down their poles.

Then things happened too quickly to be noticed individually. The vine that was fastened to the cleat just beyond the window snapped out of the water, as taut as a bowstring, and the whole house shuddered and moaned and then leaned so far to the right that all three children were more or less hanging off their trees like monkeys, and then the house snapped back to the left—port, Susan found herself thinking, left is port—and the children all swung on their branches in a shower of cracking twigs and falling leaves, and then there was a final loud shattering sound high above them, and Susan felt something harder and smaller than leaves sprinkle her hair and shoulders.

"Boys!" she called out. "Some of the windows are breaking above us! Keep your eyes closed!"

Susan sealed her eyelids shut. She could feel the floor twist beneath her feet as the mermaids turned Drift House hard to starboard. The roaring of the Great Drain receded, but then a much closer—much wetter—worry replaced it. Susan, her eyes squeezed tight against the fallen glass, suddenly felt a thick stream of water rushing over her feet and ankles. Shaking away the bits of glass in her hair, she squinted her eyes to see that a torrent of water was surging through the window she'd opened a moment ago. The mermaids had managed to turn the house against the current, and now frothy water was climbing up the sloped sides of the pyramid and rushing in a great torrent through what had become, in effect, a hole in the side of her ship.

"Susan!" she heard Charles now. There was a panicked note in his voice. "Susan, there's water rushing in!"

And then: "Susan, help!" Murray screamed in a voice of pure terror. "Susan, it's washing me away!"

Susan looked around. Charles was a few feet from her, his eyes tightly closed, his arms and legs securely wrapped around his branch. But Murray's legs had washed free, and he was stretched out on top of the water like a surfer, his hands barely able to hold on to his length of quivering trunk.

Susan was quick-witted enough to see that Murray was in no real danger of being washed away—even if he let go, he would only be swept into the house. But still, there was more than enough water rushing in through the open window to drown a five-year-old boy—and, as well, a hole in the side of a boat is hardly the best way to keep it afloat. She summoned all her strength and threw herself against the inpouring water.

"Murray, hold on! I'll have the water stopped in a moment!"

It was only a few short if arduous steps to the window. The water was rushing full into her face, making it difficult for her to find the edges of the window. But eventually her fingers grasped them beneath the sheet of water and, summoning one more breath, she heaved the window up and threw the latch tight. On the other side of the glass, the water frothed against it like suds splashing the glass door of a washing machine—a very big, very *angry* washing machine.

The house wrenched and creaked and shook as the mermaids dragged it against the powerful current. Susan heard one last pane of glass shatter above her and fall through the banyan leaves, and then a calm descended over the room.

Then:

"Good Lord and butter!"

"President Wilson!" Uncle Farley's faint voice could just be heard. "Please. There are children present. At least I hope there are. Children," he called in a louder voice, "are you all still with us?"

Charles opened his eyes, blinking repeatedly, until he seemed to realize his glasses had fallen off. Then Murray lifted his face from the messy floor. His cheeks were smudged with mud and there was a little cut on his forehead—but there was a *huge* smile on his face and a wild gleam in his eye.

"That was *fun*! Let's do it again!"

"Oh, Murray, you silly," Susan said, sighing with relief. "Come give me a hug please."

A moment later Charles, having recovered his glasses, came up and joined them.

"It was a little blurry," he said, pointing at his wet glasses, "but I saw what you did. That was very brave of you."

"Susan, Charles, Murray?" Uncle Farley's voice cut through the tranquility of the solarium, followed by the sound of cracking branches and thudding footsteps. "Are you here? Children, answer me!"

"This way, Uncle Farley. We're all fine."

A moment later he crashed into view. If anything, he was even wetter and muddier than Murray. Bits of vine and leaf clung to his hair and beard and clothes, making him look like some wild man of the forest.

"I tried to come to your aid. But the water, the lurching, the crashing . . . it was all a body could do to remain upright."

Susan could just make out a particularly big splotch of mud on Uncle Farley's bottom that suggested he hadn't remained quite as upright as he was suggesting. She wasn't sure, but she thought she heard Murray suppress a giggle.

"We're fine, Uncle Farley. We managed to survive."

A flash of color in the branches caught her eye then. It was President Wilson, hopping from branch to branch, gripping something white in his beak. He was the only dry member of the solarium.

"Er, Charlth," he said, lisping slightly because of the object in his beak—which was, Susan saw now, a handkerchief. "For your glattheth."

President Wilson hopped down to a low branch, and

Charles somewhat hesitantly took the handkerchief from him—he obviously didn't like having his hand so close to the great sharp beak or long taloned feet.

"Th-thank you, President Wilson," he stammered, and commenced to dry his glasses.

President Wilson perched on the branch, surveying the disheveled assortment of humans below him. His eyes came to Uncle Farley, and when they did he rolled them disdainfully, and then they passed on to Susan, who could have sworn the parrot winked at her.

"All I have to say is," President Wilson declared in a resigned voice, "I'm glad the mermaids put the girl in charge."

TEN

"Imitations"

The mermaids pulled the house all through that day and night and the following day and night as well. Perhaps that's a bit misleading, for it sounds as if they kept at it without pause, when in fact they took numerous breaks, some of which, to the children's minds, seemed to last an eternity.

When the mermaids were towing they were, naturally, under water, but when they took a break they anchored Drift House to the sea floor (it wasn't all that deep, Diaphone told Susan, only six or seven hundred feet!) and they came to the surface, and then Susan and Charles tried to count how many of them there were. They'd seen four at first—Diaphone, plus the three others. Susan was sure there were many more; they just didn't all come up at the same time. Charles, however, was convinced there *were* only four—they just moved around so often it was impossible to count them. If Susan looked carefully, he insisted, she would see that besides Diaphone there was one mermaid with red hair, another with blonde hair, and a third with green hair (but very lovely green hair, Charles had to admit, not at all like the color that comes from a bottle, but like sprouting grass).

Another reason why it was so hard to count the mermaids was that they kept themselves at quite a distance from the house—indeed, seemed almost to disdain Drift House and its occupants. On a couple of occasions, when Susan or Uncle Farley called out to them, they refused to answer, even though Susan was sure they could hear the humans' voices. Sound, as you know, travels great distances over water, and on several occasions Susan caught snippets of the mermaids' conversation. Of course, this was in the mermaids' own language, which sounded sometimes like music to the Oakenfelds' ears, and then other times less like music than like a record being played at the wrong speed, now too slow (so that the voices became all deep and slurry), now too fast, so the words came high-pitched and squeaky, and sounded quite funny to human ears.

"It seems to me," Susan said to Diaphone at one point, "your speech has something in common with that of whales. Is that true?"

Of all the mermaids, only Diaphone would approach Drift

House—in fact, she seemed loath to leave. At first the children thought this was because she was the official envoy of "Her Most Aqueous Empress Queen blah blah blah, Undisputed Sovereign of the blah blah blah blah blah," as President Wilson put it, until Diaphone told them that the real reason the mermaids disliked being out of the water was because they were so clumsy and slow on dry land (or, in this case, the poop deck). In fact, they hated it so much they harbored a prejudice against all beings that went about on two or four legs. Despite this, it seemed to Susan that Diaphone often cast long, curious glances through the windows of Drift House, though she never once accepted Susan's invitation to come inside.

Now, in response to Susan's question about the relationship of the language of mermaids and whales, Diaphone nodded.

"It *would* sound like that, to a being with such a poorly developed sense of hearing. In fact whales have a much more limited range of expression than mermaids—they're not very smart, the poor creatures, though they *are* fun to ride in sometimes."

"You ride in whales?" Susan said.

"You see what I mean?" Diaphone said. "I said *on*. We ride *on* whales."

"My sense of hearing is *quite* well developed," Susan said. She was quite sure Diaphone had said "in." Although it was hard to say for certain, because even when Diaphone spoke English there was still that musical undertone to her voice, the sense that she was providing a choral accompaniment to her own words.

"*Really*?" Diaphone said. "Just listen then." She and Susan and Charles were sitting on the poop deck—Uncle Farley was in the gallery beneath them, and Murray, after promising to steer clear of the dumbwaiter, had gone off by himself. Now Diaphone opened her mouth. Her throat muscles moved, her lips quivered, but Susan could hear nothing. She was just about to say as much when, out on the water, two of the mermaids who were tossing some kind of ball back and forth with their tails began laughing so hard they dropped their toy—which promptly swam away.

"Ultrasonic frequency!" Charles said. "Cool!"

"What did you say?" Susan said.

"I asked them if they'd like to join us for 'afternoon tea.'"

In fact Susan had just asked Diaphone this very thing. She thought it polite, and she very much wanted to get to know Diaphone. The dumbwaiter had produced fragrant pots of Lapsang souchong and Ceylon and a basket heaped with steaming scones and croissants, but Diaphone turned up her nose at their offering, saying that "dry crumbly things" were fit food for fish perhaps, but not for the Royal Court of You-know-who, Undisputed Sovereign of the You-know-what.

"Diaphone," Susan said, "I don't mean to give offense, but where we come from it's considered quite rude to make fun of a gift someone offers you."

At this Diaphone threw back her head and let loose peals of musical laughter. "Oh, you are a tempestuous girl-child, aren't you! I do think I could come to admire you, if we became better acquainted." She paused for a moment, her face darkening, then shook her head. "You must understand," she went on in a quieter voice, "the mermaids of the Sea of Time have devoted eternity to the pursuit of the beautiful and the graceful. By your reckoning, we have spent thousands of years refining and perfecting our speech, our manners, our way of life. How could you, who have only been alive a few moments, possibly hope to understand this?"

"I'm twelve years old," Susan said defensively.

"If you please, Miss," Charles said over his sister. "That's why we have history."

Diaphone had resigned herself to the fact that Susan's boy-slaves—especially the middle one—*would* speak to her, and now she turned to him with an impatient look in her face.

"Tell me what you mean by this, slave?"

"If you please, Miss," Charles repeated nervously, for he found Diaphone quite intimidating, "human beings don't live as long as mermaids, but we pass down the things of one generation to the next through books and stories."

Diaphone looked confused for a moment, and then her face cleared. "Ah yes. Books. The one who was here before you"—the children had talked to Diaphone enough to know that this referred to the mysterious Pierre Marin—"was quite fond of such things. They made me sneeze, and they tasted terrible."

"You don't *eat* them," Charles said. "You read them."

Diaphone sniffed, as if even the memory irritated her nose. "I took one to show Her Majesty. It didn't survive the journey."

The children looked at the deep water surrounding Drift House and imagined that, no, a book wouldn't survive a journey to the bottom of that sea. Charles thought of trying to explain this to Diaphone, then decided against it.

"What happens," he said instead, "is that someone writes something down that he doesn't want forgotten, and someone else—"

"Writes?"

Charles sighed. He suddenly realized that reading and writing were very hard to explain to someone who didn't know about them. But before the conversation could continue, a cry from one of the mermaids on the water distracted Diaphone. Though none of the children could understand it, they could all hear the greedy quality to the sound.

"Flying fish!" Diaphone exclaimed. "Now, Susan Oakenfeld and slave, I will show you what a civilized being eats." Without another word she flipped overboard.

"I am *not* a slave," Charles muttered as he joined his sister at the railing and looked in the direction Diaphone had gone. In a moment they saw it: a great number of tiny torpedoes arcing above the water looking for all the world like a swarm of flies above a picnic basket. Charles found the sight exciting (if not exactly appetizing) but Susan was caught up in the fact that this image, too, had been on the drawing room mural. The idea bothered her, because it seemed everything it had depicted was coming to pass—which meant that there was still a whale to come—a whale, and that strange island, and, most disturbing of all, an unwanted journey to the depths.

"Look!" Charles said, pulling her from her thoughts.

Susan stared out over the water, where four larger shapes had joined the flying fish in jumping through the air. The mermaids were catching their own version of afternoon tea. The Oakenfelds could see they were sporting as much as they were hunting, like cats teasing mice before finally finishing them off. They would bat the fish into the air and toss them around to each other. But soon enough all four mermaids had fish in their hands. The three more aloof ones settled down on the water, but Diaphone dove beneath it. The children could see her through the clear water, racing toward the house, and they scattered away from the edge of the deck just as she launched herself into the air in a great geyser and came to land on the railing, the fish dangling out of her mouth like a bone from a dog's.

Diaphone settled herself, then took hold of the fish's tail with one hand. The tail twitched weakly, and with a shock Susan realized it was still alive. When Diaphone ripped it from her mouth there was a big bite mark taken out of its side, and she chewed with her mouth open, the grotesque white fish flesh mashing to bits before the children's horrified eyes.

"Now *that* is real food," Diaphone said, taking another bite from the fish. She looked at Charles. "You were telling me about reading, boy-slave. And writhing."

"Er, writing," Charles said. "Writhing is what . . ." His voice trailed off, but he indicated the twitching fish in Diaphone's hand.

"Don't be presumptuous, slave. This is mine. If you want some, you'll have to catch one yourself."

"No thanks," Charles said. And then, more to distract himself from the gross sounds and sights of Diaphone eating the raw, quivering fish, he continued: "As I said, writing is when you take the things people say and turn them into letters, which make words—"

"Words are sounds, boy-slave. Even a whale knows that." A chunk of fish flew from Diaphone's mouth and landed on the deck.

"Y-yes." Charles tried not to stare at the white blob. "But the letters stand for the sounds. They make you think of the sounds in your head."

"Ah, I see!" Diaphone chewed for a moment, swallowed. "The letters are like memories!"

"Well, yes. Sort of. But they're everyone's memories, or everyone who can read anyway."

Diaphone took another big bite out of the fish, which seemed finally to have given up the ghost and hung limp in her hand. "Yes, yes, I understand," she said, waving Charles silent with the nearly eviscerated carcass in her hand. "This is one of your imitations."

Now it was Charles's turn to be confused. "Imitations?"

"These things humans make, that are meant to look like real things."

Susan, who had followed this dialogue with a mixture of confusion and fascinated horror (the latter directed at the mermaid's eating habits, the former directed at the conversation her brother was conducting), now jumped in. "I think she means art, Charles. Painting and sculpture and literature."

"Yes, that's your word," Diaphone said. " 'Art.' " The word sounded small and inadequate in the mermaid's mouth, as if there were not enough syllables for her melodious voice to play with. "The problem, you see, is that mermaids care nothing for made things. We think all beauty resides in what nature has created." And she held up the chewed-up fish as if in example.

"What's the difference?" Susan said, not quite able to look at the carcass, but not able to turn away.

"Well, you, girl-child, you are created. Clumsily created, of course, with those leg things that are like an extra set of arms, and in your case a rather minuscule amount of hair. But those shapeless hangings you drape over your body like seaweed.

Those"—Diaphone interrupted herself to take one final bite out of the flying fish, and then she tossed the skeleton overboard—"those are made."

Now that the disgusting fish was gone, Susan was able to concentrate better, and she realized Diaphone was referring to her clothes. She was wearing her favorite blouse—white, with lavender stripes and thin gold piping. It was too hot for it, but she had put it on especially to impress Diaphone, and took great offense at the latter's comment that it was "shapeless"—not to mention the mermaid's remarks about her hair.

Charles, however, had a different response.

"That doesn't make sense at all."

Diaphone seemed to take this remark about as well as Susan had taken the mermaid's comment about her blouse.

"What do you mean, boy-slave?"

"Why, to say that only created things can be beautiful is just silly. Nothing is beautiful in nature, or ugly for that matter. It just is. Only things people *make* can be beautiful, because beauty itself is an idea, is—is *made*."

"Charles," Susan said. "Stop talking gibberish."

"Don't say 'gibberish' just because you don't understand," Charles said. "Now listen. Why do we call something beautiful? Because it's special in some way—rare, but also reflecting an extra amount of care. Like Susan's blouse, for instance. That embroidery is beautiful because it took extra work to make it, and it was done carefully, and in neat, pretty lines."

Susan blushed. "Thank you, Charles." Then she thought for a moment. "But we also call people pretty, and mountain ranges and sunsets and things. And nobody made them."

"No one made them," Charles said. "But someone did decide they were beautiful. And they decided that because they had certain qualities people agreed were beautiful. For example, nobody likes a mountain when it's just a big heap of mud, but if it's covered with green trees then everyone says how lovely it is. That has nothing to do with the *mountain* and everything to do with who's looking at it. And of course everyone has their own ideas of what they think is beautiful, which is why Susan is so fond of that blouse and Diaphone prefers to wear, um, clamshells."

"Yes, yes, I *see*," Diaphone suddenly said. She picked at a bit of fish in her teeth with a fingernail, then continued. "The boy-slave is more interesting than I gave him credit for. But," she said, spitting the bit of fish over the railing, "you must admit that a mermaid is more beautiful than a human."

"Please, Miss," Charles said, looking almost as shocked by Diaphone's spitting as he had been by the way she'd flayed the fish alive with her teeth. "I don't want to be rude, but . . . but no. Mermaids are beautiful in different ways from human beings. I think Susan is just as pretty as you."

Susan blushed even more than she had before. Diaphone sat silent for a long time, staring at Charles with a thoughtful look on her face.

"Although I am inclined to disagree with what you say, boy-slave, there is a tone in your voice that impresses me. Your belief in your *sister's*"—Diaphone said the word as if it were hard for her to believe that boys and girls could be related to each other—"your belief in her beauty is compelling, and makes me want to believe in it too. I shall have to consider this more. But right now, boy-slave, I think you should leave us."

Charles blinked. "Leave?"

"I desire to speak with Susan Oakenfeld in private. It is imperative that I learn if she is capable of aiding the Court of Her Most Aqueous Empress, Queen Octavia, Undisputed Sovereign of the Sea of—"

"Leave?" Charles repeated, cutting off the litany of Queen Octavia's titles and dominions. He was indignant. He thought he'd finally managed to convince Diaphone that he was an equal member of Drift House's crew, and now it felt as if she were sending him to his room as though she were his babysitter. And it seemed to him that Susan had a bit of a smug look on her face, as though she were pleased at being singled out for special treatment. Now, though he knew it was childish, not to mention rude, he stamped his foot on the deck. "Susan has already agreed to help you. And it's not fair—"

Charles stopped speaking when, with an elegant gesture, Diaphone slipped the thin edge of her fluke beneath the piece of fish flesh that had spewed from her lips a few moments ago and, almost nonchalantly, flipped the piece of meat through the air and snapped it into her mouth.

"I would hate to have to ask you again, boy-slave."

"Run along now, boy-slave," Susan said.

To be fair to Susan, she had meant her words to be ironic, but Charles clearly didn't get the joke. He looked at Susan with an indignant expression on his red face, obviously trying to think of some suitable retort. But that was the sort of thing Susan was good at. Charles had to settle for stamping his foot again (he was barefoot, so it didn't make a lot of noise) and then he turned and stalked off into the house.

ELEVEN

Murray Version 2

"I THINK YOU SHOULD LEAVE," Charles muttered to himself as he walked down the third-floor hallway. "I would hate to have to ask you again, boy-slave."

Stupid mermaid.

"Run along now, boy-slave."

Stupid Susan.

Charles decided to look for Uncle Farley. After all, the two of them shared an aptitude for science, and perhaps without those pesky girls around (Charles was trying to convince himself that *he* was ditching Susan and Diaphone, and not the other way around) their rational male brains could figure out how to control Drift House. Along the way, he peeked into the bedroom he shared with Murray, whom he hadn't seen for some time now. The room was empty, the beds freshly made (the sheets changed no less—that Miss Applethwaite, whatever she was, was a remarkably efficient housekeeper, although someone had to tell her that Charles was a little old for sheets printed with ponies and puppies).

He headed downstairs then. When he got to the first-floor entrance of Uncle Farley's study, he found the door closed. All the other doors on the first floor of Drift House were glass, but the oversized double doors to the enormous gallery room Uncle Farley used as his study were made of thickly embossed wood painted black and edged with gilt. Altogether a majestic and forbidding portal, barring access to the secret chamber beyond. Charles knocked timidly, waited a moment, then knocked again, more loudly. Hearing nothing, he nudged the door with his hand. To his surprise, it swung wide.

The gallery was an impressive room by any standard. It had eight sides, for one thing, with floor-to-ceiling windows set in every other wall, and floor-to-ceiling bookcases set in those walls that didn't have windows. The ceiling was double height, with only a narrow balcony running around the second floor, and an enormous bell-shaped chandelier seemed to fill up the open space inside the balcony. The chandelier was made up of thousands of lines of crystals that seemed to be pouring out of a narrow fissure in the ceiling and raining down in a wide fluted shape, like a fountain turned upside down, and this inverted fountain spilled its light (it wasn't turned on, but still it seemed to leak luminescence) on a roomful of heavy, intricately carved tables and plinths and trunks scattered about the room. The surface of each of these was filled with so many arcane objects that Charles could hardly focus on one of them—vases and urns, candelabra, statues, smaller and larger boxes, bits and pieces of ancient armaments, bell jars beneath which were displayed unrecognizable items of taxidermy, and innumerable books, from tiny palm-sized volumes to leatherbound portfolios bigger than the card table his parents brought out on bridge night. As I said, impressive by any standard. But to a nine-year-old boy interested in antiquated technology it was merely—

"Cool!"

Charles's voice echoed across the vast room. He had already forgotten Diaphone and Susan's insults. And, as well, he saw on one table the Philco Tombstone that Uncle Farley had suggested controlled Drift House's course, surrounded by stacks and stacks of books. An idea occurred to him. He would figure out how the radio worked all on his own. With barely constrained eagerness, he stepped into the room. . . .

And there, I'm afraid, we shall have to leave him, for there is another Oakenfeld in Drift House who commands our attention. Or at least Uncle Farley's attention. After spending the early morning trying once again to figure out how to work that Tombstone radio, which had been stubbornly silent for more than a day now, he had decided to seek out Murray, to see if perhaps he could unlock a few of his nephew's secrets.

Farley left his study through the second-floor exit (just missing Charles as he headed downstairs on the main staircase). He looked through all three stories of the main part of the house without finding the youngest Oakenfeld, though he did see, from the third-floor windows, Susan and Diaphone engaged in what appeared to be an earnest conversation on the gallery terrace—what Murray and Charles liked to call the poop deck. It was good that the two of them were becoming friends, Farley thought. Although he found the mermaid's haughty demeanor annoying, he thought that perhaps Diaphone could be of assistance at some point. Now, panting slightly, he set off for the solarium. As he passed through the library he nearly tripped over a chest that one of the children must have moved from its normal place, and, huffing more heavily, he pushed it back against the wall and continued on his way.

The solarium was a large room. Indeed, Farley sometimes suspected it of being bigger inside than it was outside—when he'd first taken possession of Drift House, he'd spent many frustrating days trying to figure out if this was true, but to no avail. Every time he measured the room he came up with a wildly different figure (a phenomenon he attributed less to magic than to his own uselessness with a measuring tape), and so, like the Tombstone radio, his theory remained nothing more than a suspicion. The room seemed particularly large to-day, and clotted with leaves and vines and branches and other things that could trip you, and Farley was particularly tired. He considered getting a snack from the dumbwaiter to fortify himself for the search, but decided to call out first.

"Murray," he called into the green depths. "Are you in here?"

Only silence answered him. After a moment, he turned to go back to the dumbwaiter. Susan had eaten no fewer than *three* crumpets for breakfast, leaving her uncle only one. Farley thought he would rectify that now. Or perhaps he would go straight to lunch, and have Miss Applethwaite whip him up some of her delectable shepherd's pie—whoever or whatever Miss Applethwaite was, she was an expert at baking the cheese to the perfect crispy texture without drying out the potatoes.

Before he'd gone a few steps, though, he heard a rustling behind him. Farley turned and peered back into the dense green foliage.

"Murray? Is that you?"

Again, only silence greeted him—but no, there was the rustling again. Farley took a few steps back into the room.

"Murray?"

The rustling got louder as whoever or whatever was concealed in the room apparently ventured into the tangled undergrowth that grew among the banyan's many airborne roots. A moment later, there was a crack, an *oomph!*, and a loud crash.

"Who's there?" Farley called out now, his heart beating not just from exertion, but from fear. "Murray, if that's you, I demand you answer me."

"It's okay, Farley," a voice answered him weakly. It was impossible to tell if the voice was Murray's or not. "I'll be along in just a moment." If it *was* Murray, he was sounding even more grownup than he had since he'd emerged from the dumbwaiter.

"Murray," Farley said, pushing through the thick heavy roots after the sound of the voice. "What is going on here? Why are you hiding?"

"Keep back, Farley," the voice called again. It was louder now, but it still had a weak, almost faded quality to it, as if it were not Murray's voice, but a poor recording of it.

"That's quite enough, Murray," Farley said, continuing to press through the solarium's thick growth. "I insist that you come out right—*oof*!"

Farley, as we've had the opportunity to note a few times, was not nearly as athletic as his great size might have indicated, and now he'd tripped over an exposed root and fallen flat on his face. Fortunately, the spongy soil (and the padding of his stomach) softened the impact. But nothing could soften the impact of the sight that greeted him.

There, a few feet in front of him, was a very, very, *very* old man, similarly stretched out on the loamy soil of the solarium as though he'd fallen as Farley had. The old man's withered hands were busy fiddling with something at his neck, and when Farley squinted around the intervening foliage he saw that the man was trying to fasten the clasp of a necklace with his ancient gnarled fingers. The necklace was thin and made of gold and, as you've probably guessed by now, suspended a tiny heart-shaped locket from its links.

Farley peered through the whiskers and wrinkles and age spots to the soft brown eyes of the figure before him. Could it be?

"Murray?"

Smiling slightly, the old man gave up trying to fasten the locket around his neck.

"Hello, Farley. Surprised to find me like this?"

The voice was eroded by years and years of use, the same way a stone step is rubbed smooth by the passage of thousands upon thousands of feet. But it was Murray's voice.

Farley pushed himself into a sitting position.

"My word, old boy. I—I really don't know what to say."

Murray looked down at himself as if he were equally surprised at the body that housed his brain. He looked up at Farley.

"Miss Applethwaite said that any man whose favorite dish is shepherd's pie is a simple, honest soul, no matter how impressive his array of knowledge. It really *is* Applethwaite, by the way, in case Susan ever asks you again."

This statement seemed to render Farley speechless for a

moment, but at length he managed to sputter out, "The *dumbwaiter* did this to you?"

Murray used his tremulous, spindly arms to push himself into a sitting position. "Partly. I'm still not sure of the full extent of it, to be quite honest. Part of it was Miss Applethwaite—the dumbwaiter, as you say—and part of it was the sea itself, and part of it was some third factor—perhaps something that's in me."

Farley shook his head. "I'm afraid I don't understand at all, my dear b-b-b—my dear man. Can you start from the beginning and tell me exactly what happened?"

"I only wish I could." Murray gestured with the locket in his wizened hand. "I thought perhaps this was blocking my memories, so I took it off. But it turned out to have another function altogether." He indicated his aged body and laughed slightly—an old man's rheumy wet laugh. "Something else is preventing me from remembering most of what happened after I went into the dumbwaiter, although I have a general idea. The dumbwaiter is a kind of wishing well. You ask for something and Miss Applethwaite delivers it, to the best of her ability."

"But what . . . who is Miss Applethwaite? I thought it was just a name President Wilson made up."

"No, no, she's a very real person."

"But, then, where is she? I've been in the basement a thousand times. I've never seen anyone down there, ever."

Murray shook his head. "I'm afraid I don't know. My head,

my memories"—Murray laughed slightly—"they're just as fuzzy as an old man's."

Suddenly a startled expression flashed across Farley's face. "My word, Murray, what did *you* ask for?"

A rueful expression flickered over the hollow cheeks of the old man, and for a moment Farley could see the face of the five-year-old who knows he's done something he shouldn't have.

"I was very angry at Susan," Murray said. "I thought she'd played such a mean trick on me, making me hide and then not coming to look for me."

Farley nodded. "I shall have to reprimand her for that."

"I think Susan has learned her lesson on that point, Farley," Murray said, and chuckled. "At any rate," he continued in a soberer tone of voice, "what I wished for—what I *think* I wished for—it's so very hard to be sure—but what I think I asked for was to be as far from Susan as I could possibly be, so that when she finally came looking for me she'd never be able to find me. And what could be farther away than the future?" He shrugged again. "As I said, I thought that perhaps if I took the locket off, I would be able to remember more. Instead I found myself in this form. In fact, I have the nagging feeling that the locket, or at least the pictures of Charles and Susan it contains, are actually to *help* me remember. I think that before I came back from wherever or whenever I was, I knew I was going to forget everything that had happened, and I gave myself a clue to try to jog my memory. But despite my outward appearance, I am still very much a five-year-old boy, with a few extra memories thrown in just to confuse me."

Now Farley chuckled. "I'm afraid we're all a bit confused." He peered deeply into the eyes set into the faded skin of the face before him. It really was Murray. "You *sound* older," he said. "I mean, you're not talking like a five-year-old. You really don't remember anything more?"

"Studies have shown that amnesiacs who have forgotten their entire life stories nevertheless retain a complete grasp of language—and before you ask me how I learned this, I don't know. I don't remember."

Farley peered into Murray's eyes. "Do you by any chance remember a book?"

Surprise flickered over Murray's face. "A book? What kind of book?"

"Yesterday, when I found you in the dumbwaiter, you were resting on top of a book. But when I lifted you out the doors fell closed behind me, and when I opened them again the book was gone. Do you know anything about this?" From the hopeless tone in Farley's voice, it was clear he didn't expect a positive answer.

And, indeed, Murray appeared stupefied. "A book," he repeated. "Do you think Miss Applethwaite might have taken it? Did you ask her?"

"Murray," Farley said. "The extent of my communication

with Miss Applethwaite is to place my breakfast, lunch, and dinner orders."

"Yes, it's all rather mysterious, I have to admit. I've tried contacting her again, to no avail, although I have dim memories of actually seeing her, and speaking to her. She's"— Murray curved his hands, indicating a fullness of figure—"round. Like you."

If Farley had had any lingering doubts that the old man in front of him was really his five-year-old nephew, they were dispelled by his last words. They were not the sort of thing one adult could have said to another.

Murray chuckled now. "I even tried going back into the dumbwaiter, but nothing happened."

"Murray!" Uncle Farley said, then cut himself off with his own chuckle. "It feels very silly to scold someone who is so much older than I am."

Murray laughed too. "Oh, I know, Farley. Everything's topsy-turvy around here, it's hard to know where to look or who to trust. I'm afraid I don't even really trust myself."

Farley put his plump hand on Murray's thin one. "I trust you. And so do Susan and Charles." He squeezed Murray's hand, but lightly, because of the tenderness of the old man's bones.

Murray sighed. "I remember being lonely for a long time," he said in a sad voice. He held up one of his aged hands for inspection. "A very, very long time. I remember missing you, and my mother and father, and Charles. But most especially I missed Susan, because she was dead. Some terrible menace awaits her, Farley," he went on in a more urgent voice. "It's driving me crazy that I can't remember what it is, because I'm sure I came back to save her. I'm sure I *can* save her, if I can just remember. But I can't. Not for the life of me—or Susan."

Tears welled up in the old man's eyes, and once again Farley could see the face of his five-year-old nephew clearly.

"There, there, Murray. We must all have faith. You made it this far, you will make it the rest of the way. We all will." He thought hard for a moment. "Do you think it's these Time Pirates Diaphone mentioned?"

"They're certainly bound up in it. But it's bigger than that. For some reason I seem to think it has something to do with the Great Drain itself."

"But the mermaids have already saved us from that."

"I know very little," the aged Murray said then, "but one thing I'm sure of is that nothing on or in the Sea of Time is entirely safe from the Great Drain. Not even the mermaids themselves. Its power is stronger, and stranger, than we can even imagine."

A distant voice came from the direction of the main body of the house.

"Murray? Uncle Farley?"

It was Charles.

"Er, one moment, Charles," Farley called, then immediately regretted it, for he heard Charles's heavy passage as he made his way toward his uncle's voice.

"Uncle Farley? Where are you? I've been looking all over for Murray and I can't find him anywhere."

Murray began fumbling with the necklace in his hands, but his arthritic knuckles refused to cooperate. Quickly, Farley scooted across the ground to help him fasten the necklace around his throat.

"Are you sure this will make you look like a five-year-old again?" Farley said as he too fumbled with the clasp.

There was a crashing sound as Charles came looking for them.

"We'd better hope so. Hurry, Farley, he's almost here!"

As Farley finally managed to seal the clasp, he heard Charles crash through the branches just a few feet away. Farley looked up nervously, prepared to knock his elder nephew down if he had to, in order to block the view of his radically altered younger brother—he couldn't imagine anything good coming from Charles seeing Murray in this state. Even as he looked up, Charles's face appeared through the leaves.

"Uncle Farley," Charles demanded. He bobbed to the left and right as he tried to see around his uncle's body. "What's going on? Why didn't you answer me? I've been very worried. I couldn't find you or Murray anywhere."

"I'm sorry, Charles," said a voice below Farley. "I must have fallen asleep. It's very peaceful in here."

Farley looked down and saw, much to his relief, his younger nephew in his proper five-year-old body. The two exchanged a brief look, and then Uncle Farley turned back to Charles.

"I—I was looking for him myself," he said. He looked back down at Murray. "We've all been quite worried about you, young man. I think you should keep your wandering to a minimum."

"I'm sorry, Uncle Farley," Murray said in a contrite voice.

Charles looked back and forth between the two of them suspiciously.

"There's something going on here," he said. "But that's okay," he continued in a louder voice when his uncle opened his mouth to protest. "No one wants to tell *me* anything, but that's *quite* okay. I will figure it out on my own."

And, so saying, he turned around and crashed back out of the room.

Murray looked up at his uncle.

"Poor boy," the five-year-old said with the faintest trace of a smile on his face. "He'll understand more when he's older."

TWELVE

The Importance of Family

WHILE ALL THIS WAS GOING on, Susan and Diaphone were alone on the poop deck. They had chatted amicably about this and that, and at length Susan persuaded the mermaid to try one of the dumbwaiter's scones, though after nibbling at it Diaphone placed it discreetly on the railing next to her. Susan bent over the tray to get a croissant for herself; she thought she heard a faint splash, and when she looked up saw that Diaphone's scone had disappeared. The mermaid was staring at her tail as if it were the most fascinating thing in the world. Oh well, Susan thought, at least she tried.

All of a sudden Diaphone looked up at Susan with a broad smile on her face. There was something odd about her expression, Susan thought, and then she realized it was the first time Diaphone's smile had seemed genuinely *nice* to Susan, as opposed to haughty, or mysterious, or pleased with herself.

The mermaid cleared her throat. "Although we have only known each other a short time, Susan Oakenfeld," she began, "I have come to like you, and to admire you as well. You have shown a remarkable degree of fortitude under very stressful circumstances." The mermaid allowed herself a mischievous giggle. "I probably shouldn't tell you this, but Queen Octavia"—she left off the list of titles and dominions—"was quite taken aback by your counteroffer to her demands. It has been many a century since I've seen something fluster Her Majesty."

Susan, as we know, grew quite embarrassed when someone complimented her even the littlest bit—and she'd certainly never been told that she'd "flustered" a queen before. Her face went as red as President Wilson's plumage.

"Thank you," she said quietly. "I was only trying to look out for my family."

"You love your brothers very much, don't you?"

For some reason, Diaphone's question made Susan remember an incident that had occurred last year, in seventh grade. One of Susan's teachers had referred to Charles and Murray as Susan's "half brothers." The things that Susan had *actually* said to her teacher got her detention for a week; all you need to know is that Susan had insisted that not only were

Charles and Murray her *whole* brothers—her *real* brothers—they were her *little* brothers, and as their big sister she felt extremely protective of them, and she considered the teacher's comment quite hurtful. The teacher had apologized, but she'd still given Susan detention for the parts I've left out.

Now Susan told Diaphone, "My real father died when I was too young to remember him. But I do remember the two years after. It was just my mother and me, and it was very lonely. So when Mum married Dad, and Charles came along, and then Murray, I was very thankful. I knew that it didn't have to work out that way. We could have remained alone. And, I mean, I love my mother very much, but having Charles and Murray and Dad is just, well, *better*."

Throughout this speech Diaphone had a slightly thoughtful, slightly perplexed expression on her face.

"Diaphone," Susan said, studying her new friend. "I've heard you call the other mermaids your sisters. But do you have a mother and father?"

A faraway look came over the mermaid's face. Her skin softened, her eyes glazed as she stared into the empty blue sky that covered the Sea of Time. Susan thought that Diaphone was remembering something—something pleasant, apparently, for a small smile appeared on the mermaid's pink lips. But then the smile faded.

"I—I do not know," Diaphone said. "I have been alive for so very long that I no longer remember where I came from, or how I came into being." A bright smile forced itself onto the mermaid's face. "But the other members of Queen Octavia's court are like sisters to me, whether they are actually my sisters are not. And the queen herself is so much more than a mother."

She seemed about to go on when the sound of running footsteps called her attention to the house. She looked up just in time to see Charles come bursting through the doors to the poop deck.

"I am not a boy-slave and I am not a baby and I am tired of being sent away!" he declared.

Poor Charles: he had labored fruitlessly over the Tombstone radio in Uncle Farley's study for a long time, and then, when he found Murray and Uncle Farley in the solarium, he had been convinced they were hiding something from him. Now he said, "You and Diaphone and Murray and Uncle Farley are all having secret conversations, and I'm tired of being left out!"

Susan was just about to say something apologetic to Charles—he was red faced with anger, and she was afraid he might actually start crying—when she noticed that Diaphone was staring attentively across the water. Following the mermaid's gaze, she saw the head of the green-haired mermaid poking from the gentle waves. Her mouth was moving, though Susan couldn't hear a thing. Oh please, Susan thought, her stomach churning. Not more fish.

After a moment Diaphone turned to Susan and Charles.

"The Advance Guard of Her Most Aqueous Empress Queen Octavia, Undisputed Sovereign—"

"Oh, spit it out!" Charles said. He really *was* in a bad mood.

Diaphone gave Charles a look that was half angry, half respectful before turning back to Susan.

"The Time Pirates' vessel is not far now. You must make your plans."

"What plans?" Susan said.

"Your plans to rescue Ula lu la lu. You didn't think the Time Pirates would simply hand her over, did you?"

"B-but I don't know what kind of plans you make to rescue someone. Can't you give me some advice?"

A funny expression flashed over Diaphone's face. Susan could have sworn the mermaid looked guilty.

"That is not part of our bargain, I'm afraid. The Time Pirates are humans like you, so you should be able to figure them out. You are a resourceful girl, Susan Oakenfeld." The mermaid glanced at Charles. "And your brother has a most interesting mind. I am sure, between the two of you, that you will come up with something ingenious."

The mermaid prepared to flip herself into the water, but Susan grabbed her hand.

"Diaphone, please," she said. "I'm scared."

Diaphone smiled, but there was a touch of sadness to it. Her cool hand squeezed Susan's warm one tightly.

"No matter what happens, remember how much you love your family. And remember that I will try to help you in—" Diaphone's voice stumbled "—in any way I can."

And before Susan could say another word, Diaphone flexed her tail and vanished over the side of the railing.

A short while later, the family convened in the music room for a quick conference. It had taken some doing for Susan to persuade Charles to join them, since, as he said, "Everyone seems to be getting along just *fine* without me." Susan had hoped that Uncle Farley might have some advice for her, or perhaps Murray would have remembered something further, but as it turned out it was President Wilson who hatched the idea.

"You shall pretend to be Pierre Marin."

President Wilson was looking directly at Susan when he spoke, but she turned her head from one side to the other—Murray on her left, Charles somewhat farther away on her right—before looking back at the parrot.

"Me?" she said, as if he'd told her she must fly to the moon and bring him back a piece of cheese. "That's the most ridiculous thing I've ever heard."

"A hat, a wig, a thick coat to give you a bit of shoulder," President Wilson said. "It will be dark soon. They'll hardly know you from one of their own."

"I have to agree with Susan on this one," Uncle Farley said, coming into the music room with a heaping tray. "From what I've gathered, Pierre Marin and the Time Pirates knew each other quite well at one point. I can't see how they would mistake a twelve-year-old girl for a one-legged fur trader."

President Wilson seemed unbothered. "According to Diaphone, it has been hundreds of years since the Time Pirates last saw Pierre Marin." He clawed dexterously at Uncle Farley's tray to see what there was for a parrot to eat. "Diaphone says the Time Pirates are human, so it's logical to assume their memories have dimmed somewhat."

"Unlike a *parrot's* memory," Uncle Farley said acidly.

Finding a dish of unsalted cashews, President Wilson buried his beak in it.

"If it's been hundreds of years," Susan said, "won't the Time Pirates wonder what Pierre Marin is doing alive after all this time?"

"That is the one thing we *don't* have to worry about," Uncle Farley said, passing steaming bowls all around the table. "This is the Sea of Time, after all. Hmmm, paella. An interesting choice on, um, Miss Applethwaite's part." He looked at Murray. "You didn't have anything to do with this, did you?"

Murray picked up his spoon and plunged it into the dish of seafood and rice. "*No recuerdo,*" he said innocently, and began eating.

Susan was too distracted to pay attention to the fact that her five-year-old brother had just spoken Spanish. Her eyes flitted between President Wilson and Uncle Farley. "But . . . but . . . this is ridiculous!" she said.

"Exactly!" Charles suddenly exclaimed.

She turned to him, glad for the support—which made her all the more surprised at what came out of Charles's mouth next.

"I think it's the *perfect* thing to do."

Even President Wilson seemed startled. "Explain yourself, Charles."

"Well, look," Charles said. "Murray said it was Susan's involvement with the mermaids that got her k-killed." He stuttered at the last word.

All eyes turned to Murray, who nodded between bites of his paella.

"Now, we're not exactly sure what happened to Murray in the dumbwaiter, but it seems as if he got into the future somehow, where he saw what was going to happen to Susan, and then he found a way to come back to try to prevent it."

Again, everyone looked at Murray, who, after exchanging a glance with Uncle Farley, nodded again. He had put his spoon down and was rubbing the locket on his chest.

"So it seems that what we want to do is *not* do what we would have done in Murray's version of the future."

"Charles, please," Susan said. "In plain English."

"What we have to assume is that people who don't know the future go about their lives in a normal manner. They do things according to their character, without second-guessing if it's right for them."

"They live their lives," Uncle Farley said, his voice slightly confused.

"Exactly," Charles said. "But we know the future. Or, at any rate, we know enough of the future to know that whatever Susan did got her k-k-killed"—he obviously didn't like saying the word—"so we have to come up with the exact *opposite* thing that Susan would do, and have her do that. And what could be further from the mind of an intelligent girl like Susan than something as stupid as President Wilson's idea? It's the perfect plan!"

"Now, see here!" President Wilson began at hearing Charles call his idea stupid. But then, realizing Charles was backing him up, he broke off.

"I don't understand," Susan said—she felt like she'd been saying that a lot lately. "Are you saying President Wilson's plan is so stupid it's smart?"

"Can we find another word for 'stupid,' please?" President Wilson said. "Perhaps 'innovative,' or 'unconventional'?"

With a considerate nod to President Wilson, Charles said, "I'm saying that it's the last thing I can imagine *you* doing, which means it's probably what you *should* do. I mean, what *would* you do? To rescue a mermaid from a pirate ship?"

As Susan had said to Diaphone earlier, it was not the sort of question she had ever considered might be put to her.

"Well, ah, I, uh, I suppose I would ask, I mean, I imagine I would appeal to the pirates' sense of fairness. I mean, mermaids belong in the sea, don't they? Not on it, in a ship."

"Sec, yes, that's you. That's Susan. Always looking for things to be in their right place—like when we first came here, and you were so upset by the fact that the house was crooked."

"But how is doing something so outlandish as dressing up as a pirate better than doing something logical?"

"Because don't you see? We already know from Murray that doing something logical didn't work."

"But Murray doesn't remember *what* I did."

"Yes, yes, I know, that's a bit of a problem. Which is why we can only base our judgments on what seems most likely—most like you—and then try to veer as far from that as possible."

Susan sat in silence for a moment, and then she said, "Charles, I do believe you've thought so much you've burned out your brain. Of all the strange things we've seen and heard in the past three days, this is the silliest by far."

"I can see why you're worried, Susan," Uncle Farley said then. "But I think Charles is on to something. We have to make sure we don't conform to our usual behavior, so that we don't make whatever mistake it is—er, was, or will be—whatever mistake got you, got you . . ." Uncle Farley, it seemed, couldn't say the word at all. "The only thing is," he finished up, "I think I should go, rather than Susan."

"I thought of that," Charles said. "But doesn't that seem logical too? I mean, of course you'd send the grownup. So doesn't that mean we should probably reject that course as well?"

"Well, ah, hmmm," Uncle Farley said. "This does get rather difficult, doesn't it? I mean, casting about for the most unlikely possible plan? Why don't we just, I don't know, build ourselves wings and swoop down and carry this mermaid off?"

"Can we do that?" Charles asked seriously.

"Oh, well, um, no. I was trying to be silly."

"Too bad," Charles said. "That might have worked. I don't know," he said, turning back to Susan. "I think the disguise is the best way to go."

"I'm afraid I have to put my foot down on this one, Charles. If anyone's going aboard that ship, it'll be—"

Uncle Farley was cut off by the sound of scraping in the hallway outside the drawing room.

"Murray!" Susan said. "Where's Murray?"

He had disappeared again. But before anyone could panic, his voice emerged from the hallway, straining to be heard above the scraping noise.

"I'm here." He appeared in the doorway, bent over oddly, and then a moment later the object that was making the scraping noise appeared as well. It was a big leatherbound trunk. Charles recognized it as the trunk Murray had tried to hide in two days ago, during that fateful game of hide and go seek, and Uncle Farley recognized it as the trunk he'd tripped over earlier in the day, on his way to the solarium, but only Susan made the larger connection:

"Why, it's the trunk in the portrait of Pierre Marin!"

Murray lifted the lid on the trunk now, and pulled a wad of brightly colored cloth from it, as well as what looked like yards and yards of tangled black thread.

"You see, you really won't be able to go," he said, turning to Uncle Farley. "You won't fit."

"Murray," Uncle Farley said. "What do you mean? What have you got there?"

"I believe they're Pierre Marin's pirate clothes," Murray said. "And I'm afraid they're too small for you." He set everything down except a piece of bright purple velvet—a frock coat, Susan saw, when he held it up with his hands.

"Why, it's tiny!" Susan said.

"I do believe," Murray said, using that odd grownup locution yet again, "it's just your size."

"Murray," Susan said, taking the purple jacket from her youngest brother. "How did you know where these were?"

Murray shrugged, and his hand rubbed the golden locket dangling from his throat.

Susan tried the coat on, and indeed, it fit her as if it had been made for her. She was tall for her age—the tallest girl in the eighth *and* ninth grades. And the men of the past were rather shorter than your fathers and uncles, not having had the advantages of fresh vegetables and meat to build strong, long bones. But even within this context, it was pretty clear that Pierre Marin was a bit of a shrimp.

Susan sauntered around the music room in the great purple frock coat. The wide flaring lapels seemed to add width to her chest and the high collar obscured the thinness and delicacy of her neck. From the back, it was impossible to tell if the figure were boy or girl—the only thing visible was her head, hands, and ankles. There would be a wig and cap to cover the first, gloves for the second, and for the third—

"Er, President Wilson," Charles said.

President Wilson, hearing the doubtful tone in Charles's voice, turned an annoyed eye to him.

"What is it, boy?"

In answer Charles pointed at Susan's two slender ankles.

"Yes, yes, they're a little spindly for a man, but boots will cover that up."

By now, Susan was in front of a tall mirror, into which she snarled her best imitation of pirate-speak. "Abaft, maties, full fathom five and prepare to empty the head over the gunwales and—what's that? I beg your pardon, President Wilson, but my ankles are not *spindly*!"

"Yes," Charles said, "but they are *two*."

"What are you going on about, boy?" President Wilson said exasperatedly.

But Uncle Farley understood. "Oh dear, Charles, you're right. It's, it is a little difficult to cover that up, isn't it?"

This time it was Susan who exclaimed:

"What *are* you two talking about?"

Charles sighed. "You have two legs, Susan. Pierre Marin only had one."

But before Susan could do anything more than make a shocked expression, a tapping noise came from the doorway. Everyone turned to see Murray standing in front of the trunk with what looked like a banister spindle in his hand. He was tapping it against the doorframe like a professor trying to get his class' attention.

"You found the leg!" Charles said immediately. "Murray, you are amazing!"

Murray blushed slightly, and held out the leg for inspection. It was somewhat thicker than the children would have imagined it to be, tapering down to a flat base at the bottom and about as wide and hollow as a cereal bowl at the top, with a complex arrangement of leather straps that were obviously meant to fasten it to one's thigh. All in all, it was an ugly, scary-looking device, if strangely fascinating too, and at the sight of it Susan found her temporary confidence deflating. But then she happened to glance out the windows, and saw that the light was fading over the Sea of Time. The water was dark as an empty mirror, and there was something so lonely in the sight that Susan felt her resolve strengthen again. She would do just about anything to get back to dry land. To the sights and sounds of home. To her mother and father.

"Well," she sighed, "let's try it on then."

They had to go out in the hall and look at Pierre Marin to see which leg it should go on (the right), but, like the frock coat, it fit perfectly. In the end, everything fit perfectly. It was almost as if someone had known beforehand of their plan and tailored the clothes just for her. In particular, the pants. They were narrow at the waist, and wide and blousy in the thigh, so that Susan was able to fold her right leg double inside them. The top of her knee rested comfortably in the top of the peg-leg, and the straps secured it to her thigh as firmly as if it were one of her own bones, and after a little practice she was able to walk around with surprising ease.

"The wig is harder to deal with than the leg," Susan said in response to Uncle Farley's question. She combed the long curly locks out of her eyes. "All this hair!"

"Yes, I'm sure it's a bother," Uncle Farley said. "But it does help to conceal your fair skin. See, now, that won't do," Uncle Farley said when Susan blushed. He rushed out of the room and was back a moment later with something small and black in his hands.

"What's that?" Susan said. It looked like a lump of coal.

"It's a lump of coal," Uncle Farley said. And he proceeded to rub it over her cheeks and chin and throat, blacking them like a grown man's stubble.

"Uncle Farley, be careful!" Susan said when he passed the coal over her lips. "Yuck! That tastes horrible!"

When she was fully dressed—hat, wig, coal-black, ruffled shirt and frock coat, big black belt and ballooning pants, peg-leg on her right leg and silver-buckled shoe on the left—Uncle Farley led everyone out onto the poop deck to see how Susan looked in the darkness. She had trouble with the stairs at first because she couldn't bend her right leg, but she learned to take them one at a time, and then she took great pleasure in the ominous thudding her wooden leg produced as she made her way down the long hallway.

"Argh!" she said, "It's a walk down the plank for you!"

("She's certainly taking to the part, isn't she?" Charles whispered to Murray, and the two of them enjoyed a little titter behind their sister's purple-coated back.)

When they were outside, Susan walked a few feet from her family and then turned around slowly. She imagined them to be the welcoming party of a group of pirates and that she had one chance to make a formidable first impression.

"Ahoy, maties! Thar she blows!"

There was a brief silence, and then Murray giggled.

"What?" Susan demanded. "What was wrong with that?"

"Well, ah," Uncle Farley said. "Your appearance is flawless"—he rubbed a bit of coal dust off his hands—"but your *voice* leaves something to be desired."

"What's wrong with my voice?" Susan said—and, saying it, heard the tones of a twelve-year-old girl echo back in her ears. "Oh," she said, understanding. "Oh dear."

"Her voice is the least of it," President Wilson said. "Mademoiselle," he continued, "*Comment s'appelle le roi de la France?*"

"Wh-what?" Susan said.

"Just as I feared," President Wilson said. "Susan, Pierre Marin spoke French. And I take it you do not."

"I got an A in French," Susan objected, but weakly.

President Wilson sighed. "I see I shall have to come to the rescue again."

He flew at Susan then. She cringed a little, squeezed her eyes shut, but held her ground, and a moment later felt a weight on her right shoulder. Although they didn't hurt, she could feel the impression of President Wilson's claws through the thick velvet of her frock coat.

"Rak!"

President Wilson's shriek was so loud that Susan jumped and nearly lost her balance on her wooden leg. President Wilson had to flap his wings to keep from being thrown off by her jostling.

"Really, Susan! Such jumpiness befits neither a great explorer nor a spy about to enter the enemy's lair. Now then," he continued when Susan had recovered her balance. "I have an idea."

It is very weird to talk to someone sitting on your shoulder. If you've ever carried a little brother or sister or cousin around on your back, then perhaps you can imagine the experience—

although if your brother or sister or cousin were small enough to fit on your back, they probably would have been unable to say what President Wilson said next:

"I shall do all the talking."

"What? How?"

"I am a rather adept ventriloquist. When the pirates speak, I shall reply and make it seem as if the words are coming from your mouth. All you need do is move your lips."

For a moment Susan was flabbergasted—as were Charles, Murray, and Uncle Farley, at least judging from the expressions on their faces. But then, summoning her resolve, Susan said, "But how will I know when to move my mouth?"

In answer, President Wilson let loose another "Rak!" that almost sent Susan over the railing.

"I shall precede each speech with one of my more avian exclamations," President Wilson said, "and trust," he added drolly, "that you won't fall over each time I do so."

"Perhaps you could rak! a little more quietly?"

President Wilson dismissed this idea. "It wouldn't sound natural. Wild parrots have only two volumes—off and on."

Susan hadn't exactly spent enough time in the jungles of the Amazon to know if the parrot on her shoulder was making this up or not. Her brothers and uncle were still looking at her with half-shocked, half-confused expressions on their faces, but now the latter shook his head and said,

"Oh no no no."

"What is it now, Uncle Farley?" Susan said.

"This is just insane. The air here must be affecting my wits for me to have allowed it to go this far. No, I simply cannot allow this ridiculous plan to go into effect—your mother would never forgive me if something were to happen to you. I would never forgive myself. No, it just won't do."

There was a long moment of silence on the deck, and then Susan said, "Do you have a better idea?"

Uncle Farley could only look at her blankly. At length he put his hands in his pockets and turned them inside out, in imitation of the gesture she had performed—was it really only *two days ago*? A lifetime seemed to have passed since then. Feeling much older than she was, Susan said,

"Then I don't see as we have any choice. *Pierre Marin, c'est moi.*"

It was Murray, using his newly acquired grownup tone, who spoke next.

"It will be all right, Susan. I feel pretty sure of it."

Susan was about to thank him when the familiar voice came up over the railing.

"Stand back, Susan Oakenfeld."

It had grown dark while the family held their dress rehearsal, and Susan could barely make out the coppery flash as Diaphone leapt from the water. She seemed almost to appear out of thin air. The mermaid looked at Susan for a moment with a startled expression, and then a genuine smile of admiration appeared on her face as she took in the scene before her.

"I think I should have said, 'Stand back, Pierre Marin.' Bravo, Susan Oakenfeld. Once again you have shown your resourcefulness."

Susan blushed underneath her wig and painted-on beard.

"It was President Wilson's idea actually."

"An adviser's wisdom is only to the credit of his owner."

"I beg your pardon," President Wilson said. "I am *owned* by no one."

"Everyone has a master, bird-creature. Perhaps one day I can show you yours."

"Why you insolent—"

"Enough!" Diaphone said. "The time for quarreling, however amusing it may be, is passed. The Time Pirates' vessel approaches to larboard!"

("Larboard?" Charles hissed to Murray, who replied, "Larboard is the same as port. Left," he added, when Charles looked at him blankly. "Larboard is left.")

THIRTEEN

The Chronos

Everyone on the poop deck (except Diaphone, who remained right where she was) rushed to the larboard—or port, or left—rail of the poop deck. The night was so black that they couldn't see anything at first, but gradually a deeper blackness began to take outline in the sky, a long curved shape that floated on the water like a giant banana. There were a couple of lanterns aboard the ship, one forward, one aft, and the light they cast made it easier to see the ship's profile. Susan could make out the stump of a mast disappearing up into the darkness, the exterior walls of cabins, the delicate tracery of rigging. The approaching ship was ghostly and quiet, seemed almost to be deserted. Somehow that thought was even scarier than a ship full of pirates—a ghost ship, sailing itself across the Sea of Time on its own unknowable errands.

All at once a burst of light seemed to explode at the front of the ship. Sparks fired into the air and an acrid odor of smoke reached the nostrils of the observers on Drift House. When the flames had subsided a little Susan could see their glow through two rows of jagged teeth, the gaping jaws of a dragon's head leaning over the prow of the ship.

She felt something grab her arm and nearly yelped. She realized it was Charles. His hand was shaking like a leaf.

"It's all right, Charles," Susan said, understanding coming to her as soon as she spoke. "It's not a real dragon—it's a Viking ship!"

A voice came from across the dark. It sounded closer than Susan would have expected.

"Ahoy! Who goes there?"

"President Wilson!" Susan said. "They speak English!"

"*They* do," President Wilson said in her ear. "But Pierre Marin didn't."

"Oh," Susan said, a little disappointed that she would have to go through with the ventriloquist act after all. "Right."

"Are you ready?"

"Read—" Susan began, but President Wilson cut her off with a loud "Rak!"

"President Wilson, they can't even see us!"

But President Wilson was speaking over her in rapid

French. She heard a "*bonsoir*" in there, and then finally "Pierre Marin," but everything in between was a blur.

"You must enter into the spirit of the thing," the parrot whispered when he finished. "You didn't even move your mouth."

Words carried from the pirate ship—in French now, in response to President Wilson's greeting.

"Rak!"

Susan began moving her mouth as the stream of French shot out from her left ear. Even though she couldn't see the pirates yet—and, presumably, they couldn't see her—she felt extremely silly.

"What did they ask for?"

It was Uncle Farley who answered her.

"They asked for permission to come abreast. They've invited you aboard." He turned to Susan. "This is it."

The next several minutes were a bit fuzzy to Susan, who felt frozen in one spot—as if, if she took one step on her pegleg, she would fall flat on her face. As the Time Pirates' vessel approached she could see several shadowy figures moving rapidly across its decks. More lanterns appeared; more figures. In the dim light it was hard to make out shapes. One appeared to be about eight feet tall. Diaphone had indicated that at least some of the Time Pirates were human, but Susan found herself wondering if they had other creatures in their number as well—there was a talking parrot sitting on her shoulder, after all, and a mermaid on the deck behind her. She didn't know if she could talk to a giant, or, or a talking lion, or something even weirder.

Susan turned to Diaphone. In the darkness, her face was nearly invisible.

"Do you have any advice for me?"

It seemed to Susan that Diaphone was silent for an unusually long time. Then:

"Ula lu la lu is not nice like me."

"Not—"

"Treat her with respect, Susan Oakenfeld, and everything will be fine. But do not encroach upon her dignity, lest you suffer the consequences. She has been many days out of water, and her temper will be the shorter for it."

There was a crash behind her. Susan stifled a yelp and turned to see Charles and Uncle Farley securing a narrow plank bridge to the railing of the poop deck with long curls of braided rope. The bridge was about ten wobbly feet long, stretching over the dark water to the Time Pirates' ship and swaying back and forth like a jump rope. Susan realized she would have to walk across its rickety length on her pegleg. At the far side of the bridge a group of shadowy figures waited, all clustered around the tall figure—much of whose height, Susan realized, was contained in an even larger version of the hat she wore.

The tall figure called out to her now, in fluent if slightly halting French.

"What's he saying, President Wilson?"

"He says he is Captain Quoin, your servant. He asks if you require a hostage to guarantee your safe return."

Susan swallowed a lump out of her throat. "Tell him that won't be necessary," she said, then nearly jumped again when President Wilson rakked! in her ear. This was going to take some getting used to! Unsure if the pirates could see her or not, she tried to move her mouth as expressively as possible as the French streamed from beside her ear.

"Tell him I'm coming now," she whispered.

President Wilson rakked! and Susan almost felt as if she were speaking the words said from her shoulder.

Just as she reached the railing she felt a little tug on the hem of her frock coat. Murray was looking up at her with his big serious eyes—serious, but not afraid. He was holding something in his hand.

"I think it will be okay, Susan. Just remember to lean into your good leg."

"My good—" Susan began, then stopped when she realized what Murray was holding out to her.

It was a sword.

"If you should have to draw it," Murray said, "just remember to lean into your good leg."

The sword seemed very heavy as Susan fastened it to her waist. The belt went under her coat and the scabbard poked out of the vertical slit in the coat's tails. As she walked she could feel it slap lightly against her left leg. If she walked slowly it wouldn't be a problem, but if she had to run she imagined it might make it difficult to keep her balance.

She looked at her family as she hoisted her good leg over the railing to the rope bridge. It probably wouldn't do for Pierre Marin to kiss his shipmates goodbye, so she saluted them instead. She looked across the deck to nod at Diaphone, and realized with a start that the mermaid was no longer there. She must have slipped overboard when Murray gave her the sword.

Now the pegleg joined her left on the plank bridge. The balance was tricky, but if she held on to the bridge's rope handles she thought she would be okay. At the far end of the bridge, on the deck of the second vessel, the group of pirates awaited her in a huddled mass.

As she moved across the bridge she could see the shadowy water between its planks. The glow from the dragon's mouth on the far side made evil-looking shadows. She peered at the dragon for a moment, thinking there was something off about it. Was it—yes, it was! It wasn't a dragon's head, but a parrot's, like the masthead that appeared on Drift House's coat of arms.

"President Wilson!" she hissed as quietly she could. "The masthead!"

"Quiet!" President Wilson hissed back. "They're all watching you!"

Susan bit her tongue. There really *were* a lot of connections between the Time Pirates and Drift House, and she found herself wondering why the relationship had deteriorated so much. But she couldn't devote too much energy to the idea, as she had to concentrate on making it across the bridge without falling into the Sea of Time. Its water seemed black and cold far below her, and Susan tried not to imagine how she'd swim if she fell, with her heavy clothes and wig and right leg all bound up in the pegleg's harness. Instead she concentrated on standing as straight and manly as possible as she inched toward the Time Pirates' vessel.

As she drew closer she realized that several dark bumps on the side of the ship that she'd taken for barnacles were actually letters. They came into view slowly, like a camera coming into focus: C—H—R—O—N—O—S.

Chronos. It must be the name of the ship.

Suddenly a rak! burst from her shoulder and she almost fell off the bridge. She barely remembered to move her lips as President Wilson began saying something in a fierce-sounding voice. Even as she recovered her balance she realized that the parrot, at least, was playing his part beautifully. His words sent all the pirates except the tall one scrambling back several feet. This had the unfortunate effect of making Captain Quoin seem even taller and more forbidding, and from the folds of his frock coat she could see a sword that seemed at least three times as long as hers.

Finally her good leg was on the deck of the *Chronos*. Then, a thump later, her pegleg joined it. Brushing herself off for some reason (why, Susan had no idea) she raised herself to her full height of five feet six inches.

"Rak!"

Susan began moving her mouth as President Wilson spoke from her shoulder, wondering what he was saying—although, among other things, she heard the words "President Wilson" and realized he must be introducing himself! But halfway through the parrot's speech she realized that, beneath his great frock coat and sword belt, the long shirt worn by the fearsome captain of the Time Pirates was in fact an old-fashioned nightdress. She almost laughed then, but managed to hold it back.

Captain Quoin made a long speech in response to President Wilson's. He spoke very expressively, waving his hands around in the air, and it was disconcerting to have to look up at his eyes and pretend to understand what he was saying. Every once in a while she nodded her head gravely, and once when the captain laughed, she did too—or, rather, she pretended to, as President Wilson supplied a surprisingly deep belly laugh from her shoulder.

When he was finished speaking, the pirate took his hat in his hand, sank to one knee, and bent forward. President Wilson used this opportunity to whisper in her ear.

"He's renewing his pledge of loyalty. You must touch him on the shoulder with the flat of your sword."

"I thought that's what you did when you knighted someone," Susan whispered.

"In this case it symbolizes your authority to cut his head off, and your benevolence in allowing him to live."

Susan gulped, and drew her sword from its scabbard. Then, realizing the crew of the *Chronos* could still see her, she straightened her face. The sword was very heavy, and she had to hold it in both hands to keep it steady. As she leaned forward she instinctively led with her right leg—her pegleg—and almost lost her balance. All at once she remembered Murray's advice to keep her good leg forward when she drew her weapon, and shifted her position. Good old Murray!

As the flat part of the weapon came down, President Wilson rakked! and Susan heard herself say something in a formal-sounding voice. When the words stopped she pulled the sword back and, with some difficulty, managed to slide it back into its sheath. She had to use her fingers to guide the tip in, and as she did they brushed up against the sword's edge. It was disconcertingly sharp.

Captain Quoin stood up, hastily pulling his coat closed around his nightdress, and somewhat crookedly replacing his hat on his head. He turned slightly, indicating his crew with one hand. He said something, and President Wilson took advantage of the captain's averted eyes to translate.

"He says there have been many changes in the crew since you were last aboard. He's going to introduce them."

The crew stepped forward and lined themselves up. Their stiff military posture was at odds with their pajamas and nightdresses. Captain Quoin began speaking, gesticulating broadly, and as he spoke President Wilson whispered in Susan's ear. "He's introducing the first mate . . . Lieutenant Cosmo . . . listing his various accomplishments . . . singlehandedly rescued the phoenix . . . cut off the giant squid's ninth tentacle . . ."

"Squids only have eight tentacles!" Susan hissed.

"Thanks to Cosmo, apparently," President Wilson whispered. "And, ah, what was that? He says the two of you have met."

"Oh great," Susan whispered, bracing herself for more ventriloquist pantomime. Instead Captain Quoin stopped speaking abruptly. There was a long pause, and then he mumbled something in a sheepish voice.

President Wilson laughed quietly.

"False alarm," he whispered. "It would seem our phoenix rescuer and squid tamer is still asleep."

Apparently hearing the laugh, Captain Quoin flashed an inquisitive look in Susan's direction. Before he could say anything there was a loud rak! in her ear, and she could tell from the tone of President Wilson's voice that he was scolding the captain for his first mate's laziness.

"*Desolé, desolé,*" the captain said over and over again. After apologizing several more times, he introduced the ship's helmsman, Zhi Wo, a Chinese man with long thin mustaches hanging down on either side of his mouth, and then the navigator, an Irishman named Jonno who had an unlit pipe clamped firmly between his lips, and then the ship's gunners, a pair of tall thin Italian twins named Pierro and Pietro. The crew stepped forward one by one to kneel before her and shake her hand, and these expressions of subservience calmed Susan a bit—for pirates, they certainly were a meek bunch!—until Captain Quoin introduced the lowest-ranking member of the crew, the ship's "boy." He was an enormous hulking Russian named Konstantin Dimitrivich, and even after he kneeled down he was still nearly as tall as Susan. When she looked into his round, good-natured face she could see that he was, despite his great size, only a few years older than she was, but even so, her knuckles squeezed together painfully when she shook his hand. She hoped he was as good-natured as his face suggested. She would certainly hate to get in a sword fight with *him*.

There was a moment of silence after the introductions were completed. Now that her eyes were used to the light Susan could see the bleary expression on Captain Quoin's face more clearly. He was still half asleep. Even as she looked at him, he tried without much success to stifle a yawn, and to her horror she found herself yawning in response. She and the captain just looked at each other for a moment, and then all at once he started laughing, and after a moment Susan relaxed and laughed too.

"Pah, and what's all the nighttime hilarity?" said a sleepy, grumpy voice farther down the deck. "A body can't catch a moment of shut-eye with all the ruckus and—winds and waves! It can't be!"

Susan turned and peered up the dark passage between the railing and the cabin amidships. A small figure emerged from the shadows. It was so small that she thought it was a large hairy dog at first, but then she realized it was in fact a tiny man wearing a fur vest that seemed easily as big as he was. Over the front of the vest hung a matted growth of beard as thick as a sheepskin. If Susan barely came up to Captain Quoin's sternum, this figure barely came up to her belt buckle.

"As I live and breathe," the figure said. "Pierre Marin! Is it really you?"

There was a rak! from Susan's shoulder, but she was so caught up in the figure approaching her that she forgot to start moving her mouth until President Wilson had been speaking for several moments. President Wilson's scolding speech tapered off as the tiny man peered at her suspiciously.

"It *is* you, Monsieur Marin?" the figure said in a voice that sounded significantly more awake than it had a moment ago. "Isn't it?"

There was a long uneasy silence, but before Susan could

say something a voice barked from over her shoulder. She started to move her mouth, then realized it wasn't President Wilson who was speaking, but Captain Quoin. He was obviously chastising his first mate for sleeping through such an auspicious occasion—the fact that he was doing it in French suggested he was doing it for her benefit, and as his speech tapered off President Wilson rakked! and added a few choice words of his own. Grumbling slightly, the tiny figure in front of her dropped to one knee.

"Begging your pardon, Captain Quoin, and, er, *pardonnez-moi*, Monsieur Marin." The little man extended his hand. "Lieutenant Cosmo, at your service."

Susan had to bend down to shake his hand, which, despite its small size, squeezed hers as firmly as Konstantin Dimitrivich's had. There was a rak! from her shoulder, and President Wilson said something in a tone that was stern but slightly forgiving. Still, she wished he hadn't spoken so close to Lieutenant Cosmo's eyes—she could see him peering at her intently as her mouth tried to keep up with President Wilson's fluent French.

She stood up, and as she did a hand fell on her shoulder. She nearly jumped, then had to bite back a sigh of relief not befitting her station. Captain Quoin was merely giving her a comradely squeeze. He said something in a tone of voice that Susan had used herself to describe her little brothers to her girlfriends, and she could guess that he was saying something to the effect that one's crew, like little brothers, were sweet, but a bit of a pain too. He laughed when he finished speaking, his hand still on her shoulder, and Susan was about to laugh back when President Wilson flapped his wings and snapped at his hand. Captain Quoin dodged President Wilson's beak, and then, with a deft gesture, chucked him under the beak!

"Watch it!" President Wilson said in English, and Captain Quoin chuckled.

"He's a feisty parrot, no?"

President Wilson rakked! once more, and said something in a rather severe tone. Susan saw Captain Quoin's eyes flicker between hers and President Wilson's, as if he wondered who was speaking. Susan tried to make her face look as serious as President Wilson's voice sounded. She heard the words *"le perroquet"* several times.

When "Pierre Marin" had finished speaking, Captain Quoin's eyes continued to move between hers and President Wilson's for a moment, and then he threw back his head and laughed wildly. Behind him (or below him, in the case of Lieutenant Cosmo), his crew laughed as well. When he'd finally regained his composure, the captain began speaking very rapidly, turning as he spoke and walking down the long narrow galley between the main cabin and the rail.

"What's he saying, President Wilson? Where's he going?"

"He says he's got a little captive we might be interested in seeing!"

"Oh, excellent, President Wilson! Well done!" She hadn't dreamed rescuing the mermaid would be this easy!

As Susan followed Captain Quoin down the deck, she heard Lieutenant Cosmo's voice behind her.

"Pah, what's this? A sorrier bunch of sailors I couldn't imagine! Pietro, tuck in that shirt! Button that coat, Zhi Wo! And you, Konstantin Dimitrivich, put those elephant arms of yours to good work and swab the deck! You want our guests to have to walk on two centuries of grime! Hop to it, boys, or it'll be nothing but porridge in the mess for the next two weeks!"

Captain Quoin walked rapidly, and Susan found it hard to keep up on her wooden leg. But she sobered quickly as the comedy behind her faded, remembering her task. She was here to rescue a mermaid. However genial this crew seemed, they were still kidnappers. Still pirates, a sword dangling from each and every belt.

Halfway down the deck, Captain Quoin stopped at a door. Fumbling in his pocket, he produced a great roll of keys, one of which he slipped into the door's lock. As he opened the door Susan noticed one key in particular. It was larger than the others, and a deep silver in color—not like the polished silver used to make cutlery, but like some slightly glowing metal. She stared at it for a long moment before she remembered what she was on board the *Chronos* to do. Using her body to shield her hand from any Time Pirates who might be watching behind her, Susan took hold of her sword. As she did so, she suddenly contemplated stabbing a man, and gulped. She would go for the leg, Susan told herself. No higher than the thigh. It would hurt, but at least it wouldn't kill.

The key turned in the lock. Susan put her good foot forward. The heavy wooden door creaked open on unoiled iron hinges and a spill of light flooded out onto the deck. Susan remembered Diaphone's parting words: *Ula lu la lu is not nice like me.*

"Rak!"

Susan's mouth had been moving for several seconds when she realized it wasn't President Wilson who had rakked! There on a high wooden perch stood a rather dainty-looking blue and red parrot. It half covered its face with one wing when it caught Susan looking, occasionally lowering its feathers just enough to glance at her with a large dark eye—no, Susan realized, not at her. At President Wilson.

Addressing President Wilson directly, Captain Quoin said, *"Monsieur le perroquet, je présente Mademoiselle Marie-Antoinette des Oiseaux."*

A wolf whistle burst from her shoulder, and Susan blushed on behalf of her parrot. She wanted to stammer an apology, but President Wilson wasn't obliging.

"Let's go say hello, Susan!" he hissed in her ear.

It was very frustrating having Captain Quoin staring at her with his wide smile. Susan wanted to tell President Wilson that now wasn't the time for flirting, but couldn't. Instead she smiled at the captain and walked toward the perch. As she approached, the blue and red parrot edged shyly toward the farther end of its stand and covered its face. Susan could feel President Wilson's claws biting into her shoulder through the thick fabric of her coat.

"Marie-Antoinette!" she heard Captain Quoin bellow behind her. "*Dit bonsoir à Pierre Marin et—et—*"

Even Susan could understand this French, though President Wilson seemed too preoccupied to answer.

"*Et?*"

Deepening her voice as best she could, she said, "Président Wilson."

"Eh, hmmm, what?" President Wilson said, though not very loudly. He was leaning his head and body all the way forward in an attempt to get Marie-Antoinette's attention. Susan, afraid he was actually going to jump on her perch, took advantage of the fact that Captain Quoin was behind her.

"Ask him where Ula lu la lu is!" she whispered, then whirled so suddenly toward the captain that President Wilson nearly fell off her shoulder.

President Wilson cast a longing glance behind them, and Susan stamped her pegleg. The thud made Captain Quoin jump a little, and the wide smile on his face faltered slightly.

"Rak!"

All business now, President Wilson commenced a long speech that sounded more and more angry as it went on. Several times Susan heard the name Ula lu la lu and the phrase "*les mademoiselles de la mer.*" He got so into his speech that he began to gesture with his wings, and Susan could see Captain Quoin's eyes move confusedly between hers and President Wilson's. She waved her arms fiercely to bring his attention back to her.

At last President Wilson seemed to finish. Captain Quoin hung his head, a look of shame on his face.

"*Oui, Monsieur Marin. Je comprends.*"

Bowing slightly, he backed from the little room—right into the group of pirates who had clustered just outside the door. Blustering some, the captain rebuffed his crew for a moment then led the way around the corner to a dark companionway that led belowdecks just behind the mast. Ducking, he plunged heavily into the dark hole. Stealing a look over her left shoulder at the bright lights of Drift House—Susan could make out the drawing room just a dozen yards away—she followed him belowdecks.

The companionway was so narrow that she had to angle her body into it, which caused her sword to slap against her leg, nearly upsetting her balance. She'd made it only a few steps when she walked straight into something. With a start she realized it was the captain.

"*Pardon, pardon,*" she said herself, even as President Wilson said something that sounded significantly less polite from her shoulder.

Captain Quoin mumbled an apology. A moment later a match struck. The captain lit a pair of candles and held one out to Susan.

Susan took the candle with a wan smile—one more thing to throw her off balance. President Wilson on her left shoulder, the candle in her right hand, the heavy sword in its scabbard slapping against her leg, and of course that pegleg. To top it off, the narrow passageway was crowded with all manner of chests and wooden barrels and piles of things covered with sheets. As she followed Captain Quoin's big broad back she weaved with difficulty in and out of these obstacles, the flickering shadow of her candle causing them to move about as if they were wiggling in the dark and the smoke stinging her eyes. The only thing that fortified her was the sudden memory of Murray's delight at the "hop-stickle course" back in Canada. How long ago and far away that afternoon seemed now!

Captain Quoin spoke as they walked, and President Wilson took advantage of his turned back to translate.

"He apologizes for the clutter," President Wilson whispered, "but says they've been on a number of successful missions lately and haven't had the opportunity to deposit their goods on Crescent Island." President Wilson was silent a moment while the captain droned on. Then: "Fascinating!"

"What?" Susan whispered. "Tell me, President Wilson!"

"He's saying they've been to 1912 . . . 1848 . . . 1776 . . . 1492 . . . and, and 1066! Really spectacular results. Gold bullion, diamonds . . . and, ah, what's this!"

Captain Quoin had stopped in front of a cloth-draped mound, and with a flourish, pulled back the cover. There, in a large cage made of wire and wood, sat the oddest-looking bird Susan had ever seen. It had obviously been sleeping, but now it opened its eyes groggily, then stood up and pressed its fat curved beak through the bars. Susan couldn't resist reaching a finger and scratching it between the eyes. The bird cooed happily.

"Elle est un oiseau de dodo," Captain Quoin said.

A dodo! Susan remembered learning about them in school—they had been extinct for hundreds of years!

"Elle est pour L'Isle du Passé—de sort que le monde n'oublie jamais."

L'Isle du Passé? The Island of the Past? Susan's curiosity was so piqued that she couldn't resist asking about the bird.

"Pardonnez-moi, Monsieur de Capitaine," she said in as deep a voice as she could muster. "*Quelle est L'Isle du*—ow!" Susan's question was cut off by a sharp cry as President Wilson bit her ear. Fortunately the crunchy strands of her wig softened the bite, or she was sure he'd have drawn blood.

Captain Quoin looked confusedly between Susan and President Wilson for a moment, then chuckled and began a long speech that Susan couldn't really follow. As he spoke he covered up the dodo again and resumed walking down the hall.

"That was very foolish, Susan," President Wilson hissed. "Your French is execrable!"

"Well, you needn't have bit me!" Susan whispered back. Then: "How *do* you say ow! in French?"

"Ow!" President Wilson said.

Captain Quoin walked a little ways farther down the passage, then stopped before a closed door.

"*Nous voici,*" he said, pulling out his big roll of keys.

As the keys jangled from his pocket, there was a rustle from within the room, and then a thud as something struck the door from the inside.

"*La sirêne n'aime pas être mise en cage.*"

En cage: in a cage! Susan shivered at the thought of being imprisoned like the dodo she'd just seen. Again, she reminded herself that these pirates weren't as jolly or harmless as they seemed.

Captain Quoin seemed finally to have found the right key in the dim light. As he slotted the big iron instrument into the door's slightly rusty lock, Susan glanced back down the passage to see if the crew had followed them belowdecks. But the passage was dark and silent.

Captain Quoin eased the door open slightly. Several thuds in quick succession caused it to rock on its hinges.

"Nous devions l'avoir mise dans une salle vide," he said. *"Nous n'avions pas pu prendre les pâtes ou une salade en deux semaines!"*

Susan caught thc word *"pâte,"* which she knew meant "pasta," and *"salade,"* which is basically the same in English. But she didn't understand what the captain was talking about until he pushed the door open a little farther and a big red tomato caught him right in the face, accompanied by a voice that had all the strange power of Diaphone's voice, but none of its beauty or charm.

"Foul surface skimmer! Release me before I bring down the full fury of the Royal Guard of Her Most Aqueous Empress Queen Octavia, Undisputed Sovereign—"

"—of the Sea of Time and All Its Tributaries and Headlands," Captain Quoin finished for her, in English. He wiped a bit of tomato from his eyes. "We have heard your threats before, and still there is no sign of your mermaid army. Just one poor pathetic creature who is wasting a perfectly good supply of tomatoes."

At these words, a fresh round of tomatoes thudded against the door, walls, and the captain's face.

Captain Quoin smiled wryly at Susan. *"A votre proper risqué,"* he said, and pushed open the door.

At her first glance of the room, Susan let out a gasp of horror—it was covered in blood and bones! But then she realized the blood was really tomato goo, which covered every available surface in the room, garishly lit by a smoky lantern hanging overhead. Several barrels of tomatoes were pushed up against the room's far wall—Ula lu la lu had evidently been imprisoned in a storeroom. The bones, however, really *were* bones, the slightly smelly remains of the fish the pirates had evidently fed their captive.

Susan saw her now. She lay supine near the barrels of tomatoes at the back of the wall. Her long hair had a dull silver color, and was tangled and knotted from her time spent in this room, and her tail was silver as well. Its scales weren't shiny like Diaphone's but dull, almost dirty, as if they had been scraped off. Susan imagined that mermaids didn't fare well when they were kept out of the water day after day after day.

Around her waist was a thick cruel belt of iron, fastened by a small chain to a bolt in the floor.

While Susan had been examining the mermaid, the mermaid had been staring at her with slitted eyes. Now she barked in a voice that made Susan's hair stand on end (underneath her wig of course):

"You, girl! Why do you not draw your sword and strike down this man creature and release me?"

Girl! The mermaid had seen right through her disguise!

Susan was so flustered by this that she was about to protest in her own voice, when all at once Captain Quoin threw back his head and laughed so hard his hat fell off.

"Your time out of the water has dulled more than your hair and tail, my little mistress. This is no girl, but Pierre Marin, the legendary founder of the Time Pirates, come at last to lead us again."

Susan was struck by Captain Quoin's words. *Founder* of the Time Pirates? She thought Pierre Marin had merely fought them in battle. But if he had actually founded the Time Pirates, he must have had a far closer relationship to them than the mermaids had let on. All at once an image of the Time Pirates' parrot figurehead flashed in her brain. With a start, she realized the *Chronos* didn't merely resemble the ship in Drift House's coat of arms—it was the *same* ship!

Susan's mind was racing, but she couldn't make sense of all her thoughts because Ula lu la lu was still threatening to expose Susan's true identity. Seeming more hurt by the reference to her disheveled and worn appearance than by the insult to her mental prowess, she made a halfhearted attempt to comb her hair with her fingers, and ran a hand over the scales of her tail as if they were wrinkles in a dress.

"My shame at being captured by surface skimmers is only compounded by the fact that I have been caught by one as foolish as you! And to think you are the captain of this vessel! Have you not eyes, man? Can you not tell girl from man?"

To be fair to Captain Quoin, mermaids have exceptionally good eyesight, for making their way around the bottom of the ocean, which is very dark. The captain bent over now, peering at Susan—and, though his eyes were squinting and confusion clouded his face, Susan noticed that one hand had gone to the hilt of his sword.

"RAK!"

President Wilson's shrill cry was so loud that everyone in the room jumped. He let loose with such a torrent of abuse that Susan could hardly move her mouth fast enough to keep up. She had no idea what he was saying, but she could guess from the look on Captain Quoin's face that if *she* were caught saying these things she would very promptly get her mouth washed out with soap!

At length President Wilson finished, and the captain, a thoroughly chastened expression having replaced the confusion on his face, bowed low to the ground as he had in their first meeting. Susan tried to take advantage of his averted gaze to tell the mermaid that she had come to save her, but she wasn't sure how. Then, spying the key ring on Captain Quoin's belt, she made the universal sign for opening a lock. She pointed to the belt around Ula lu la lu's waist.

"What do you mean, girl," the mermaid exclaimed, "have I got the key to my chain? If I had the key, do you think I would still be here? Now lop off this four-legged oaf's head and have done with it!"

"Eh?" Captain Quoin looked up slightly from his prostrate position, but a fresh torrent of abuse from President Wilson brought his nose nearly to Susan's shoe buckle.

Susan was at a loss. How to tell Ula lu la lu to grab Captain Quoin's keys without the mermaid giving away her charade? She looked down at the bent head of the man before her, her fingers tentatively touching her own sword handle. Could she . . . was it possible . . . could she *lop off his head*?

Fortunately for Susan, she was spared the choice. A great commotion suddenly broke out on the deck above them. There was the sound of running feet and shouting and something that Susan instinctively recognized as the clang of one sword striking another. Then an explosion sounded from somewhere close by, so loud that the entire ship vibrated as though it had struck a rock. There was an ominous silence for a moment, and then the sound of running footsteps in the hallway outside the room. Or, rather, stumbling footsteps, for the person running was obviously tripping over all the booty that lined the passage. One more crash, accompanied by an "Awk!" (the dodo, Susan thought) and then the runner appeared in the door, crying, "Mutiny! Mutiny!" The voice was Lieutenant Cosmo's, and his naked sword gleamed in the thin light belowdecks.

At these words, Captain Quoin leapt to his feet and drew his sword, and, almost unconsciously, Susan did the same. She could feel President Wilson's claws on her shoulder. "Steady now, Susan," he whispered into her ear. "Steady, steady."

Lieutenant Cosmo's eyes glared fiercely at Susan out of his

dark beard, and she prepared herself for his attack. But the words that came out of his mouth surprised her.

"Mr. Marin, sir!" he said, so excited that he'd apparently forgotten Pierre Marin didn't speak English. "Pardon the interruption, sir, but your crew! They've mutinied and are attacking the *Chronos*!"

For a moment everyone in the room looked at Lieutenant Cosmo in silence, and then Ula lu la lu said, "Don't be a fool, girl! Strike him down!"

Susan responded instinctively. She stepped forward with her sword straight out in front of her. But unfortunately she stepped with her pegleg, the ball of which landed right on a split tomato. It slid out from under her in one long smooth motion. As Susan fell to the floor the harness twisted free from her thigh, and she felt a rush of air around her face as President Wilson flapped off her shoulder. A moment later she was sitting on her bottom in an inch of slimy tomato paste, with two perfectly normal legs sticking out from her body, and two very confused human faces looking down at her, one through a thick matting of hair.

"I can explain this," she said. "Um, *je peux*—"

Suddenly there was a crash and the room went dark—Ula lu la lu had evidently hurled one of the tomatoes at the hanging lantern. There was a strange sliding squelching sound, and Susan heard Captain Quoin "Oof!" followed by a tumbling sound as he fell to the ground. Susan had dropped her sword when she slipped, and now she groped about wildly for some weapon. Her fingers closed over something hard and wooden—the pegleg!—and she swung it wildly in the darkness and felt it strike something solid with a loud clunk. In the other part of the room she could hear squirming flesh and flapping feathers and then the distinct sound of rattling keys. A moment later she was tossed across the room as some immensely powerful form swished past her. There was the sound of running footsteps and then, a moment later, silence.

Susan sat in the darkness a moment. Then:

"H-hello?"

Silence.

"President Wilson? Ula—ula u?"

This time, she did hear sounds—men's shouts, moving feet, swords clashing and clanging. But they all came from above.

Susan inched forward with the pegleg still in her hands. A moment later she touched something warm and furry—Lieutenant Cosmo, she realized. Something must have knocked him out cold. She wondered if *she'd* done it when she swung the pegleg, and felt half guilty, half excited at the idea. She had never knocked anyone out before!

Susan skirted the first mate, heading in what she thought

was the direction of the door. Her eyes had adjusted a bit to the darkness and she could discern the outlines of things: the sacks of tomatoes, the wooden barrels, the fur-covered bundle on the floor (which seemed to be snoring peacefully, she noted with some relief).

The mermaid's chain hung from its ring, but there was, obviously, no sign of her, or Captain Quoin, or President Wilson. "President Wilson?" she called, just to be sure, but the only sounds that came to her ears were all abovedecks.

A dull gleam caught her eye then. Her sword. She grabbed it and, blade in one hand, pegleg in the other, stepped into the darkened hallway. It was good to be walking on both legs again, and she stepped nimbly around the piles and jars and trunks in the passage. The cloth had been knocked from the dodo's cage, and the bird stared at her in confusion as she hurried past. As she grew closer to the open rectangle of the companionway she saw the flickers of lights and shadows, as if men with torches were running past the opening. Well, she reflected, there probably *were* men with torches running past the opening.

She paused at the foot of the stairs. Then, just as she set her foot on the first step, she heard a bloodcurdling cry.

"Back!" the voice screamed. "Back, I tell you! Argh!"

It was Charles!

"Charles!" Susan screamed, and ran up the stairs.

A terrible sight greeted her: Charles lay on his back, tangled in a coil of rope. One of the twins, Pietro or Pierro, stood over him with his sword at Charles's throat!

Susan didn't think. She raced forward with her sword stretched straight out in front of her. There was no other place for it to go but right in Pietro or Pierro's bottom.

"Ouch!" he cried, standing up straight and pressing both hands to the struck place. As he did, Susan swung the pegleg with her other arm and felt it impact solidly with Pietro or Pierro's head.

For just one moment, Pietro or Pierro seemed confused by what had happened. A crooked smile spread awkwardly over his face and he said, "Begging your pardon, sir." And then he fell to the deck.

"Susan!" Charles said.

"Charles!" Susan said.

"Pietro!" said a voice to Susan's right.

"Susan!" Charles said again. "Look out!"

Again, Susan acted on instinct. She swung to her right, her sword coming with her (if you have ever taken tennis lessons—which Susan had—it was a little like a backhand). Even before she saw Pierro she felt a vibration up and down her right arm as her sword struck his. There was a snarl on his face and the pegleg seemed to go for it of its own accord, smashing against his cheekbone with a loud *crack!*

Pierro wavered back and forth, a smile that was the mirror

of his brother's on his face. "If you will excuse me a moment, sir," he said, and then he fell atop his twin.

Susan stood there a moment, feeling more powerful than she'd ever felt in her life. She had the sword in one hand, the pegleg in the other, and she brandished them at whatever enemy might come at her. When Charles said her name, she whirled, and if he'd been standing up she might very well have lopped off his head. Fortunately he was still lying on the coil of rope.

"Susan," he said again. And then: "Wow."

"Come on," Susan said. "We've got to get out of here. Did you see what happened to Ula, U, lulu—"

"She went that way." Charles nodded toward the forward deck.

"We'd better go help her."

Charles gulped. "She, um, she didn't look like she needed any help."

They had sneaked up the side of the cabin, and as the forward deck came into view Susan saw what Charles meant.

Ula lu la lu sat on the flat lid of a crate on the deck. She was swinging a length of chain around her in the air, at the end of which Susan could just make out a wet-looking anchor. It wasn't a *big* anchor, but it was steel, and its two prongs looked sharp and heavy.

Just out of reach of the whirling weapon stood three of the crew—Zhi Wo, Jonno, and Konstantin Dimitrivich. Their swords were drawn, but they obviously had no desire to press their attack. Even the enormous Russian seemed cowed by the ease with which the dainty mermaid swung the steel anchor on its chain.

"Come forward, dogs!" Ula lu la lu taunted them. "Who is next to taste my steel? You shall see what it means to chain a member of the Royal Guard of Her Most Aqueous Empress Queen Octavia, Undisputed—"

"—Sovereign of the Sea of Time," Charles whispered, and both he and Susan giggled a little, despite the seriousness of the situation.

"Oh Charles!" Susan said. "I don't know whether I'm glad to be here or not, but I'm glad you're with me." And she gave him a quick hug.

"Ow, stabbing," Charles said, trying to hide the sudden mistiness in his eyes by wiping at his glasses. He indicated Susan's sword. "Careful with that thing."

Susan looked all around the deck. Ula lu la lu was still taunting the pirates. Captain Quoin, Uncle Farley, Murray, and President Wilson were nowhere to be seen.

"Charles," she said, suddenly remembering. "What was that explosion?"

"It was Murray! He fired the cannon!"

He pointed with his sword toward Drift House, and Susan followed the sharp silver line up to the roof, where she saw the snout of one of the cannons poking through a gap in the balustrade. As she peered up at it she saw the dark shape of a bird flying off to the left. She thought it might be President Wilson, but then realized it was just the weathervane mounted on the roof of the solarium.

Unfortunately, someone else saw Charles's sword as well.

"There they are!"

Susan looked behind her. Captain Quoin and Lieutenant Cosmo were advancing. The captain's wig and hat were missing, and his bare head was nearly bald and glistened in the torchlight.

"Uncle Farley got the captain good!" Charles said now. "You should have seen him!"

Susan nodded at Charles without answering. She tried to gauge the distance to the rope bridge connecting the *Chronos* to Drift House. Could she and Charles make it before the pirates around Ula lu la lu saw them and cut them down?

"I'll carve steaks from their stomachs and have em for dinner!" she heard Cosmo say behind her.

The first mate's words made up Susan's mind.

"We're going to have to run for it," she said to Charles.

"When—"

"Now!" Susan said, dropping the pegleg, grabbing Charles's hand, and jerking him toward the bridge.

She heard an uproar to her right as the pirates surrounding Ula lu la lu caught sight of her, followed immediately by a terrible cracking noise. Despite herself, she glanced over her shoulder and saw that Ula lu la lu had released her whirling anchor, leveling the cluster of pirates like a set of bowling pins. She caught an undulating flash of white and silver on the other side of the pirates, and then heard a splash as Ula lu la lu dove into the water on the far side of the *Chronos*.

"Susan, come on!" Charles said, yanking her onto the bridge so roughly that she nearly fell into the water. She and Charles dashed for the poop deck of Drift House. The clatter of pirate steps on the bridge behind them was both audible and palpable.

Charles leapt over the railing and Susan jumped after him. She turned and saw Cosmo's bearded face only a few feet from hers. Immediately behind him was Captain Quoin's bald head, glowing like the full moon. In one motion she raised her sword and brought it down with all her strength—not on the matted head of the miniature pirate, but on the ropes that held the bridge to the poop deck's railing. Though it was more than three hundred years old, Pierre Marin's sword was, as she'd noted earlier, still razor sharp: it sliced right through the two ropes. The bridge fell back and slapped against the Time Pirates' vessel. Lieutenant Cosmo and Captain Quoin dropped out of sight in a near-instantaneous sequence, and the splash they made was big enough to wet Susan's face.

FOURTEEN

The Prisoner

IMMEDIATELY, DRIFT HOUSE LURCHED BENEATH Susan's feet so sharply that she almost fell. At first she thought it was some new attack of the Time Pirates, but Charles was quicker to understand.

"The mermaids are pulling us away!"

"Oh, thank goodness!" Susan shouted—though if she'd taken the time to study Charles's face she might not have spoken so quickly. Charles's expression was clouded behind his glasses, his brow wrinkled with concern.

In moments the *Chronos* had faded into the darkness, though for quite some time Susan and Charles could hear the pirates' confused cries as they tried to fish their captain and first mate from the water. It wasn't until the sound had finally faded away that Susan relaxed her vigilance. She relaxed so abruptly that Pierre Marin's sword slipped from her fingers and clattered to the deck, and as it thudded on the wooden planks she realized how exhausted she was. She half felt she might follow the sword to the floor, but just then a sound pierced the night air.

"Rak!"

"Rak, rak!"

With a loud flutter, President Wilson swooped out of the dark sky and landed on the railing—followed, a moment later, by the more petite form of Marie-Antoinette.

"President Wilson!" Susan exclaimed.

"A moment to . . . catch my . . . breath," President Wilson panted. "Not used to . . . to such . . . long-distance . . . flights."

On the railing beside him, Marie-Antoinette nudged President Wilson with a solicitous beak. She seemed as fresh as the moment the captain had first shone a light on her.

"Charles," Susan said then. "Where are Murray and Uncle Farley?"

"Inside. They're all right, but Uncle Farley has a pretty nasty cut on his head."

"What?" Susan said, turning to run inside.

"It's okay, it's okay. Uncle Farley said that Miss Applethwaite—I mean, the dumbwaiter—would fix him right up."

"Right," Susan said, turning back to President Wilson. "What happened to you back there? You disappeared."

"Disappeared?" President Wilson said in an affronted manner. "Why, I harried that accursed captain the length and breadth of that foul ship!"

"He did," Charles said. "You should have seen him!"

"I seem to have missed *everything,*" Susan said.

"Not true," Charles said. "You saved me."

Susan was about to hug her brother again, but a bloodthirsty roar cut her off. A moment later a furry form had hold of her legs and was rolling her to the deck. Susan's wig fell in front of her face, blinding her. For a single tense moment all was darkness and confusion, and then, just as quickly, it was over. Susan pushed the wig off her head and looked up to see Charles holding his sword at the bearded neck of the shortest of all the Time Pirates.

"Cosmo!"

The dwarf drew himself up to his full height, which was only a little higher than Susan's waist. "I'll thank you to address me properly, missy. My proper title is *Lieutenant* Cosmo."

"Well, if you're going to be such a stickler about names," Charles cut in, "then I imagine you'd better be calling my sister Sus—Captain Susan. And I'm Lieutenant Charles!"

"Rak!"

"And that's President Wilson," Charles added sheepishly.

"President, huh?" Lieutenant Cosmo said. "So your leader's a little green bird?"

"Why you miserable half-sized excuse for a human—"

"Enough!" Susan said, cutting off President Wilson before a fresh scuffle could ensue. She took off her sword belt and handed it to Charles. "Here," she said. "Use this to bind his hands. And if he gives you any more trouble, *lop off his head*."

Even Charles seemed a little cowed by the determined tone in Susan—Captain Susan's—voice, and he took the belt and tied Cosmo's hands. Then, leading their captive, they went in to find Murray and Uncle Farley.

"President Wilson," Susan said at the door. "Are you coming?"

"One moment, Susan. I just have to coax my little beauty inside."

"Rak," Marie-Antoinette croaked. "My little beauty."

"Mark my words," Susan whispered to Charles as they went inside. "*That* is going to end badly."

"Oh, Sus—Captain Susan," Charles said as they walked down the second-floor hallway. "I think I should tell you that Uncle Farley wasn't injured by the, ah, Time Pirates."

"What? What happened?"

"Well, it was the, that is, he was attacked by—"

"Pah, spit it out, boy," Cosmo said. "It was them stinking fish-girls." He laughed a little. "Looks like they played you for suckers all along."

"It was Diaphone," Charles said lamely.

By this time the party was coming down the stairs, and a terrible sight greeted Susan's eyes. The first thing she saw was that the front door lay in splinters on the floor of the hall, as if it had been stove in by a battering ram. And then she saw that one of the doors to the drawing room was simply gone, as if it had been ripped right off its hinges.

Susan felt as if she'd taken a blow to the stomach. All the air seemed pushed out of her body, and for a moment she couldn't speak, couldn't even breathe. The dark water of the Sea of Time stretched beyond the front entrance like an endless expanse of black polished marble.

"What—what happened?"

"Murray said it was the key," Charles answered his sister. "Apparently, that's what she wanted all along."

"What key?" Susan said. But somehow she knew already. The glowing silver key she'd seen on Captain Quoin's key ring flashed in her mind, and she was surprised when Charles said:

"The one that was stuck in the drawing room door."

"What's this about a key?" Cosmo said. "You didn't let one of the Keys to the Great Drain fall into the mermaids' hands, did you? Winds and waves, tell me that's not what you've gone and done!"

"What are you both talking about?" Susan said impatiently, and then, when both Charles and Cosmo started talking at once, she cut them off and said, "Charles, you first."

"Murray suddenly remembered," Charles said. "He said it wasn't the Time Pirates who killed—who did what they, you know, did to you. He said it was the mermaids. They were just using you to get to the key. That's when he fired the cannon. He was aiming at the mermaids, not the Time Pirates."

"And what would we want with killing little girls and babes-in-swaddling like yourselves?" Cosmo said. "Them mermaids was after the three keys all along."

"*What* keys?" Susan demanded.

"Are you daft? The three Keys to the Great Drain. We're done for if they've got them all."

"Cos—"

"*Lieutenant* Cosmo."

"Lieutenant Cosmo," Susan said, "I'm afraid you're going to have to start from the beginning. Charles and I are new on the Sea of Time, and not familiar with its history."

"History's something for your world, not ours," Cosmo scoffed. "Here there's just what is and what ain't and what could be. Pah!" he said, when the faces of the children showed they were no closer to understanding him. "You've seen the Great Drain?"

Susan and Charles nodded.

"And the Great Drain, as its name says, empties the Sea of Time. Without it the waters would stop up and time would cease its flow—in this world and yours."

Charles, ever the scientist, said, "But where does the water go?"

"That's something what no one knows 'cept them who's gone there. And them who's gone there never come back to tell what they've seen. Aye, sure as I'm a normal-size man 'n' you two are gangling giants, there's no return to this world from the Great Drain."

Charles's gulp was audible in the silent house.

"For as long as there's been the Great Drain," Cosmo continued, "there's been the three keys. The three keys, turned in concert, are the only thing what can close the drain. The mermaids have one, the Time Pirates have another, and the third was said to be in Pierre Marin's possession—but no one's seen hide nor hair of Pierre Marin in three of your centuries." Cosmo paused and glared at his captives. "If that *was* a real drain key what the fish-girl's took from you, then we're in for deep trouble. No pun intended."

Once again the silver key Susan had seen on Captain Quoin's key ring flashed in Susan's mind. "But . . . but you said the third key is in the Time Pirates' possession. Aren't we safe as long as that's the case?"

Cosmo frowned at Susan. "Normally the captain kept his key hidden from prying eyes. But having a mermaid aboard made him fearful for its safety, so he took to wearing it on his person, thinking he could better defend it with his sword should your Miss Ula lu la lu somehow get free. But it's hard to draw your sword when you're bowing before your supposed Admiral Marin. With the captain so incommoded, and right in her cell no less, it wasn't all that hard for the mermaid to seize his keyring and liberate herself." Cosmo shook his head so violently that Susan imagined it spinning off his neck and rolling down the hall. "Pah," he said. "We're in for deep trouble, I tell you. *Deep* trouble."

He seemed about to go on, but another voice cut in.

"Susan?"

"Murray!" Susan almost yelled. She skipped the last few steps down the stairs and swept him up in her arms. "Thank goodness you're okay!"

Murray bore the hug as one who feels he doesn't deserve it.

"I'm afraid I messed it all up, Susan. I thought the mermaids were our friends and I was all wrong, and now they've got all three keys."

"You're sure of that? The key you found in the umbrella stand and gave to Charles—the one that got stuck in the drawing room door. It really was a drain key?"

Murray nodded. His eyes were as big and solemn as she'd ever seen them.

Susan tried very hard to keep her face impassive, though something about Murray's nod, so matter-of-fact, sent chills up and down her spine.

"Well, it can't be helped," she said at last. "Indeed, I daresay this is all how it was supposed to work out. There was nothing you could have done."

Murray continued to stare at Susan, his lower lip trembling ever so slightly, his dark eyes filling now, as if he might burst into tears at any moment.

"Don't say 'I daresay,'" Charles said then. "It's affected."

"It is not—" Susan began, then suddenly realized Charles had spoken only to distract Murray. "Lieutenant Charles," Susan said then, drawing herself up to her full height. "As your captain as well as your elder sister, I order you not to question my utterances again."

"Don't say 'utterances.' It's affected."

"Charles Oakenfeld!" Susan said, her outrage only half fake. "You insolent—"

"Don't say 'insolent.' It's—"

"Pah!" Cosmo cut in. "And will the two of you still be prattling when the sea nymphs halt time's flow forever? There's work to be done, children! Let's get a move on!"

Even though Cosmo was half as tall as she was, and bound as well, he was still a grownup, and spoke with a grownup's authority. Almost meekly, Susan turned to Murray. She was pleased to see he was clear eyed again. "Where's Uncle Farley?"

"In the music room."

As Susan followed Murray down the hall, she glanced up at the picture of Pierre Marin. He seemed very neat in his frock coat and wig and pegleg, and she was acutely conscious of the disheveledness of her own appearance. The expression in his face seemed almost scolding.

"Oh, what do *you* know?" she said under her breath. "You're just a painting." But she stood up straighter and neatened her coat.

Once she entered the music room, however, she immediately lost her composure again, for there was Uncle Farley stretched out on the tufted leather couch, a white bandage obscuring half his face. A big red dot showed up in the bandage like a bull's-eye.

"Uncle Farley!" Susan exclaimed, and rushed to his side.

Uncle Farley's right eye was covered by the bandages, and his left was closed. It fluttered open now—no, not fluttered, but sort of wrinkled. It was apparent he was in a good deal of pain.

"Captain Oakenfeld," he said in a quiet voice, "welcome back aboard, ma'am."

Susan felt she could burst into tears. Two days ago, Uncle Farley had been a jolly, bearded, slightly clownlike figure. Now he was stretched out like a victim of war, and she felt she should have been able to prevent this. Though she knew he was still the same man he'd always been, he seemed tiny and insubstantial in his horizontal position, and she fell to her knees so she wouldn't tower over him. Captain Susan? She felt like a great big faker!

"Oh, Uncle Farley, I'm sorry, I—"

"Shhh," Uncle Farley said, raising his hand slightly. "It's not your fault. And I'm not as bad as I look. An aspirin would be nice, but it turns out Miss Applethwaite"—he smiled here, thinly—"doesn't dispense medicine without a prescription."

Susan felt simultaneously helpless and that she had to do something. She plumped ineffectually at the pillows beneath Uncle Farley's head and noticed that his bandage had been rather sloppily applied.

"Charles," she said, "get me a fresh bandage, please. And soak some rags in cold water—Uncle Farley feels quite hot."

She barely heard Charles's "Yes, Susan," because her attention had returned to her uncle. "Uncle Farley," she said, "the mermaids have all three Keys to the Great Drain. Do you know what that means?"

Uncle Farley nodded slightly, then winced.

"Ouch. Remind me not to do that again."

"Just be still," Susan said, and then, after a moment, "Do you have any idea what we should do?"

Uncle Farley was silent a moment, and when the answer came it was from behind her.

"What should we do?" Cosmo said. "Pah! We should cut the hearts outta them fish-girls and mount em on spears to serve as an example, that's what we should do!"

Uncle Farley's eyes opened again. Without turning his

head, he tried to make out who had spoken.

"Pardon me. I don't believe we've met. My name is Farley Richardson."

"Cosmo's me name, lieutenant and first mate of the *Chronos*, and luckless captive of these two deceiving children here"—for Charles had reentered the room, laden with wet and dry cloths and a large cup of water.

"Pah!" Charles said, liking the sound of the exclamation on his tongue. "I'll untie you and take you captive all over again." He gave his bundle to Susan. "It's seawater," he said as he handed her the cup. "But it smells fresh."

"Aye, boy, untie me and let me carve chunks out of you to use as fish bait for them what's towing us!"

"Boys!" Susan said sharply. But Cosmo's words had reminded her of something. "Uncle Farley," she said, "do you think we should cut the mermaids' ropes? If they pull us back to the Bay of Eternity, won't we be powerless to stop them?"

Uncle Farley's eyes had closed again, and Susan was afraid he'd fallen asleep. She beckoned Charles over and prepared to change her uncle's bandages. As she unwound the one from his head he began to speak slowly.

"I don't see how that would help us. We'd just be drifting aimlessly. At least when they're pulling us we know where they are."

"The ones that are pulling us anyway," Charles said.

"There could be others who are already on their way to the Great Drain."

"Do you know nothing at all boy? Nearly all the fish-girls have been eaten up by the drain. I'd wager my beard what's left of em is all below us now."

Charles, who evidently enjoyed having a sword in his hand, now held it up to Cosmo's bushy chin, and said, "This beard's not exactly yours to wager, little man."

"Charles!" Susan said. "Cosmo may be our prisoner, but he's also our guest. Have you forgotten all your manners?" She glared at him sharply, then turned back to her patient. "Here, Uncle Farley, drink this," and she held up the cup of water Charles had brought with the rags.

Murray, who had retreated into a staring silence since they'd entered the room, now screamed "No!" and leapt forward. Before Susan quite knew what was happening, he'd dashed the cup out of her hands, sending cold water splashing all over Uncle Farley's stomach.

"Murray!" Susan exclaimed. "What on earth—"

"You can't drink it!" Murray said, still practically screaming. "You can't, you can't!"

"Don't be silly, Murray," Susan said, dabbing at the wet spots on Uncle Farley's big round belly. "It's just water."

"Unicorns' horns and whales' tails!" Cosmo exclaimed. "A dafter bunch I've never seen. The weanling's the only one with any sense a-tall." All eyes were on Cosmo. "Everyone knows that drinking the water of the Sea of Time will subject you to the sickness of temp'rality."

It took Susan a moment to decode this last as "temporality," and when she had, she glanced at Charles. There was that word again. But then she looked at her youngest brother.

"Murray," she said, "did you know this?"

Murray nodded. He didn't speak, but still Susan felt she understood.

"Did you—did you drink it, Murray? When you were in the dumbwaiter?"

Murray nodded, and Cosmo made a strange noise.

"Winds and waves," Cosmo whispered, and with his hands made a convoluted gesture. Susan thought he might be crossing himself, like a Catholic. "Never would I have thought I'd live to meet one."

"Meet what?" Susan said.

"One of the Accursed Returners."

At Cosmo's words, Charles lifted his sword again. "You take that back! That's my brother you're talking about!"

"It's true, Charles," Murray said.

To Susan's ears, his voice sounded older than an old man's—older than her grandfather's, or the president's when he'd spoken on TV the day after the city had been attacked. Still, she was his older sister, and their mother had charged her with looking after him.

"Murray," she said, "do you know what it means? To be a, a Returner?" She couldn't bring herself to say "Accursed."

"I'm not sure. But I . . ." His voice trailed off. He shrugged weakly, then said, "I think it means I . . . I can't die."

"Can't die?" Cosmo said. "Die a hundred times is more like it. Die over and over again!"

"I'm warning you," Charles said. "Another word—"

"Charles," Uncle Farley said weakly, "I think you'd better let Lieutenant Cosmo speak."

Once again all eyes turned to the bound dwarf, who looked back at his audience as if they were his captives, and not the other way around.

"It's said that them what drinks of the Sea of Time become subject to its current. All of us what comes here from the other world, we step out of time, see, and as long as we remain here we remain out of time, not aging, aye, and not changing save as what's done to us. Them bandages you're pressing to the giant's head"—for Susan was busy rewrapping Uncle Farley's wound—"won't do no good. Only a curing potion will fix him, or a return to your world. As long as he's here, he'll stay as he is—unless of course someone were to give him another, worser cut. But them what drinks the sea's water *do* change. They grow, they age, and eventually they die. But then they come back again, and again, and again. I can't say as I know how that happens, seeing as I've never met one of the Accursed myself, but what I've heard is there's a forgetting that happens, in that the Accursed come back to the moment when they first drank the water nigh as ignorant as they were then. And then it begins to come back to them: the future what they've got to live again and again."

"But is it the *same* future?" Murray said. "I can't remember. I don't think I did any of this before, but I'm not sure."

"Pah. What I want to know is, what made you drink the water in the first place?"

Susan had finished wrapping Uncle Farley's head—even if it was futile, she was going to make her best effort—and now she said to Cosmo, a bit defensively, "I would imagine he was thirsty."

"There's no thirst on the Sea of Time, and even if there was, it'd take more than thirst to drive you to drink it."

Even as he spoke, Susan realized it was true. Despite the sweetness of its odor, there was something alien in the sea's water. Not scary or repellant or anything like that. It was just something she would never want to drink. She had only offered it to Uncle Farley unconsciously, out of concern for his wounds.

She looked at Murray. It was strange, how he could look like a little boy and an old man at the same time.

"Murray?"

"Oh, I was mad, Susan! I was being contrary, like Mum would say. I just, I . . . I wasn't thirsty, but when I looked at the sea I just *knew* I shouldn't drink it, and so that's exactly what I did!"

"Locks and clocks!" Cosmo said. "And what could make you so angry, weanling?"

Murray dropped his face and muttered something into his chest—into his locket, Susan saw, as if he were too embarrassed to speak it aloud.

"What's that, child?"

But before Murray could speak, Susan said, "He was mad at me."

"Whatever are you talking of, girl?" Cosmo demanded.

But Susan had had enough of talking and she ignored him. She stood up and walked over to Murray. She was going to hug him but something about the set of his face changed her mind. So young, but so old too. She stuck out her arm for a handshake.

"What I did was wrong," she said. "I promise not to trick you, or let you down again."

Murray took her hand. "And I promise not to be so willful," he said solemnly, and they shook once on the point.

There was a moment of silence in the room, and then Cosmo's voice cut into the silence.

"Pah. Is there no one willing to take charge of this drifting vessel? There's fish-girls what's got all three drain keys, and here you all are, hugging and shaking hands like a kiddies' birthday party."

"We *are* children," Charles said.

"Well, you won't be much longer, if them mermaids close the Great Drain. You'll be fish bait for that what swims in the deepest water. C'mon, kiddies! We need a plan."

"What we need," said Uncle Farley from his supine position, "is some sleep."

"Eh—"

"It's dark and we have no way of powering ourselves. There's nothing we can do until the mermaids present themselves. I suggest that we all get a good night's sleep so that we're fresh and ready when that happens."

"Sleep! Did them girls cut away your brain with their—"

"Quiet!" Charles said, more crossly than he'd intended, for Uncle Farley's words had made him see how tired he was. Not physically tired—there was no fatigue on the Sea of Time—but his mind was *exhausted*.

"Uncle Farley's right," Susan said. "There's nothing we can do now but get some rest."

Cosmo just stared dumbfounded as the children had a short private conference with Uncle Farley, at the end of which Susan said, "Lieutenant Cosmo, I'm afraid we're going to have to lock you in a closet for the night, just in case." Cosmo swore up and down at this, but seemed a little mollified when Charles brought him a steaming tray from the dumbwaiter. To Charles it smelled foul, but he could see

Cosmo's nostrils quiver. As soon as they'd locked the door of the first-floor closet they heard the clang of silver as the lid was knocked off the tureen, followed by the near-silent exclamation "Tails 'n' sails! It's fish-head slew"—or "stew" perhaps; the children couldn't tell because the last word was drowned out by slurping noises. They checked in on Uncle Farley one last time—he was snoring peacefully on the couch—then headed upstairs.

"Good night, Charles," Susan said outside his bedroom door.

"Good night, Susan," Charles said.

"Good night, Mur—" Susan began, then saw that Murray wasn't around.

"He's already in bed," Charles said. He put his hand on his sister's shoulder. "You'll be okay, Captain Susan. I believe in you."

Susan could only smile at her brother, and then she made her way to bed. As she pulled off Pierre Marin's clothes and pulled on her own pajamas, she felt herself becoming younger again, younger and younger, until all at once she was what she was: a twelve-year-old girl who'd stayed up way past her bedtime. She climbed into her crisp, warm sheets and was asleep as soon as her head hit the soft enveloping comfort of her pillow.

If she'd stayed awake a moment longer, and if she'd peeked out her window, she might have noticed a face bobbing on the vast undulating surface of the Sea of Time. It gazed at the darkened silhouette of the floating house until the last light had been extinguished within its windows—and then it gazed a moment longer, as if . . . as if making sure. And then, so quick you'd've doubted it had been there, the head disappeared. A fanned fluke appeared briefly in its stead, and then with the tiniest splash Diaphone disappeared beneath the black surface of the Sea of Time.

FIFTEEN

The Island of the Past

THE NEXT MORNING THE LIGHT was as bright as it had ever been, the water as blue and vast and calm, the house softly creaking like a cradle whose endless rocking had started before you were even born. How alike were days at sea! Susan thought, sipping a cup of hot chocolate at the table in the music room opposite Charles, who was reading a book while he munched his cereal just as he would have at home. Uncle Farley had eaten a little bit—"just a crumpet, if you please, or perhaps two"—and was dozing again, while Murray, who'd finished his breakfast, stood at the window gazing out with the spyglass. The only odd element was the bearded dwarf, who stood on a chair at the end of the table, trying with some difficulty to eat a bowl of porridge with his bound hands. His beard had caught more of the gruel than his mouth, but luckily the dumbwaiter was generous with its portions. Yes, Susan thought, it's all so calm and peaceful and—

"LAND HO!"

At Murray's shout, Cosmo dumped an entire spoonful of porridge on his beard. He seemed about to start cursing when his mind caught up with his ears, and he realized what Murray had said.

"Land, boy? Did you say you've sighted *land*?"

Nimbly hopping off the chair, he followed Charles to the window, and Susan, glancing at Uncle Farley, followed after. Uncle Farley's eyes were still closed, his breath moving evenly in and out of his crumb-spotted stomach.

At the window, Susan pulled back the gauze curtain and glanced out. As she did she thought back to those images of islands glimpsed from the water that she had seen in movies, recalling thin strips of sandy beach or the thick green band of a tropical forest. Those images had nothing to do with what she saw now. A series of smooth, deep green blobs rose out of the sea. At first Susan thought each mound was an island—together they formed an archipelago, she reminded herself (geography: A+)—but now she saw they were all connected by a flat plain that rose only a foot or two above the surface of the water. The mounds were sort of like hills, but they were so smooth and regular, their sides so vertical, that they seemed more manmade than natural. The tallest was as tall as an apartment building on Park Avenue, and as big around too, while the smallest was the size of a little house. Indeed, Susan thought, they reminded her of houses, perhaps because many of them had chimneys—the bigger ones had several—from which thin streams of smoke emerged and dissolved quickly in the sea breeze. There was no mistaking the island in the distance for a tropical paradise: it could only have been the island she'd seen in the drawing room mural her first night here, and no other.

"By the Sphinx of Minsk!" Cosmo said from her waist. "I can't believe my eyes!"

"Cosmo," Susan said now, "do you know what this place is?"

"Know what it is?" Cosmo said, as if she'd asked him if he knew his captors were gangling freakish giants. "And what else *could* it be, girl?"

Susan waited for him to answer his own question, and when he didn't she said, "Perhaps you could offer us the benefit of your experience."

Cosmo's chest, smaller than hers, but big for a man of his size, swelled up in pride beneath his beard, the impressiveness of the gesture only slightly diminished by the fact that the beard in question was covered in drips and globs of spilled porridge.

"Aye, and who should be the captain here, and who the

captive?" he said. Charles seemed about to say something to this when Cosmo spat out, "Pah, girl. Anyone with eyes can see it's the Island of the Past that lies there on the water."

Susan suddenly remembered what Captain Quoin had said last night: that the dodo was destined for *L'isle du Passé*—the Island of the Past. She tried to remember if he'd said anything else about it, but couldn't.

Meanwhile Cosmo was looking at the children expectantly. When their faces remained blank he rolled his eyes and said, "The depths of your ignorance! I'm amazed you didn't build your ship of stones and mortar and expect it to float above the waves like a water glider. Do you not even know about the Island of the Past?"

Charles, in his driest possible voice, said, "I think that's obvious by now, isn't it?"

"What's obvious is the bad end awaiting anyone entrusted to your care. All the rest is to be inferred."

"Now see here—"

"Lieutenant Cosmo," Susan said, interrupting before the two first mates lapsed into another squabble. "I understand our ignorance of the Sea of Time must seem shocking to you, but we are strangers here, after all. Imagine how you might feel if you were in our world."

"What, New York City?" The bound, porridge-stained dwarf seemed to fight back a scoff. "As sure as I could best you in a race from the Mouth of Yesterday around the rim of the Great Drain, with stop-offs at Scrying Island, Air-Heart's Rookery, and the Dry Well From Which All Thirsts Are Quenched, why, as sure as that I could beat you in a race from the Brooklyn Bridge to Grant's Tomb with a climb up the Empire State Building and a ride on the Staten Island Ferry thrown in to boot."

It took the children a moment to process all this, at which point Susan said, "You—you've been to New York?"

"Been? Been when it was an island with no name, just ash and maple trees and mosquitoes thick as fire smoke in the air, been when Wall Street was just a wall and Canal Street just a canal. Aye, and other times too—and each time I went I checked my maps and legends, as any good sailor does before setting out for a new destination."

"But how?" Charles said.

"Always with the wrong questions at the wrong time" was Cosmo's reply. "Look out the window, children. Does yonder hills look like Manhattan to you? Then why're you asking me questions about a place you're not likely to see again, unless you survive what's coming at you?"

And all at once Susan realized: the Island of the Past *was* approaching them: it had grown rapidly in the past several minutes, and the white-capped water Susan had at first taken for waves breaking on the shore was in fact the froth of the island's progress.

"Aye, she's a floater, is the Island of the Past. Though 'floater' seems a little inadequate to appellate something what can outrace anything in the sea's water. We'd not be getting this close unless the fish-girls was in possession of some powerful stuff indeed."

And, as if they'd been waiting for someone to mention them, there came the familiar sound at the door.

THUD—THUD—THUD.

The first thing Susan noticed when she went into the hallway was that there *was* a door—two actually, on the drawing room, and on the front of the house. Mr. Zenubian, whoever or whatever he was, was certainly an efficient handyman.

When Susan pulled open the front door, she knew to look down at the water. But even so she was surprised, for she had expected to see Diaphone, and instead she beheld the crystalline features of Ula lu la lu. In the bright light of day she could see that the silver-haired mermaid was every bit as beautiful as the copper-haired Diaphone, but where the latter's expression was haughty, Ula lu la lu's was clearly, simply, cruel. Her smile, when it came, revealed a row of sharp white teeth that made Susan wrap one hand protectively around her throat.

"Susan Oakenfeld."

For some reason, the mermaid's words left Susan speechless. Her name in the mermaid's mouth seemed as strange—and wrong—as if Susan had tried to breathe underwater.

Susan's silence seemed to annoy Ula lu la lu, and she went on quickly.

"Susan Oakenfeld, I come to you as an emissary from Her Most Aqueous Empress Queen Octavia, Undisputed Sovereign of the Sea of Time and All Its Tributaries and Headlands, and All the Vessels That Sail upon It. You are honored and blessed by a royal summons, Susan Oakenfeld. The queen has need of your legs."

Perhaps it was the British part of Susan, but her initial impulse was to say, "Yes ma'am" and curtsy. But she was Captain Susan now—Charles had named her so—and the well-being of Drift House and its crew were her responsibility, now more than ever. And look how she'd handled that so far: in a mere two days Murray had drunk the forbidden water of the Sea of Time and been subjected to some kind of reincarnation she didn't really understand, Uncle Farley had suffered a nasty wound that had left him incapacitated, and, worst of all, she had helped the mermaids come into possession of all three Keys to the Great Drain, with which they could apparently, if incomprehensibly, stop time.

She steeled herself.

"You broke your bargain," she said in the most condemning voice she could muster.

"Queen Octavia is not in the habit of recognizing bargains struck with slaves, child!" Ula lu la lu barked in such a harsh voice that Susan winced. "She is in the habit of compelling their service and punishing them for failure!"

"I *didn't* fail," Susan said. "I—I rescued you. I wish I *had* failed—I wish you were still locked up in that room with all of those smelly tomatoes. But I didn't. And now I want to go home!"

When Susan finished speaking Ula lu la lu stroked a lock of her hair the way a wise but slightly sinister-looking man strokes his beard. At length she smiled again, and the sight of her gleaming teeth sent a chill down Susan's spine.

"Diaphone said you had genuine courage," Ula lu la lu said. "But you sound like a whining child to me. Nevertheless, Her Royal Highness has said that you are the one we need, so I am here."

Susan felt like a child. She stomped her foot and said, "I demand that you take me home!"

"You will never see your home again, Susan Oakenfeld. The queen has an errand for you from which you will not return. But if you complete it successfully then at least your boy-slaves—your brothers—will survive. In a few moments we will be beaching your vessel on the Island of the Past, and after that my sisters and I will take you to the queen's court. I give you this time"—Ula lu la lu could not resist an ironic smirk here—"to make your final farewells and settle yourself. Do not waste it."

"But—"

But it was too late. Ula lu la lu had disappeared beneath the water.

A moment later, Susan had awakened Uncle Farley and sat with him and Charles and Murray and Cosmo. The first thing she'd done was untie Cosmo's hands, because it had become suddenly clear that he really wasn't their enemy, but she'd yet to make him her friend. And she sensed that she needed every friend she could get right now, if she was to survive.

She explained everything to Uncle Farley, and then said, "What do I do?"

Uncle Farley looked at her with cloudy eyes. "I don't know, Susan. I just don't know."

Cosmo's freed hands seemed to have freed his feet as well. He paced up and down the room.

"Aye, and to be sure they need you to work the keys," he said at length.

Everyone in the room looked at Cosmo as if they were drowning and he was a life preserver tossed in their midst.

"Lieutenant Cosmo?" Susan said. "They need me to work the keys?"

"Aye. They need you to close the lock."

"But I don't understand, Lieutenant. The mermaids have hands. Why can't they close the lock themselves?"

"Hands, yes. But it's feet they don't have, isn't it?"

"I'm afraid I still don't understand, Lieutenant."

"Ice, lice, mice, and gice!" Cosmo said, clapping a hand to

his forehead (Susan could only guess that the last word was meant to be "geese"). "It's like lecturing to a nursery!" He clambered onto a chair, facing the children like a professor on a podium. "Now listen," he said. "What's known of the lock to the Great Drain is that it lies beyond the vortex that empties the Sea of Time. Now, no one what's ever been to the bottom of the whirlpool has come back to tell what they've seen, but it's generally believed that there's a place of no water down there. And where water can't go, the fish-girls won't go neither. And that's where you come in."

"No water?" Charles said now, his fingers drumming on the handle of his sword. "At the bottom of the sea? That's preposterous!"

"He's talking to me from a floating house, and he carries on about what's preposterable. It's a good thing *he's* not in charge," he said to Susan. "Your inexperience may not serve us, but his daftness'd get us killed as sure as sure is sure."

"Why you—!"

"Charles!" Susan said, for her brother had half drawn his sword. "We've got to work together here! Put that away!"

Charles pushed his sword into its scabbard with a disgusted motion. "This is nonsense, Susan. He's done nothing but insult us since the moment he snuck aboard like, like, like some little wharf rat."

"Wharf rat! Why, you giant of a thing, I'll make you pay for that!" And Cosmo jumped down off the chair and made for Charles with his hands outstretched.

"Boys!" Susan yelled, so loudly that everyone in the room jumped. "Stop your squabbling this instant!"

Silence filled the room, punctuated after a moment by Charles muttering under his breath.

"Charles, if you have something to say, say it so that everyone can hear."

Charles looked up at his sister with an annoyed look on his face.

"I said, don't say 'squabbling.' It's affected."

"Charles, this is hardly the time for jokes."

"Who's joking, Susan? Since you've put yourself in charge, there's been nothing but disaster after disaster. And now . . . and now . . . and now you're going to get yourself killed just like Murray said you would!"

Charles's face was so red Susan was afraid he was going to cry in front of everyone, and she wanted to save him that embarrassment if she could. She knew he was only lashing out because he was scared, and she tried to think what her mother would say in a similar situation, to calm him down.

"Charles," she said, searching her mind. "You . . . I . . . you'll understand when you're older," she said finally, knowing her words were inadequate, but unable to come up with anything better.

For a moment Charles looked as if he would actually draw his sword on her. His face went even redder than it had been and he stamped his foot on the floor, and then, a moment later, he stamped his foot a second time and ran from the room.

Cosmo broke the silence with a chuckle. "Aye, it's a temper in that one, it is. He'll make a fine warrior one of these days."

"Why couldn't you have said something nice when he was *in* the room?" Susan said exasperatedly. "What am I going to do now?"

No one in the room answered her until the house, as if falling into the gap, provided its own response. It gave a series of lurches accompanied by a long scraping noise, and then, with a moan and a shudder like you let out to shake the pain of a toothache or a stubbed toe, it stopped.

Murray turned from the window to Cosmo, Susan, and Uncle Farley.

"We're here," he said.

"We're stuck is more like it," Cosmo said.

"Stuck on the Island of the Past," Susan finished. "Just as the mural predicted."

And the familiar sound of knocking filled the house.

PART THREE

A Race Against Time

SIXTEEN

The Golden Bubble

THE FIRST THING SUSAN SAW when she opened the front door was a gigantic glass bubble floating in water so clear she could see the white sand through the transparent curve of glass. The bubble was about five feet in diameter and girded by golden bands like lines of longitude on a globe. Or was it latitude? She could never remember which ones traveled north and south, even though she'd gotten the question right on her geography exam. It was one of those things you forgot as soon as you'd provided the answer on the test—like, say, the difference between port and starboard.

The bands segmented the glass of the bubble like an orange that's been cut into wedges, and one of these segments stood open, like a door. Seeing the opening, Susan knew immediately what the bubble was for.

"Susan Oakenfeld!"

Susan started at the sound of Ula lu la lu's voice; she had been so caught up in the sight of the globe that she hadn't noticed the mermaid. Now she looked down at her silver hair, her two rows of sharp white teeth, shining hungrily from her smirking mouth.

"It is time."

"But I haven't packed, or bathed even."

"It is time, Susan Oakenfeld."

"But I need to say goodbye to my brothers."

"If you value your 'brothers' "—Ula lu la lu grimaced when she said the word, as if it were distasteful to her—"then you will get into the transport pod."

This raised a different question. The globe was a dozen feet from the front door, floating delicately atop eight or nine feet of water.

"How am I meant to do that?"

Susan's request appeared to take Ula lu la lu by surprise. She looked at the globe, then back at Susan—at her legs, as if they were some kind of deformity. For the first time since Susan had met her, she seemed at a loss. Then, almost snarling, she said,

"I shall have to carry you."

The words struck Susan like a spear thrust. The idea of having Ula lu la lu's dead white arms wrapped around her made her want to run upstairs and hide under the covers. She even turned, and looked down the front hall, but there was no one there save the painting of Pierre Marin, which from this angle seemed *not* to be looking at her, as if even Drift House's founder had abandoned her to her wits.

Susan looked back at Ula lu la lu. She was about to ask her if she would honor her side of the bargain if she did the mermaids' bidding, but one look at that proud cruel face told her she would be wasting her time. So instead she said:

"If you harm Charles or Murray or Uncle Farley or President Wilson or Lieutenant Cosmo, then I promise you I will drink the water of the Sea of Time and come back as many times as it takes to kill every last one of you."

If it was possible, Ula lu la lu's milky skin went a shade paler. Then, shaking herself slightly, the mermaid held out her arms and said, "It is time" once again. But her voice was less confident than it had been before.

Susan walked down the front steps until she was ankle deep in the water. It was the first time she had actually touched the Sea of Time, and it was surprisingly, pleasantly warm. She was barefoot, and still in her pajamas, and, reflexively, she bent over to roll up her cuffs. As she did she saw a flash of movement out of the corner of her eye and felt two cold smooth things wind themselves around her like a pair of snakes. Susan just had time to realize the snakes were Ula lu la lu's arms and then she was aloft, held above the water and feeling the immense power of the mermaid's churning tail as she was carried to the transport pod. Ula lu la lu's tail was beating so rapidly it was just a silver blur, and the transport pod was a golden blur at the edge of her vision. Susan felt as if things were losing their sharp edges—a sensation that was reversed when, a moment later, Ula lu la lu heaved her through the air and she landed heavily on her hip inside the pod. The curved gold-flecked glass looked fragile, but felt solid beneath her body; there was an equally fragile click when Ula lu la lu sealed the open segment back in place, but when Susan pushed against it the door felt solid as rock. On the far side of the glass, Ula lu la lu smiled wickedly, and then she dove beneath the water.

Susan felt a moment of panic. The inside of the globe was so quiet she felt she could hear her own heart beating, and there was an airless quality to it as well, though she could in fact breathe freely. But then there was a tug, and she saw (below her, which was quite disconcerting) that Ula lu la lu had taken the end of a golden braided chain and was pulling her out of the shallows into open water. The house retreated quickly, and then the strange island appeared on either side of it, the great flat plain with the weird green hills that seemed to spring from it like blobs from a lava lamp. She peered at the windows of Drift House hoping for a sight of Charles or Murray, but all the windows were empty of everything except their curtains, pale green ones in the dining room, dark burgundy-tinted brown ones in the library, baby blue ones in the bedroom above the library—and then, with a tug so violent Susan fell backward, the transport pod suddenly plunged downward, and the waters of the Sea of Time closed all around her.

SEVENTEEN

Murray Version 3

In fact, there was someone at one of the windows. At the window with the baby blue curtains, in fact, which was the room Charles had been forced to share with Murray three days ago. It was *always* that way, Charles thought, fingering the pastel fabric disdainfully. He was always treated like a baby, as if he were no older than Murray, when in fact he was practically as old as Susan—and a great deal smarter than she was, at least in science.

He watched the golden-veined globe containing his sister retreat across the water with a mixture of anger and some other emotion he couldn't quite identify. That emotion was fear, and even deeper down there was sadness as well, but he was still too angry to realize it. Stupid Susan, was all he could think. Always treating him like a baby, desperate to keep the privileges of being the eldest child all to herself.

Charles blinked.

The golden globe containing his sister had disappeared.

He rubbed his eyes then, then used an edge of the curtain to rub at the crystal-clear window, but Susan was gone. Though it was logical to assume Ula lu la lu had pulled her beneath the water, the surface of the Sea of Time was so calm it seemed almost as likely that the golden globe, like the soap bubble it so resembled, had burst into thin air, taking his sister with it.

A funny feeling, like a great number of bursting bubbles going off in his stomach all at once, made Charles feel as though he hadn't eaten anything in a long time, and for the next several moments he busied himself cleaning his glasses, for they had suddenly misted over. But then he steeled himself and turned from the window.

After a bit he had calmed down enough to change out of his pajamas. He knew he should take a bath but decided against it—he was the oldest child in Drift House now, and with Uncle Farley injured he was practically running the show. He even put on yesterday's clothes, which were somewhat grimy from the battle aboard the *Chronos,* and over his jeans—which had a fearsome-looking rip in one thigh, almost as if he'd been stabbed—he buckled his sword belt. He would show everyone who was the most resourceful child in this family!

Drift House was quiet as he made his way down the hall. Murray was off somewhere—he seemed always to want to be by himself now, since he'd emerged from the dumbwaiter—and of course Uncle Farley was asleep on the music room couch. Lieutenant Cosmo appeared to be hiding too. Probably looking for something to steal, Charles thought, or to use as a weapon against the Oakenfelds. Charles didn't approve of Susan's decision to untie him, but it was too late to do anything about it now. But let the dirty little man try some funny business and . . . Charles's fingers fumbled for the security of his sword hilt.

As he descended the stairs, Charles heard voices coming from the drawing room.

"Oh, my coy, coy mistress—why *do* you resist me?"

The voice was President Wilson's, and it had a plaintive edge to it. Charles was too young to have chased after a girl yet, so he didn't recognize the sound of someone hopelessly besotted.

"Why *do* you resist me?"

This time the speaker was the pirates' red and blue parrot, Marie-Antoinette.

"Oh, madam, you tease me. You bounce my words back to me as Echo did the hapless Narcissus."

"Hapless Narcissus," Marie-Antoinette cooed quietly.

"Oh, you are *wicked*, madam! Wicked, I tell you! In your wings, my heart is a harp from which pulses the music of love."

"Music," Marie-Antoinette repeated. "The music of love."

Charles almost felt sorry for President Wilson. Drift House's mascot didn't seem to realize that Marie-Antoinette was just a regular parrot, echoing the last thing she heard with no more inkling of what she was saying than a tape recorder. But as amusing as their comedy might have been to him under different circumstances, he didn't have time to dally now. He was going to show everyone what the Oakenfelds' middle child was capable of.

The front of Drift House faced the open sea, the rear was lodged on the shore of the Island of the Past, and Charles had just turned toward the back door when the familiar sound of knocking came from behind him—from the front door. Thinking the mermaids must have forgotten something—or Susan, he thought, she probably wanted her *toothbrush* or something like that—he turned exasperatedly back to the door and pulled it open. His eyes were trained on the water at his feet from the habit of dealing with the mermaids, but the water was empty of everything save the purple slivers of darting minnows no bigger than his finger. Then a voice made him look up. The voice was familiar, and yet there was something about it Charles couldn't quite place.

"By the books! Is it—can it be—*Charles*?"

Charles looked up. There, floating five or six feet above the water, was a thick multicolored carpet, rippling ever so slightly, like the water below it. And there, sitting cross-legged in the center of the carpet in a pair of blousy brown trousers and a purple vest, sat a boy around Charles's age, but athletic and tough-looking. A thick gold turban was wound about the boy's head and a tiny heart-shaped locket dangled against the bare skin of his chest, which was brown from long exposure to the sun, the muscles underneath visibly corded like those of a much older youth.

It was Murray.

Murray's mouth hung open in amazement, and then, slowly, it closed.

"Charles! I can't believe it's you!" And then, when Charles still didn't say anything, Murray said, "What's the matter? Surprised to see your little brother after all these years?"

Charles looked behind him into the empty hall, then turned back to the figure on the carpet.

"But, but—"

"What's-a-matter, Charles? Don't you recognize me?"

Charles could only nod and stammer. "But . . . but . . . but I don't *understand*."

Murray's smile grew wider, but softer too, more sympathetic to his brother's confusion. "What's not to understand, Charles? It's me, Murray."

Charles rubbed his eyes. There was no doubt about that: it was Murray.

Suddenly the boy on the carpet broke into laughter.

"Oh, I shouldn't yank your chain. Of course you don't understand. I bet it ain't been but a day or two since I took off. But I been on my own for, I dunno, years and years now."

"But—but you're *inside*," Charles insisted. "I—I just had breakfast with you."

Now it was the turn of the boy on the carpet to look confused.

"What?" he said, his manner suddenly imperious and threatening in a way that the Murray Charles knew had never been. "What're you talking about, Charles? I swear, if you're trying to play a trick on me—" He reached for the handle of the sword at his waist.

"But I did. You had pancakes cooked with some kind of mushy strawberries—"

"Compote? Strawberry compote?"

"Yes, that's what you called it. Strawberry compote. And then after breakfast you wandered off like you always do, and—" Charles broke off suddenly. "Did you go back in the dumbwaiter?"

"No, Charles, I'm right here."

The voice came from the hall behind Charles, and he whirled around in a panic.

There, in a pair of pale brown corduroys and a wrinkled blue T-shirt, stood five-year-old Murray. His arms were crossed behind his back in a gesture that looked very adult to

Charles, and he regarded his older brother with sympathetic mournful eyes.

"What, what?"

Charles whirled back to the boy on the carpet. No, not "the boy": it was clearly, unmistakably—undeniably—Murray. He was spluttering in confusion.

"But how did you? I mean, how did I? I mean . . ." He shook his head. "I don't know what I mean. What the heck is going on here?"

"It's quite simple," the Murray in the hall said to the Murray on the carpet. "You made it back. I made it back. We," he said in his slightly sad voice, "made it back." He sighed then, like an old, old man. "You'd better come inside," he said to the bigger version of himself.

Sometime later, the three boys sat in the library, having elected not to disturb either Uncle Farley or President Wilson. The two Murrays had spoken very rapidly to each other, and Charles had had a great deal of difficulty following their conversation, but he thought he had finally figured it out.

A few days ago, his little brother had climbed into the dumbwaiter with the desire to get as far away from Susan as possible. He'd fallen asleep in the dumbwaiter, and when he awakened no one was around. He had wandered the rooms of Drift House looking for his brother and sister and uncle, but they were long gone. He had tried climbing back in the dumbwaiter and wishing himself back to them, but to no avail. He had thought to run away, but when he opened the door there was only the Sea of Time all around him. He was thirsty, and even though he sensed he shouldn't, he took a drink of the sea's water. For several days he had floated on the open sea, drinking copious amounts of water to slake his thirst and hunger, until finally the house washed up on an island—not the Island of the Past, but some other. Murray had left the house in search of his family, but instead he nearly fell into the hands of a band of brigands. He ran from them, but when he returned to the place where Drift House had washed up he found that it was gone. He had hidden from the brigands for several weeks in the treasure-laden caves on the island, living off the scraps he could steal from their camp when they were sleeping or away, until finally they noticed the missing food and laid a trap for him. It had taken him some time to convince the brigands he wasn't an agent of the mermaids or the disgruntled creatures of the Island of the Past—which, at that point, Murray hadn't even heard of yet—and the brigands, confronted with what was obviously nothing more than a lost little boy, had taken him under their wing. They had taught him to navigate the Sea of Time, to fight with a sword and fire a cannon and board another vessel racing at top speed on the open water, and even taken him on several raids to all sorts of exotic places, but all the while Murray was looking for a way to get back to his family. But as the weeks turned into months and the months turned into years, that seemed more and more impossible. How could he explain to his mother and father that he was no longer the five-year-old they had left at Uncle Farley's house? And so, after several years, he had left the brigands peaceably. He had acquired his carpet by then, and he used it to search the scattered islands of the Sea of Time in search of some greater magic than mere time travel: something that would make him five years old again. Something that would allow him to go home. He had, after nearly two years of searching, finally managed to track down the Island of the Past—which, apparently, was a floating island, and quite good at evading detection—and he was just about to leave when he'd seen Drift House stuck on its beach.

When the ten-year-old Murray had finished telling his story, the five-year-old Murray looked at him silently for a moment, and then he said, "You found what you were looking for."

An expression of joy such as Charles had never seen flashed over his ten-year-old brother's face, but it was gone just as soon as it had come. "Here?" the elder Murray said. "There wasn't nothing here—nothing I could use, anyway."

"Not here," the five-year-old Murray said. "Not now."

"Then when?" the ten-year-old Murray practically shouted. "Where? Oh, tell me, please! You don't know how much I've missed everyone! You don't know how lonely I've been!"

The five-year-old Murray's face went very stiff for a

moment. Though Charles couldn't know it, he was thinking about the experience he had had in the solarium with Uncle Farley, when he had taken off the locket and found himself transformed into an old man. In a very quiet voice, he said, "What you need is still a long way from now."

"But you must know where it is!" the elder Murray insisted. "Tell me where it is and I'll go there right now!"

Five-year-old Murray was shaking his head.

"I'm sorry. I've tried to remember, but I just can't. I—I realized last night that we were going to meet you here, and I stayed awake all night trying to remember. But I couldn't. I just . . . couldn't."

"You're lying!" the ten-year-old Murray yelled. "I know you're lying!" In a flash his sword was out of its scabbard, its shiny sharp tip just inches from Murray's throat. "Tell me where it is or I'll have your head!"

"Pah! It's not one of *my* charges you'll be harming today. Do anything but lower your blade to the floor and it's *your* head that'll be rolling across this fine old nonflying carpet."

The words were Lieutenant Cosmo's, of course. He had heard the loud voices and snuck into the library, and now he stood behind ten-year-old Murray, his sword tickling the bare skin between the boy's turban and the embroidered collar of his vest.

Moving slowly, the older Murray lowered his sword to the ground, and then, hands in the air, he turned slowly to face his assailant. But even before he'd finished his rotation, Cosmo had dropped his sword and Murray had thrown his upraised arms around the dwarf's bearded head.

"Cosmo!"

"Mario!"

For a moment, Charles was confused—or, rather, even more confused than he already was—but then he understood. The "brigands" who had sheltered his brother were none other than Captain Quoin's crew of Time Pirates. As for the "Mario" business, Susan had mentioned Zhi Wo, and Jonno, and Pierro and Pietro; Charles guessed Murray had wanted to fit in with Cosmo and Co.

There was much embracing and exclaiming before the dwarf and ten-year-old Murray separated from each other.

"Serpents and portents," Cosmo said, jerking a thumb at five-year-old Murray. "This little un really is you?"

" 'Fraid so," the elder Murray said, in the salty tone Charles found a little hard to reconcile with the memory of his soft-spoken, sweet-voiced brother. "Puny little stripling, wasn't I?"

"It's a wonder you survived, really. Aye, but the captain could make a man of anything with two legs. Even this un here, eh, Charlie-o?" Roughly, but not unkindly, Cosmo thumped Charles on the chest. Through his thick beard Charles could see the faint glimmer of what would have been a face-splitting grin on a clean-shaven man. "So Mario," he said, "you made it back after all. Aye, and if anyone was gonna make it back, it was you."

"Looks like I did," the elder Murray said. "But this little un won't tell me how, nor where, nor when."

"Aye, that sounds about right," Cosmo nodded. "You know as well as I do the sea don't make its mysteries known nor free nor easy. Especially not to one what's drunk of its water."

"Aye indeed," the elder Murray said, his voice taking on an even saltier brogue in dialogue with Cosmo. "It ain't never easy, is it?" He turned to Charles then. "Ah, Charles," he said, grasping his older and yet slightly smaller brother by the wrist. "You don't know how much I've missed you. It's been . . . it's been . . ." Murray's voice choked up for a moment, and then he regained his composure. "It's been five years since I seen you, or Sus—or Mother and Father."

Charles barely had time to wonder why Murray hadn't said their sister's name out loud when he said, "Mother and Father." When Charles had known Murray, his little brother still called their parents Mummy and Daddy. Charles didn't know what to say.

"I—I'm sure you'll get back," he said. He pointed to five-year-old Murray. "See? You do. You do get back."

All eyes turned to the somber face of the littlest boy in the room. His sad eyes made it seem like getting back almost hadn't been worth it. "I think you should go," he said then.

"But I only just got here," the elder Murray protested.

"You should go before Uncle Farley sees you. I think that would be a mistake. And besides, we still have to figure out a way to rescue Susan."

"Susan?" The elder Murray's face clouded with confusion. "Whaddaya mean, rescue her? Isn't she . . . I mean, the mermaids told me . . ."

"Spit it out, seedling," Cosmo said. "What've them fish-girls been telling you?"

"Why, they told me she . . . Susan . . . she died."

"What?" Charles said. "No! Not after all we've done!"

"They said she drowned."

Suddenly the younger Murray's face grew excited for the first time. "Drowned, you say? But how? When?"

"It was years ago, they said. They said she tried to get away—they wanted her to do something and before they could pull her down to their city she jumped out of their breathing bubble and tried to swim away and, and drowned."

"But she didn't!" Charles said. "I saw her. She didn't. She went down below the water with them."

"Yes, yes!" the younger Murray said. "Don't you see, Charles? They lied! They must have been afraid I'd come back, and so they told me she died, knowing I would tell her, and that would make her do what they told her to without putting up a fight!"

"Pah, my head, child," Cosmo cut in. "What you say is

giving it a mighty pain. Speak more clearly afore my brain explodes."

"Listen," the younger Murray said. "The mermaids know more about the Sea of Time than anyone—even the Time Pirates," he said firmly, when Cosmo looked about to disagree with him. "Somehow they must have known that I'd drunk the sea's water, and that eventually I'd return to the moment I drank it. So they planted an idea in my head. They told me Susan died."

"They did," the elder Murray agreed.

"But why would they do that?" Charles said.

"Carpets and car parts, I think I understand now," Cosmo said. "They wanted to break your spirit."

"Yes, that's it," the younger Murray said. "They thought that if Susan believed it was hopeless then she'd just give up. She'd just do what they said. We all would. But it's not, don't you see? It's *not* hopeless! We can save her!"

"But how?" the elder Murray said. "How do we do that?"

"I don't understand it fully yet," the younger Murray said. "But . . . but I'm starting to think I *know*. I just have to try to remember." A look of intense concentration clouded his face for a moment, and then he shook his head. "I need to go off and think. And in the meantime *you*"—he pointed at the bigger version of himself—"had better go before Uncle Farley sees you. I think the fewer people who know about your existence the better."

A look of sadness crossed the ten-year-old Murray's face. "You're probably right," he said after a moment. "Oh, but Charles, there's so much I want to tell you! So much I want you to tell me!"

Charles looked at the older version of his younger brother. He was so plaintive, so . . . so hungry.

Before Charles could think of something to say, the younger Murray said, "You'll get the chance to talk to Charles all you want. It may take longer than you realize, but one day you will." His voice was compassionate but also firm, and for the first time Charles heard a trace of hope in it as well. "*We* will. We'll get back to them. We'll all be together again."

Before he left, Murray pulled Charles aside. "I'll leave you my carpet," he said in a conspiratorial whisper.

Charles looked at the rolled-up bundle that leaned—no, floated actually, a few inches off the floor—in the hall. At the thought of flying it he felt bubbles and butterflies in his stomach. He knew this was hardly the time to get excited, but he could feel his fingertips tremble with eagerness to seize hold of the carpet and unroll it.

"You can use it to explore the island," the elder Murray was saying. "Perhaps there's someone"—he winked slyly—"that'll help you save Susan."

"But how will you travel?" Charles said.

"Ah, don't worry about me, old boy. I've got ways of getting around."

Now Charles glanced out the back windows at the wide expanse of the Island of the Past.

"I'm not sure—"

"Aw, c'mon, Charlie-o. I know there's a little pirate in you, waiting to come out."

He hugged Charles then, a strong bearhug that squeezed the breath from Charles's chest.

"Pah, and what's with all the hugs and tears," said a voice behind them. Lieutenant Cosmo was emerging from the library with the younger Murray. "You'd think we was a bunch of fish-girls and not the Time Pirates we—ah, what? Put me down, you daft boy! Put me down!"

Murray had swept up the first mate of the *Chronos* in a hug so big it lifted him off his feet. After setting him down, he stepped toward the younger Murray, and then he stopped.

"Better not," the five-year-old Murray said in a voice much older than the older Murray had yet acquired.

"No, better not," his ten-year-old version agreed.

In the doorway, he hugged Charles one more time, and Charles could feel in the press of his brother's strong body the urgent beating of his heart.

"We'll save her, Murray," Charles said. "Susan will be here for you when you come back."

There were tears in Murray's eyes, tears of sadness mixed with joy. "It's—oh, Charles, you don't know the half of it." He seemed about to say something else, then shook his head. "I know you'll save her, Charles. You were always the smartest of all of us."

He turned then, and before Charles knew quite what was happening Murray had jumped into the Sea of Time. For a moment his purple and gold form hung in the water and then, quick as a fish—or a mermaid—he was gone.

EIGHTEEN

Queen Octavia

Undisputed Sovereign of the Sea of Time
and All Its Tributaries and Headlands
and All the Vessels That Sail upon It

SUSAN HAD PREPARED HERSELF FOR the idea that Ula lu la lu was going to pull her under water, so she didn't yelp when the sky vanished—she didn't know if the mermaid could hear her through the glass, but she wasn't going to give her the satisfaction if she could. But she did fall down at the jolting movement, and so she was lying on her back as a film of blue water spread between her and the blue sky. The sky and the water were both blue, both nearly the same shade, but the one seemed the most lovely color in the world as it faded away, while the other was horrible, horrible, horrible as it swallowed her up.

Ula lu la lu swam with amazing rapidity. Susan marveled at

how strong the mermaids were. It had only taken four to pull Drift House clear of the Great Drain, and now this single mermaid was pulling the great buoyant bubble that held Susan's body under the water as if the gold chain she held in one hand was a silk ribbon attached to nothing at all.

As they went farther down, the light changed. The pale blue color deepened, and darkened, yet also seemed strangely luminescent, like black light almost (Susan had a friend back in New York, Charlotte Jackson, and Charlotte had a black light set up in the closet of her bedroom, and the two girls liked to sit on the floor and look at the way the white clothes glowed, and when they laughed their teeth glowed too, just like their clothes). Now, almost imperceptibly, the light thickened. Ula lu la lu's form, only a few feet away, grew fainter and fainter, until all Susan could see was a faint and possibly imaginary flicking of her tail. But then even that was gone. Pure unadulterated darkness surrounded the cocoon that held her, and Susan realized with a start that she could see nothing inside the globe as well. Once, in Charlotte's closet, one of the girls had kicked the black light's plug out of the socket and the two had been plunged into a similar darkness. They had screamed and giggled for a moment and then Charlotte found the door handle and opened it and light had flooded in.

But there was no door handle here. It was completely dark, inside and out, and Susan wondered if she would ever see anything again. Perhaps the mermaids lived in total darkness at the bottom of the sea, and all she would know would be their hollow voices and the feel of their cold hands on her shoulders. Despite herself, she shivered.

"Buck up, old girl," she said to herself, giving herself a little hug, and then she said, "Don't say 'buck up,' it's affected," in her best imitation of Charles's American accent, and she laughed at herself. Oh, she *did* wish she hadn't left things so badly with him. She hoped he wouldn't only remember her in that way.

Just then Susan spied a glow in the distance. She supposed it was below her, but in the vastness of underwater that word didn't seem to mean anything different from "above," or "left" or "right" for that matter, or "diagonally." There seemed no inevitability of gravity or anything like that—no confident "down" against which all other directions were measured. But there it was, off in the distance, a smear of light, blue green in color—aquamarine, Susan thought. Sea blue. What other color could it be?

The glow grew with the same swiftness that the sky's light had receded. Within a few moments of her noticing it, it had spread itself out in a lumpy kind of way, like radioactive jam—blueberry jam, Susan thought—smeared thickly on a vast piece of bread. And as she got closer Susan could see that the lumps and bumps were hardly inconsequential, but, like the Island of the Past, big as hills on the sea floor, great round smooth glowing domes that Susan instinctively knew were hollow. No chimneys poked from them, but even so, Susan understood that these hills were the buildings of the mermaids' underwater city.

A few heartbeats later and Susan—and Ula lu la lu, who had become visible again—were among these great hills. The mermaid city stretched out for miles all around her, and its many hills, some taller, some shorter, some narrow like columns and some wide as pyramids, couldn't help but remind Susan of the city she had left only a few days ago. How many worlds away Manhattan seemed now!

Ula lu la lu slowed once she reached the city. It was as if she wanted Susan to be able to see this magnificent thing the mermaids inhabited. And it *was* magnificent. So simple in its undulating outline, and yet so beautiful in the interplay of colors that swirled over them. Susan felt as if she were swimming inside a painting—one of those dappled kaleidoscopic impressionist paintings she'd seen on a class trip to the Museum of Modern Art. The peaks and valleys of the mermaids' city, which had seemed uniformly aquamarine at a distance, were up close covered in great rolling fields of luminescent blue-green seaweed, but sprinkled and swirled liberally throughout were dots of pink and purple, spirals of yellow and orange, bursts of red, muted brown-orange patches, dribbles of emerald teardrops, and clouds of translucent white pearliness. And this was just the city: in the water all around were a million more miracles. School after school of fish, each more dazzling than the last. Amber pin-striped creatures whose bodies were as paper-thin as an upright sail, and then another school, equally thin but horizontal this time, their broad backs a deep rich grayish purple and their bellies as white as the white of an eye or an egg. There were fat-bodied fish as thick as her waist, their bodies reddish, brownish, bluish, greenish, depending on how the light struck them, and then there were fish so small that she could only see them collectively, like a shimmering light. An eel as long as her bedroom in Drift House curled quickly around the curve of one hill, followed a moment later by a second—or was it the first, swimming in a complete revolution even faster than she'd realized? And then she saw a great form that seemed heavy even as it floated in the water, a pale blue whale whose flukes wagged up and down as slowly as a palm frond wafted over an ancient Pharaoh. And a million more fish, far too many for her to see, let alone examine or remember. And she *did* want to remember them, she realized. She wanted to tell Charles and Murray all about them, when she got back to the surface.

But what she didn't see were mermaids. Though the city that stretched all around her could have housed thousands upon thousands of inhabitants, and though she had by now noticed many great openings in the sides of the hills through which she had glimpses of green-gold glittering caverns filled with water, Susan had not seen a single mermaid besides Ula lu la lu. And all at once the great beautiful painted city of the mermaids acquired a tinge of sadness to her, because she realized it was empty. This city, which was so obviously meant to be the seat of an undersea kingdom, had somehow been turned into a ghost town.

Ula lu la lu had slowed even more by then. Susan saw that they were approaching a great hill near what must have been the center of the city. The hill was as broad as a football field and tall as a ten-story building, with sides that sloped as gently as a circus tent. Directly in front of her a great arched doorway opened up. The arch was upheld by two carved mermaids standing on their tails, their arms raised above their heads as if they'd been caught dancing and frozen that way. Then the mermaids were gone and Susan was in an enormous domed chamber whose curved roof was banded in red and white stripes, each edged with the thinnest gold piping. It made her dizzy to look at it, so she looked down. There, below her, stood a great throne carved in the shape of an openmouthed fish, its lolling tongue looking plush and pink despite the fact that it was obviously carved from a single piece of stone. Susan half expected Queen Octavia (Undisputed blah blah blah of the blahbetty blah blah and all the blah blah blahs That Sail upon It) to appear from the fish's bowels, but the tongue remained empty, stuck out at her in an almost impudent leer.

Susan realized then that Ula lu la lu had stopped swimming, and the pair of them were rising slowly, lifted only by the buoyancy of the golden-veined bubble she was trapped inside. Susan thought they would bang into the striped ceiling and was a little afraid of the bubble cracking, but then she looked up and saw a hole in the center of the dome, and a moment later the bubble passed through it—and, with a little jump, as when a buoy you've been holding under a pool bursts from your hands—Susan's bubble popped out of the water into the open air of an undersea cave.

This room was much smaller than the throne room below, and not nearly so grand. Not grand at all, in fact. Its unfinished walls were brown-blue rock and flecked here and there with natural veins of ruby and amethyst, and the dais at the end of the room held not a throne but a couch with a single gilt arm and a tufted red seat. But this chair, unlike the throne in the room below, held a mermaid.

Susan didn't need anyone to tell her this was Her Most Aqueous Empress Queen Octavia, Undisputed Sovereign of the Sea of Time, yes, and All Its Tributaries and Headlands too, and, of course, All the Vessels That Sail upon It. Even the thin gold circlet that banded her forehead and held her great heavy masses of black hair off her face was an unnecessary symbol. Her face itself was her crown. The superiority in her expression made Diaphone and even Ula lu la lu seem like a pair of schoolgirls in comparison. It was impossible to imagine anyone disputing the sovereignty of those dark piercing eyes, that green-white skin that seemed as flawless and hard as glazed porcelain, that wide pink-lipped mouth that seemed to smile and frown at the same time. The queen's face was so compelling, so commanding, that Susan could hardly look away, but she had to, for fear that she would be mesmerized by its beauty and cruelty.

When Susan did manage to look away, she saw that the edge of the pool she floated in was ringed by mermaids—nine of them, she counted, including Diaphone, who refused to meet her eyes as Susan turned around inside her glass bubble to look at the fish-girls who were looking at her. With the exception of Diaphone, all of the mermaids were pointing at her and giggling, and though Susan couldn't understand a word of their melodious language, she knew they were making fun of her short hair, her shapeless clothing, her legs. And, angered by the mermaids' mocking, she suddenly said out loud:

"Well, at least I'm not a slave!"

There was a moment of stunned silence on the mermaids' part, and then, louder than ever, they burst into laughter, pointing at the girl in the glass cage who was, to their eyes, very much a slave. *Their* slave. But then the mermaids fell silent again as another voice shook the air of the room—indeed, very nearly shattered the glass of the bubble Susan floated in, which vibrated wildly as the imperious words flowed over it in an angry wave of sound:

"Bring her to me!"

Ula lu la lu had by that time taken a seat on the edge of the rim of the pool, and now she hauled Susan's bubble to her by the gold chain, hand over hand, like a fisherman pulling in a load. The smile on her face was more malicious and gloating than ever, and Susan felt real hot anger in herself for the first time. What did *she* have to smile like that about? She was a liar and a thief and, and *rude* too. But Susan didn't say any of this, because she was so angry she was afraid she would start crying.

Ula lu la lu pressed on the side of the bubble and one of the wedge-shaped pieces of glass slid open with a little hiss of air. The first thing Susan noticed was that the mermaids' cavern *reeked* of dead fish. Early on a Saturday morning last summer, Mr. Oakenfeld had taken Susan to the fish market, and the smell had been similar but somehow fresher—it wasn't particularly pleasant, but it didn't make you not want to eat fish ever again. But this smelled more like trash, like refuse left too long in the bin instead of being thrown out—and indeed, as Susan climbed awkwardly from the wobbling bubble she felt and heard the crunch of brittle bones beneath her foot, and saw that the floor of the cavern was littered with fish skeletons, some still with bits of flesh on them. The sight was disgusting, but it also undercut the royal bearing of the figure reclining on the couch, and, steeling her shoulders, Susan walked to her, the bones snapping like toothpicks beneath her feet with every step.

Queen Octavia said nothing as Susan approached. She half sat, half lay with one of her arms propped on the arm of the couch—the couch's arm was fish shaped, Susan saw now, a miniature replica of the throne downstairs—and the queen's long supple tail stretched over the far edge of the couch and trailed to the floor. As Susan got closer she could see that Queen Octavia's tail was a little different from the other mermaids'. Their tails were fishlike, broad and solid and covered in scales and ending in wide flat iridescent flukes. But Queen Octavia's tail was as smooth as the skin of a shark save for several long dark lines that stretched from the place where the tail joined her body all the way down to the tip, which was fluted in little strands—a bit like a peeled green banana. There was something *weird* about that tail, and Susan was trying to figure out what it was when Queen Octavia shifted slightly, and a thick plait of black hair fell away from her chest. There, hanging from a leathery bit of seaweed, hung the three Keys to the Great Drain: a gold one, a silver one, and the rusty one from Drift House. Queen Octavia touched them slightly, as if making sure Susan saw them, and then she stretched her hand and beckoned her human captive forward with one hooked and waggling finger.

"Come closer."

Her voice had quieted, but lost none of its edge. It was like some great unpleasant recording of a thousand bells ringing cacophonously, but played back at a very low volume. Susan took three more steps on the crackling bone floor.

"You have done well, Susan Oakenfeld."

The compliment took Susan by surprise, and she looked down at her feet.

"We have waited a long time for you, Susan Oakenfeld. I almost thought you wouldn't come in time."

Susan looked up in confusion.

"Your Majesty?"

Queen Octavia was stroking the keys on her breast lightly, as if they were a small pet. The gesture was rhythmic, captivating, and it took an effort on Susan's part to shift her gaze. She saw then that the plush of the couch's cushion was actually threadbare, ripped open in a few places and missing several buttons. The fish-shaped arm of the couch was missing many of its golden scales and one of its ruby eyes was gone too, leaving a gaping empty socket.

"Yes, Susan," Queen Octavia said now. "You are beginning to see, aren't you?"

But Susan didn't see—or didn't understand what she saw.

"Your Majesty? I'm sorry . . . I don't think I *do* see."

"You saw our great beautiful city, didn't you? Ula lu la lu showed it to you from your transport bubble?"

Susan nodded.

"And you noted that it was empty, didn't you? Our great beautiful glowing domes, each more beautifully planted than the last with the rarest and most delectable of phosphorescent sea flowers—every last one of them empty."

Again, Susan nodded. She had noticed that the city was empty. Deserted. *Lonely.*

"And this room: littered with bones. And my couch: tattered, like a peasant's."

Susan swallowed slightly.

"You, you could sweep up the bones," she stammered. "And, and sew the buttons back on your couch."

"SWEEP?!?"

The word shot out of Queen Octavia's mouth like a cannonball, and Susan fell back several steps.

"SEW?!?"

Something Susan didn't quite understand happened then. A piece of Queen Octavia's tail seemed to detach itself and stretch out toward her with a whiplike movement. In a moment it had wrapped itself around her waist, fastening her arms to her side in an inescapable embrace, and a moment later Susan felt herself lifted in the air. She would have cried out, but her stomach was so constricted she couldn't suck in a breath. Queen Octavia held her up in the air for just a moment and then Susan felt the awful squeezing around her waist relax, and she fell a few feet to the floor. The fish bones pricked her hands and knees, but it took a moment before she could catch her breath and stand up.

"Why, you're not a mermaid at all," she said when she had. "You—you're an octopus!"

And it was true. Or at least it was from the waist down.

Queen Octavia's "tail" was actually a set of tentacles, long, sinuous, and whirling all about her now like a nest of hissing serpents. Apparently she was still upset over Susan's remark that she clean up after herself.

"I have not endured *three thousand years* of powerlessness to push a broom over the floor like some charwoman!" she was saying now. "I have not watched the teeming population of my kingdom dwindle to this pathetic remnant of itself only to pull a needle through a piece of tattered sacking like a seamstress! I am Her Most Aqueous Empress Queen Octavia, Undisputed Sovereign of the Sea of Time and All Its Tributaries and Headlands, and All the Vessels That Sail upon It, and I will seal my borders and protect what is left of my court with the dignity and majesty that attaches to that office!"

Susan had a hard time following Queen Octavia's speech because her voice was so loud that it echoed frighteningly around the undersea chamber, and also because she was mesmerized—and terrified—by the flapping flailing tentacles that grew from the lower half of the queen's body. But a few words had made an impression.

"Your Majesty? Seal your borders?"

"For three thousand years the Great Drain has consumed the past. Swallowed it and all that comes from it right out from under me. And now, once and for all, I will stop it. I have all three keys now, and no force on or under or above the earth shall stop me from closing the drain."

Susan had seen the Great Drain up close and been terrified of it. But even as she had been overawed by its great and terrible power she had also sensed the rightness of that power, of its ability to consume anything that came too close. There were just some things like that, and the idea of doing away with them—of closing the Great Drain—seemed to Susan indescribably more horrifying than the drain itself.

Stumbling backward from the writhing tentacles and vicious sharp-toothed mouth of the figure on the couch, Susan heard herself shouting, "You can't! You won't! I won't let you!"

At her words a silence fell over the cave. The queen's limbs settled suddenly, as if she were a toy that had been switched off. Her mouth closed, but her lips remained parted in a smile that made Susan's skin crawl, as if those teeth were already nibbling at her.

And then, from behind her, came a titter. And then another, and another, and soon all of the mermaids were laughing out loud, their rude chortling echoing through the cavern and making it sound as though there weren't nine but nine hundred spectators laughing at her. Susan turned and looked at them—at Diaphone, really, and when she met her former friend's eyes the mermaid's merriment seemed to fade, and she looked away.

Susan turned back to Queen Octavia. The mermaids were still laughing, but instead of making her feel bad their mockery strengthened Susan's resolve. She would show them what an Oakenfeld could do! And, though she couldn't know that Charles had had almost exactly the same thought not so very long before, she took great comfort in being part of such a resourceful, strong family.

She looked down at the floor and found a fish skeleton that was mostly unbroken, and she brought her foot down on it in a loud crunch that cut through the mermaids' laughter. When the cave had gone quiet again, she looked Queen Octavia full in the face and said, "I won't do it. And you can't make me."

Queen Octavia attempted to smile innocently, but the razor's edge of her teeth made that impossible. Her left hand stroked the keys at her chest reflexively.

"Why, whatever do you mean, Susan? What won't you do?"

"I don't know exactly why you brought me down here," Susan said, "but I do know it wasn't just to make me feel sorry for you and your empty city. You need me to do something."

Queen Octavia's tentacles had settled back into the illusion of a tail again. Now she separated one from the bunch. It stretched up and away from her, impossibly long, supple, curling and swaying through the air like a dancing cobra. Slowly, hypnotically, it reached toward Susan, its tip thin and delicate and twitching a little bit, like the tip of an elephant's trunk, and the small white suckers on its underside pulsed ever so slightly, as if eager for something to grab on to, and stick to. With the delicacy and lightness of a spider, the tentacle separated a single strand—just one—from Susan's hair, and snapped it from her scalp. The queen brought the hair to her face, looked at it a moment, sniffed it as if to see if it were edible, then dropped it to the dirty floor.

"Why, Susan," she said, using her tentacle to comb a lock of her own hair from in front of the Keys to the Great Drain. "Whatever would we need *you* to do for *us*?"

Susan felt her stomach churning in fear, but she kept her voice calm.

"You need me to close the drain."

Queen Octavia stroked the keys at her breast, first with a hand, then with a tentacle.

"*We?* Need *you*? To *close*? The *drain*?" She lifted the rusty key to her nose—the key that had come from Drift House—and sniffed it too. "Why, whatever *for*?"

"I—I don't know," Susan said. "Maybe there's some spell or something that keeps evil mermaids like you away."

"But Susan." The tentacle dropped the key back to Queen Octavia's chest. "As you yourself pointed out, I am not a true mermaid."

"You're worse than a mermaid. You're something that shouldn't be."

Suddenly Queen Octavia leaned forward, far enough that the three keys dangled off her neck over the bone-covered floor.

"Soon I will be all there is. I and my nine daughters will

have the Sea of Time all to ourselves, and the Sea of Time will cover all there is, as soon as you close it for me."

Susan stamped her foot again.

"I won't."

"You will," Queen Octavia said. "You will, or your brothers will die."

NINETEEN

Charles Goes It Alone

CHARLES WAS AFRAID HE WOULDN'T be able to lift Murray's carpet, which, rolled up, was three times taller than he was, and thicker around than his shoulders. But it was surprisingly light. Well, not light exactly: it was actually weightless. Charles found he could lay the carpet in the air like a log floating in water and push it down the hall as easily as his mother had once pushed Murray in his stroller.

As he passed the music room, he peeked in and saw Uncle Farley on the couch, the white bandage covering one eye, his round stomach rising and falling evenly in his sleep. Charles was a little afraid to take the carpet outside lest it float into the sky, but when he edged it through the back door the roll showed no sign of rising or lowering. He pushed it out over the back steps, turned it sideways, and then, like a kitten unrolling a tube of paper on the floor, he held onto the tasseled end of the carpet and gave the big roll a push. It unfurled smoothly, evenly, until a dark patterned rectangle—a rich burgundy field shot through with subtle colors and patterns—floated above the half dozen feet of knee-deep water separating Drift House from the dry beach of the Island of the Past.

All he had to do was get on it.

He put his hands on the edge of the carpet and pushed down. It gave slightly but also held firm, a bit like a trampoline, but not so springy. He jumped a little, twisting as he did so, and then he was on the carpet, his bent legs dangling about a foot off the step. He saw then that he'd left the back door open, and, being a responsible child, he jumped off the carpet and closed the door, and then he jumped back on.

He stood up this time. Walking on the carpet was a bit like walking on a bed, although, again, not so bouncy (not that Charles would *ever* bounce on his bed, especially not so hard and high that his head nearly touched the ceiling and the babysitter had to threaten to call his parents on their cell phone—no, not Charles). He walked to the center of the carpet and looked at the funny hills of the Island of the Past, each of them as separate as lumps of cookie dough on a tin before they've been put in an oven and melted into each other.

"Well," he said, walking toward the front of the carpet, "how do I get there?"

And, with a little forward lurch that knocked Charles on his back, the carpet set off in the direction of the hills.

Charles lay on his back for a moment, staring up at the sky and feeling as if he were floating on an impossibly thin raft through which he could feel every ripple of the water beneath him. And then, unable to contain himself, he let out a whoop of exhilaration as loud as any Murray had ever let out in his whooping days.

"*Yeah!*"

He rolled over on his stomach and fixed his glasses on his face, which had been knocked slightly askew by his fall. He was facing the back of the carpet and could see Drift House, one corner edged up on the beach, the rest of the building rocking slightly in the shallow coastal water. It was surprisingly far away—the carpet was skimming over the ground at a rapid clip. From this distance Drift House looked more like a ship than ever. The pair of cannons gleamed on the flat roof, and the rooster-shaped weathervane jutted from the solarium's apex, stiff and unmoving in the windless air. Then, as Susan had done three days ago, Charles realized that the bird was actually a parrot. And then, unlike Susan, he made the connection between the parrot silhouette on the weathervane and the identical shape on the knobs of Uncle Farley's Tombstone radios. They were exactly the same, he realized, and even though he didn't know what the connection meant, he sensed that it was crucial to figuring out how the radios controlled Drift House's movements. Perhaps the weathervane served as an antenna for the radios?

But all that was moving farther and farther away as the carpet continued to glide across the plain, and Charles figured he'd better look where he was going. He rolled over on his back and sat up, facing forward. He saw that the carpet was making more or less for the space between two hills, one about the size of a typical five- or six-story building in New York City, the other much, much larger, but not so high, an enormous warehouse-sized hill. This second hill was closer than the first, and as Charles looked at it he saw what appeared to be openings in its side at regular intervals. He stood up and walked toward the front of the carpet to see them better—he was almost positive they were windows—but before he got there he put his right foot down and the carpet jerked so sharply in that direction that he fell on his side, and this time his glasses fell off his face entirely.

He sat up slowly, reaching for his glasses, which had fallen nearly to the edge of the carpet. The carpet was soft and buoyant, so of course he wasn't hurt, but he was still confused about what had just happened. For some reason he found himself thinking of Sir Isaac Newton's famous line, and as he put his eyeglasses back on he said it out loud: "For every action, there is an equal and opposite reaction." The carpet had veered right, and he had fallen to the left.

But as soon as he thought that, he sat up and looked at the woven pattern beneath him. Now he saw that a long, thin green pole ran up the center of the carpet—no, not a pole, but a slender tree trunk, and at regularly spaced intervals branches spoked out of this trunk, each as thin as a golf club, if not quite as straight. At the end of every branch a symbol hung like a leaf. Toward the tip of the tree were several he recognized—a star, a crescent moon, a sun rising above the horizon, two clutches of arrows, one pointing right and the other pointing left, a slithering snake, a falcon with its wings folded in full dive, while lower down were symbols he didn't recognize interspersed with what looked like leaves covered in swirly writing. And then down at the bottom there were a few more familiar symbols: a bell, a trumpet, a sword, a shield. At the base of the trunk was a clock, and a few notches above that was something Charles thought was a cone, but then he realized it was probably a magician's hat—it had several stars embroidered on it.

Charles looked up then. He saw that the carpet's turn had pointed him straight for the squat hill—yes, they were definitely windows, with stone sills and thick braces supporting the tops and sides, and closed wooden shutters bolted with heavy iron hardware. He looked down at the tree again, and all at once he understood. There were three arrows in each clutch, crossed exactly like the three arrows in the weathervane mounted atop Drift House. Now, crawling forward, he tapped the leftward-pointing arrows and the carpet jerked slightly in that direction. When he pressed it and held it, the carpet turned smoothly to the left; in a moment it was aimed between the two hills again.

Charles could feel his heart beating wildly. He looked at the various symbols dangling from the tree branches. What did they all *do*? The arrows were easy, but what would the bell do? The snake? He knew he shouldn't play around, but keep pressing forward in search of something useful, but at the same time he told himself that it would be prudent to learn all the carpet's features. What if he was set upon by an army of man-eating lions or, or *dwarves* or something—how would he know how to get away? And what if the carpet had special defensive capabilities, like one of James Bond's cars? Maybe the carpet could shoot poison darts at the army of bearded, dirty dwarves who might jump out at him from behind the next hill, or spew an oil slick that would cause them all to fall down when they tried to pursue him? Shouldn't he be prepared?

Yes, he told himself, pushing up his glasses unconsciously. Yes, he should.

So reasoning, he lifted his left hand high in the air like a magician—look at me, nothing up my sleeve—and then slammed it down on the sun.

"AAAHHH!!!"

The carpet shot straight into the air like a bottle rocket. Its momentum was so great that Charles was pressed flat against its surface. It was several long seconds before he thought to pull his hand from the sun, and when he did the carpet stopped abruptly. Unfortunately Charles didn't. The carpet stopped but Charles continued to rise. His body flew above the carpet another five or six feet—enough for him to see that he'd shot hundreds and hundreds of feet up in the air. The carpet was a dark solid rectangle below him, but all around the rectangle was a border of empty sky through which he could see the faint green plain of the Island of the Past far below, and at the edges of the green he could see the blue of the Sea of Time. He was up in the air just long enough to remember another of Newton's laws—bodies in motion tend to stay in motion—and then he fell back onto the carpet's soft surface.

"Oh my good gravy," Charles heard himself say, touching the carpet with shaking hands, half afraid it was going to collapse under him. "Oh my. Oh—oh my."

When he was back on the carpet Charles couldn't see the land beyond its edges. He couldn't see anything but sky. He was higher than the highest hills of the Island of the Past.

He nudged his glasses up his nose and looked down at the tree beneath him. The left and right arrows were on opposite sides of the same branch; the symbol opposite the sun was the moon.

"Rising sun," Charles said out loud. "Rising sun, setting

moon, for every action . . . oh, pretty please," he said, and pressed on the moon.

He pressed lighter than he'd pressed on the sun, and felt the carpet sink beneath him swiftly but not dangerously, like an old, somewhat rickety elevator that lurches a bit when it first starts moving. In a few moments the tips of the tallest hills were visible; soon enough Charles could see the flat plain of the island itself. When he was a few feet above the ground he took his hand off the moon and the carpet came to a smooth stop.

It took Charles another moment or two to get the carpet going again, but soon he realized that if he pressed on the star at the top of the tree the carpet began moving—the longer he pressed, the faster he went. He set himself at a smooth pace aimed across the broad plain of the island, steering here and there around a hill, and then resolved to pay more attention to the landscape around him than the carpet beneath his splayed legs.

The first thing he noticed was that nearly every hill had windows and doors cut into it—some smaller than human-sized, others enormous enough to accommodate a city bus or elephants or creatures Charles couldn't or didn't want to imagine. But all the windows and doors were uniformly closed against the morning light. Indeed, Charles got the sense that inside these hills the most fantastic array of creatures must be sleeping. Ogres maybe, or—or dragons! Charles imagined rounding a hill and coming upon a herd of snow-white unicorns grazing on the rich green grass; or maybe he would see a phoenix, fiery gold and shedding flames like feathers as it soared over the plain! He had met mermaids after all, and, and cannibalistic dwarves, and an older version of his younger brother—why not unicorns and phoenixes?

But he saw nothing. Not unicorns, not phoenixes, not mermaids or dwarves or herds of Murrays of all ages. Just the endless plain of close-cropped thick green grass and those funny dumplinglike hills with their closed doors and windows, here and there a smokeless chimney. Just about the same time Susan was noticing that the mermaids' city was deserted, Charles found himself wondering if the inhabitants of these hollow hills were sleeping, or if they were in fact missing.

The light was full and hot and Charles was feeling a bit thirsty and hungry. He knew that there was no such thing as thirst or hunger on the Sea of Time and that his cravings were just the force of habit, but even so, he wished he'd thought to get the dumbwaiter to whip up a lunch for him before he'd set out—at this rate it would be hours, days before he found anything. He was just starting to think he should turn back for that lunch (and starting to wonder exactly which way was "back" on the roadless green plain) when, for the first time since he'd set out, he saw movement. A large shimmering blob was bouncing in the distance. The blob had an irregular motion—not like a ball bouncing at an even pace, but like something alive, jumping high up in the air and hanging there for a moment, and then falling back down to the ground, only to jump up a moment later.

Charles felt a little lump in his throat at the idea of actually encountering another fantastical creature, but he swallowed it down and pressed on the left arrows to point the carpet in that direction, then pressed on the star to speed up a bit—the bouncing blob was moving at a fairly rapid clip across the land.

As he got closer, Charles could see the blob more clearly—saw that it was, in fact, a giant frog, normal looking in every way save for the fact that it was about the size of a Shetland pony and sported two enormous multicolored wings growing from its back—oh, and two antennas as well. Other than that, it was a totally normal giant frog.

Somewhat nervously, Charles pressed on the star to catch up with it.

"Excuse me," he said when he was just behind it. "Mr., um, Miss"—he really wasn't sure—"MisterMiss Frog? Could I talk to you?"

The frog had great dark purple eyes the size of soup bowls on either side of its head, and it rolled one of these toward Charles languidly. But as soon as it saw him, it stopped short and burbled,

"Dinosaurs and trolley cars!"

Charles had found that if he pressed on the left and right

arrows at the same time—he could use his feet for this—the carpet would slow down, and he slammed on the brakes as it were, then turned to face the winged frog.

"I'm sorry. I didn't mean to startle you."

The frog's antennas twitched excitedly.

"Are you a man?"

Charles wasn't sure how to answer the frog's question.

"Well, er, that is, I'm a boy."

"Yes, yes," said the frog, blinking one of its great purple eyes. "I can see that. But are you a human boy?"

"Er, well, yes," Charles said, although he suddenly found himself wondering if he really was. It was a very strange question to be asked. When the frog spoke, Charles seemed to hear two voices—the one he understood, speaking English, and another behind it with a distinctly froglike sound that gave the creature's speech its burbling quality. Susan had described a similar effect when she first spoke to Diaphone, though Charles himself had not heard it, and he was about to ask the frog about it when the latter croaked:

"Unbelievable! A human boy, on the Island of the Past! Well, not *on* it," the frog added, half to itself, "but *over* it." The frog looked him up and down. Its eyes were so big that Charles had the idea it could look right through him. "No horns?" the frog said now. "No hooves? Maybe a little corkscrew tail? I can see that you don't have any wings." And the creature's great wings, which arched over its body like a brightly colored parachute, rippled in a proud gesture.

"Um, no," Charles said, a little defensively. "I'm just a regular boy."

"Hmmm," the winged frog said, and as it continued to speak the undertone to its voice faded away. "I am afraid I shall have to ask you to prove it."

"Prove that I'm a boy?" Charles said. "But—but I am!"

"In my experience, which is admittedly quite limited, boys are known to carry boylike things on their person."

"Boylike things?" Charles repeated, completely confused. But he dug in his pockets nonetheless, and was surprised when his fingers touched something small and hard. It was the Canadian quarter he'd found under the rug two days before. He held it out before him. "I, um, I only have a quarter. A Canadian quarter, actually." He had no idea if that could be considered boylike or not.

The frog's big nostrils quivered. "It's the Sea of Time, boy, if boy you are. I'm not hungry."

"It's not food," Charles said. "It's money."

"Money? Well, why didn't you say so?" In a movement quicker than Charles's eye could follow, the frog's tongue shot from its mouth and caught the quarter on its sticky tip and pulled it into the gaping, moist cavern—and just as quickly spat the money back out. One of its great eyes stared at the tiny disk on the grass, while the other rolled toward Charles. Slowly, wonderingly, the eye looking at Charles blinked. "Whirling drains and whooping cranes," the creature said in a breathy, astonished voice. "Has it happened then? Are there no more men? Are you the last?"

"I—what?" Charles stammered. "No, I'm not the last man. There are millions more. Billions." But he was starting to understand something about this strange place. "Are you the last of, well, whatever you are?"

"Ah, well," the frog said, and Charles thought he detected a sad note enter the creature's funny echoing voice. "Of course you would ask that question. Why would a human boy have ever heard of a butterfrog?"

"A butter—"

"—frog," the butterfrog said. "My name is Spiralla Plop."

"Spiralla—"

"Plop," the butterfrog said. "But you can just think of me as the butterfrog, since I am, indeed, the last of my kind."

"Oh. I'm so sorry."

Spiralla Plop examined Charles a moment, then said, "It is apparent that you are new here, or you would have known that already."

"Yes," Charles said, and he pointed back in the general direction he'd come from—or thought he'd come from. It was a little hard to tell with all those unmarked hills. "We, that is, my brother and sister and uncle and I, and President Wilson, and a dwarf from the Time Pirates, we came in Drift House. Or, I mean, the mermaids towed us here, and then they took my sister with them."

"The mermaids?" Spiralla Plop said. "The Time Pirates? Drift House! This is getting curiouser and curiouser!"

"If you please," Charles said, "I was hoping someone could help me get Susan back."

"What is a Susan?" Spiralla Plop said.

"It's a sister. My sister. The mermaids have taken her."

"Ah, yes, I see." One of Spiralla Plop's eyes rolled left, one right, as though she could in fact see everything around her.

"Can you help me?" Charles said.

"Me? I am just a butterfrog. The last of the butterfrogs. We were a peaceful race, and as the last of them I intend to remain a peaceful person."

"Yes, you already—"

"The butterfrogs hail from a certain moment in history," Spiralla Plop spoke smoothly over Charles. "The time of the sphinx, half lion, half man, and the hippocampus, half horse, half fish, and, yes, the mermaids, also half fish, this time joined to the body of a woman." (Spiralla Plop obviously hadn't seen Queen Octavia, or she would have said half octopus, half virago.) "We were creatures born of simpler imaginations, when the world wanted to fuse two principles—the ferocity of the lion, for example, coupled with the riddling speech of men—into a single form. The speed of the horse was joined with the dexterity of a warrior to make the centaur, or the wings of a quetzal bird affixed to the body of an anaconda to make the dreaded, revered feathered serpent. Or, in my case, the delicate beauty of a butterfly coupled with the more quotidian aspect of a frog."

"Butterfrog," Charles said, understanding now.

"So I am," Spiralla Plop said. "The last of the butterfrogs." And she sighed heavily.

"But what happened to the others?"

Spiralla Plop sighed. "Down the drain, of course."

"The Great Drain?"

"Is there another?" Spiralla Plop didn't wait for Charles to answer. "Yes, they all vanished down the drain. History evolved into something more subtle, and we did not, and one by one the drain found ways of tricking us too close to its vortex—as sometime water creatures it wasn't too difficult. The second-to-last of my kind, my dear cousin Heavenly Splat, disappeared more than two thousand years ago as time is reckoned in places where the clock still ticks. Shortly after that, the Time Pirates came for me and brought me here, so that all trace of the butterfrogs shouldn't vanish from the earth."

"So you live here?" Charles said. "All alone?"

"Alone of the butterfrogs, yes. But there are other creatures from the past here. There is a lovely unicorn I sometimes come across, and the gryphon occasionally blackens the sky with its clawed shadow. Sometimes I see the kangalope bouncing along on his little legs and, let me see, yes, the ele-phish has his own lake just a few miles from here. Of course there are many, many others—a special section is reserved just for the dinosaurs, lest they trample the smaller creatures underfoot—but most of the inhabitants of the island tend to keep to themselves." And here one of Spiralla Plop's antenna waved at the nearest hills, at the barricaded windows and door. "The phoenix once told me that time is losing its belief in life, and that one day the world will be left bare and smooth, just open empty expanses of water, land, and sky. But the great sphinx told me the phoenix had it backward—that life is losing its belief in time, and soon enough all creatures will be identical, and exist in a never-ending present." Spiralla Plop did something with her shoulders that Charles thought might have been the butterfrog equivalent of a shrug. "I myself do not see the difference between the two visions, really. And as I said, the time of the butterfrogs and all the other conjoined creatures is long past, so it does not concern me."

"But is there really only one of each kind of creature here? Isn't that"—Charles had difficulty thinking of an adequate word—"*lonely*?"

"Yes, yes. Only one," Spiralla Plop replied, pointedly

ignoring the second half of Charles's question. "Which is why I was so surprised to see you."

"Me? Why?"

"Because there's already another man here."

"Another man! But where? Who?"

"Who is an easy question," the butterfrog said. "Where is somewhat more difficult." The butterfrog's antennas twitched as if she were trying to receive a signal from a satellite. "Where . . . where . . ."

"Well, at least tell me *who*," Charles said, consumed with impatience. "Who is the other man?"

"Why, I thought you'd have known. It's that famous sailor, Peter, Pater, Prior, Prayer . . . Prayer Something?"

"Pierre Marin?" Charles nearly screamed.

"That's the one," Spiralla Plop said, and her antennas twitched one more time, then pointed in a straight line off to her left. "Pierre Marin. He lives just past the fourteenth hill to the left. Look for the horse," Spiralla Plop added. "You won't be able to miss it."

TWENTY

The Proposition (part 2)

After her initial audience with Queen Octavia, the mermaids put Susan back in the gold-veined air bubble and Ula lu la lu pulled her to another of their many abandoned underwater hills. Like the palace, this hill had an air-filled chamber inside, but it was much larger, and much, much plainer. Susan thought it might have been a loading dock, judging by the remains of dozens of packing crates and loose piles of straw that littered the vast shadowy cavern. Queen Octavia traveled to the loading dock by a separate route and was waiting for them when they arrived. Ula lu la lu herself sat at the edge of the water, but Queen Octavia used her long powerful tentacles to walk on dry ground. She towered above Susan like a person walking on stilts, and moved with slow slightly jerky steps about the littered room. The three Keys of the Great Drain hung around her neck, and as she walked she toyed with them absently with a hand, or the tip of one of her tentacles.

"Once we had a thousand slaves to work this warehouse. The tribute of the world came to us, and was unloaded here and sent throughout our great city. But for ages almost beyond reckoning this room has been empty, the slaves sucked away by the Great Drain, the mermaids who kept them in line lured off by the same force. Though we are the strongest swimmers in the sea, yet not even we can fight the current if we get too close. Our human aspect makes us susceptible to the curse of curiosity, while our fishlike part leaves us innocent of suspicion. It took an eternity, but the drain is as patient as all forces of nature, and one by one it carted my subjects away. You saw my court. Nine of us—all that's left of a kingdom that stretched from Tenochtitlán to Rome to Shanghai, from the utopia of Atlantis a thousand years before Socrates to the Viking settlements in Greenland a thousand years before your time."

Now Queen Octavia laid one of her tentacles atop a crate and pressed it flat. Susan could almost see the suckers get their grip, and with a terrifyingly light motion Queen Octavia pulled her tentacle away and the thick wood of the box splintered and ripped like wet cardboard. The queen sifted through straw and pulled out a gleaming gold plate, as big as a shield, heavily embossed with the shapes of fruit: pineapples, bananas, mangos.

Queen Octavia smiled, and for a moment her haughty cruelty was softened by nostalgia.

"I mentioned Tenochtitlán. This is from the Royal Court of one of the greatest of the Aztec rulers, Tenoch himself. His people would journey to the edge of the sea several hundred miles from their inland capital and leave these offerings for us, so great was their fear of our power. But it has sat here unused, the memory of its giving erased as, soon enough, will be the memory of its giver. And do you know why, Susan?"

Susan, who was busy looking about the room for some object she could use to brain Queen Octavia and Ula lu la lu, hardly heard the question. What she would do after that was a little vague, since she was still several hundred feet underwater.

"Huh?"

"Time, Susan! Time is the greatest of all destroyers. Not war, not disease, but time. Against it the largest army and the tiniest microbes are less than a single match dropped into Mount Vesuvius—invisible, useless. Time has wiped away the greatest of cultures and peoples from the world with no more care for their accomplishments than a wave has for the beach it washes over endlessly, wiping away all its distinguishing marks and features."

"But, but *memory*," Susan said, distracted despite herself

from the search for a weapon. "I mean, our minds. Things can live on there."

"Spoken so like a human, you shortest-lived of all sentient beings. As if your memory will not vanish with that pathetic four-limbed tree you call a body." And Queen Octavia lashed at one of Susan's legs with the tip of a tentacle, which stung like a whip.

The spot where Queen Octavia had flicked her thigh burned like a bee sting, but Susan refused to rub it. Suddenly she remembered what Charles had told Diaphone on the poop deck. About reading, and books, and the way people keep track of things.

"But we pass our memories on," she insisted. "I mean, from one generation to the next. Parents teach their children. They write things down, tell them stories—"

"Stories!" Queen Octavia's laugh was short, but the echoing cave stretched it out for her. "What are stories next to gold and diamonds, to swords and cities and slaves? Bits of empty air measured against the greatest monuments." She grabbed up something from another crate now, a great purple cloak edged with white fur and feathers, and draped it over her shoulders. "What is a story compared with this? Can it keep you as warm? Convey the royal magnificence of your character? Conceal a weapon in its folds which you can pull out at just the right moment and slip between your enemy's ribs?"

"But that cape," Susan said. "It can't tell you how to live. Only a story can do that."

Queen Octavia slitted her eyes. "Your life *is* precious to you, is it not?"

Susan was taken aback by the question. "Of course it is."

"Yours—and your family's?"

Susan blinked back a sudden burning in her eyes. She would *not* cry in front of this octo-woman. She was trying to think of something to say when she saw Queen Octavia's eyes shift slightly, looking over her shoulder, and a moment later she heard a light gurgling sound and a long soft hiss of air, like a great patient sigh. A mist of warm, pleasantly scented water settled over her shoulders like a cloak.

Trembling slightly, Susan turned toward the pool of water. For a moment she thought it had disappeared—had been stoppered up by a gigantic rock. But then a great smooth black circle in the rock blinked once. It was an eye as big as her body, set in a dark gray-blue whale as big as a yacht.

A line along the bottom edge of the whale's body quivered now, and even as Susan realized it was the line of the whale's lips an ancient deep voice that she felt as much as heard rumbled into the room.

"Hello, Susan Oakenfeld."

Though there was no trace of threat in the whale's voice, and though it came on the most pleasantly perfumed breath she'd ever smelled, it was still the most terrifying thing Susan had ever experienced. She wanted to run to the farthest reaches of the cave, but knew that Queen Octavia was right behind her with those awful squeezing sucking tentacles. The only thing that tempered Susan's misery was the sight of Ula lu la lu off to one side of the whale. She'd been forced to roll out of the water to avoid getting crushed, and now she lay pouting on the dirty floor of the warehouse looking a bit like an oversized fly-fishing bait.

"H-hello," Susan said, and then, remembering her manners, she said, "Wh-what's your name?"

"My name is Frejo," the whale said, and the vibration from his sweet-smelling voice was so heavy it made her knees knock. She didn't know if the whale could see her this close—she was as close to his eye as a fly buzzing an inch in front of your face is to a human—but whether he could see her or not, Frejo seemed to sense her discomfort. "Don't be afraid of me, Susan. I haven't come to harm you."

There was a rustle behind Susan, a slither of tentacles.

"I have summoned Frejo—"

A hiss escaped the whale's lips, a current so strong that Susan's clothes whipped about her body.

"Mind your speech, sea woman. Not even at the height of your power could you have presumed to summon me, and your mermaid armies are a thing of memory long since faded."

For the first time Susan saw Queen Octavia's composure

break. A look of pure helpless malevolence flashed across her face. But then, a moment later, she smiled it away with a mouth that could have flayed living flesh from bone.

"Frejo answered my call," she said now. "He has come to carry you to the bottom of the vortex that empties into the Great Drain."

"I have *come,*" Frejo said in his hissing, rumbling voice, "to take my leave of this world, and I have accepted to carry a passenger with me, rather than have you toss her to the waves' mercy in that silly glass egg of yours."

It took Susan a moment for her brain to make sense of this exchange, and when she had she said, "Carry me? Passenger?"

Queen Octavia nodded. "In his mouth. No doubt you have heard *stories*"—Queen Octavia nearly spat the word—"about such a thing. Jonah. Pinocchio. You will be safer in Frejo's care than you would be in our transport vessel, and be able to breathe until you pass through the bottom of the vortex and can close the drain."

Susan glanced back at the whale, who sat silently, his dark hide beginning to steam slightly in the cavern's air.

"But I don't understand," Susan said. "Why don't *you* go? I mean, at first we thought you needed me because of my legs. But you—I mean, you have eight legs. Why don't you do it yourself?"

"But, my dear, that would defeat the purpose, wouldn't it?"

"I don't understand," Susan said.

"My goal is to save my kingdom. Not to lose it in the process."

"L-lose it?" Susan stammered, for she sensed what Queen Octavia was going to say next.

"There is no return from the Great Drain, Susan Oakenfeld. Not even Frejo here could swim against the full storm of the vortex's current. And it is said that the locks these keys close lie on the opposite side of the drain, in the world that comes after this one."

"After?"

"In the same way that one card comes after another in a deck, there are worlds that come after each other in the great sequence of things. And once the door between this world and the next is closed, whatever is on that side of it will remain on that side forever."

"But, but." Susan waited for Queen Octavia to interrupt her, but the octopus woman said nothing. She didn't have to: Susan knew that if she wanted Charles and Murray and Uncle Farley to live, she had no choice but to do her bidding. Then another thought occurred to her. "Queen Octavia, what happens if I *do* close the drain? To the rest of the world, I mean? Will there be a . . . a flood or something? Will everyone drown?"

"Foolish child," the queen responded. "It amazes me how ignorant the people of your era have grown about time. There will be no flood. No one will drown. Indeed, no one will ever die again. Time will never again destroy anything or anyone, ever." Susan's face must have betrayed her lack of belief, because the queen nodded at the whale. "If you don't believe me, ask him."

Susan turned back to the whale. She had no reason to trust him, yet she did. She could not imagine a creature of such obvious and enormous power would ever need to lie.

"Queen Octavia does not deceive you, Susan Oakenfeld," the whale said now. "Nor does she tell you the whole truth. For, though no one will die as a consequence of your actions, nor will anyone ever be born again, or grow, or change. Life in your world will flow by in the same continual present that exists on the Sea of Time." Something in Frejo's voice did not make this sound like the happy, peaceful outcome it seemed like, and Susan was about to ask him about it when the queen spoke behind her.

"Enough," she said. "You have my word, and the whale's, that no one will be killed by the closing of the drain. But if you do not close it, then you and all the inhabitants of Drift House *will* die. Make your decision now."

Susan looked at the queen for a moment. There was no doubting her either, not on this point. And, without speaking, she held out her hand. Her lips trembled like Frejo's with the desire to just, just *blow* this creature away, but her hand was still and calm in front of her.

With a jerk, Queen Octavia snapped the seaweed garland

from her neck and gave the keys to Susan. She used one of her human hands rather than a tentacle, which relieved Susan slightly. She hated having those snakelike things come close to her body.

"A kingdom founded on such injustice will never thrive," Susan said now. She had worked the sentence out in her head and was glad it rolled smoothly from her lips. "Though I close the Great Drain, still will your reign come to an end one day."

Queen Octavia leaned over until her face was so close to Susan's that she could smell the fish on the monarch's breath. She stared into Susan's eyes for a long time, then nodded.

"Yes," she said, and Susan wasn't sure if she was speaking aloud or to herself. "You are a remarkable girl. I almost wish things could have been different. There is much I could have taught you."

Susan steeled herself, lest she undo her former speech with a childish outburst. She pivoted on her heel, the three keys clutched in her hand.

"I'm ready, Frejo."

"My tongue will be softer than your own bed, Susan Oakenfeld, and my breath will fill your lungs with its sweetness. I will make your journey pleasant."

"Thank you, Frejo," Susan said, grateful for the whale's kindness.

The great mouth opened, just a fraction of its height. Inside it was all wet pinkness stretching back into black.

Susan took a step toward it, then felt the awful sticky touch of one of Queen Octavia's tentacles and heard the queen's voice hiss into her ear.

"The black, and then the silver, and then the gold. Turn them in that order, Susan, or your brothers die." She squeezed slightly with those awful suckers. "If you fail, I will crush their bones myself."

Susan shrugged the queen's tentacle off and climbed over Frejo's lip. It was soft and rubbery and a little bit slippery, and she slid the last few feet into the warm mouth, and then it sealed closed behind her, swallowing her into complete perfumed darkness.

TWENTY-ONE

Pierre Marin
Captain Emeritus of the Time Pirates
Human Representative to the Island of the Past
and Its Resident Historian

PIERRE MARIN'S HOUSE WAS A well-kept if plain-looking shack. It had a roof that slanted only one way, so that it looked as though half of it were missing, and it stood upon the open grassy plain like something grown from a seed dropped by a high-flying bird. Charles would have had a hard time believing the itty-bitty thing held the former lord of Drift House had it not been for the crest visible above the front door, the single-masted Viking ship and the red and green parrot floating on its sail. The parrot reminded Charles he wanted to investigate the connection between the bird on Uncle Farley's Tombstone radio and the identically shaped weathervane on the solarium's apex, but for the moment such concerns pressed less heavily on his mind than the "big horse" Spiralla Plop had mentioned.

The horse straddled the house as if the shack were its offspring. Its four legs were thick as redwood trunks or the chimneys of a factory, and at the very top of their immense length they supported a curved belly as big around as the hull of the *Chronos*, and a little longer as well—Charles thought the entirety of the Time Pirates' ship could have fit inside the gigantic torso, if you took the mast off. The horse's head stretched so high its ears would have pricked the clouds if the sky above the Sea of Time had ever been anything but clear, and as Charles stared up at the one dark eye and flared nostril he saw a black shape detach itself from the pale wooden tresses of the horse's mane and glide out over the island. From this distance Charles couldn't make out any features of the winged creature (if meeting Spiralla Plop had taught Charles anything, it was not to assume that everything that had wings was a bird). He half hoped it was the gryphon, although Spiralla Plop had made it sound a fearsome thing—if it *was* the gryphon, Charles hoped it wouldn't come too close. But in fact it appeared pretty birdlike. Its great rectilinear wingspread looked like a wooden plank soaring through the sky, and only the slight tilting from left to right as it manipulated the currents gave any hint that it was alive.

"Be-yootiful, no?" a voice said from much closer to the ground, and even as he heard the words Charles detected a whiff of pipe smoke. "Eet ees ze famous condor of Californ-ee-ya."

Charles looked down. The man in the doorway of the shack was unrecognizable as Pierre Marin, at least as Charles had seen him in the picture in the front hall of Drift House. He was older, for one thing, and wigless, and his few remaining strands of gray hair were pulled back in a little ponytail that left large tracts of shiny skull exposed. He wore a long-sleeved white shirt and gold-piped vest that stretched across his soft belly, tan calfskin breeches buttoned at the knee, and long white socks. Or sock rather, for there was only one leg of course, and one black shoe with one shiny silver buckle. The other ankle was a stick of polished black wood, and Pierre Marin leaned his weight on this peg as if it were the stronger of his two limbs, and when he smiled at Charles he gripped his pipe between even rows of square white teeth.

"Welcome, Charles Oakenfeld. Eet ees good to finally meet you."

Charles couldn't tell if Pierre Marin's funny pronunciation was a French accent, or if it was caused by the fact that he had to speak through clenched teeth to hold his pipe in his mouth.

"Good afternoon, Monsieur Marin." He looked up at the sky, at the distant dark shape of the condor. "I thought the California condor was extinct," he said, looking back at Pierre Marin.

Pierre Marin took his pipe from his mouth. "It is in your world," he said, his accent gone now, although the quietest trace of something else lingered behind his voice. "But here there will always be the one, a living memory, lest our own species live so long it forgets what it destroyed."

Charles nodded. There was something calming in Pierre Marin's manner, despite his slightly obscure way of greeting. Charles felt as if he were speaking for the first time with someone who understood him.

"Is this the Trojan Horse?" he said.

Pierre Marin nodded.

"Why do you keep it here? It was never alive, was it?"

"Not technically. But the people who made it were, and the people it killed."

Charles nodded. If he squinted he could see the trapdoor in the horse's belly through which the Greek soldiers must have come out to burn Troy to the ground.

Pierre Marin followed Charles's gaze. "You will note the horse is female, which should have tipped the Trojans off that something gestated in its belly. Most equine statues were of stallions."

Charles looked at the horse a moment longer, then down at Pierre Marin.

"I've come to ask for your help."

For the first time, Pierre Marin's smile faded.

"You want me to save Susan."

Charles laughed slightly. "Is there anything I need to say out loud? Or do you already know it all?"

Pierre Marin returned Charles's dry laugh.

"Until very recently, if you had asked me that question I would have responded by saying no one truly knows the future. The past is the only thing that can be known, and the more we know about it the more accurately we can make predictions about what is going to happen—the drawing room in Drift House is a fine example of just how accurate those predictions can be. But then I met your remarkable younger brother. Now I would say that I don't know the future, but I do know what Mario told me."

It took Charles a moment to remember that Mario was the name the Time Pirates had given Murray. "But Mur—Mario thought Susan would die. So how did you know I would ask for your help in saving her?"

"Because Mario thought you were dead as well," Pierre Marin said. "And, seeing that you are very much alive, it is logical to conclude that Mario's other information had been transmitted falsely as well. *Les sirênes* are great deceivers."

At Pierre Marin's strange pronouncement, Murray's last words flashed through his mind: "you don't know the half of it," Murray had said, and now Charles realized *he* was the other half. He was surprised: when Pierre Marin had said that Charles was supposed to be dead, he hadn't felt frightened.

There was something so matter-of-fact about the way this man spoke, something that rose above emotion, something scientific and clear that made death seem like just another datum. Of course, Charles wasn't dead, so that helped. He wasn't sure of many things, but he was sure of that.

"You speak English," Charles said then.

"I speak some English, yes. The accursed British brought it to Quebec with their armies. But I am not speaking English now."

"But I can understand you perfectly. And I don't speak French."

"Has there been any creature you *haven't* understood on the Sea of Time?" Pierre Marin said. "Mermaid, pirate, or butterfrog? Did it never strike you as odd they all spoke 'English'?"

For the first time since the conversation had begun, Charles felt a twinge of emotion, and it was embarrassment. It did seem an oversight on his part, and to be called out on it by so great a personage—a man whom he wanted to impress—caused him to blush so hotly his glasses fogged.

"Like everything else on the Sea of Time," Pierre Marin said as Charles rubbed at his glasses with the tail of his shirt, "language exists apart from history. What is spoken here is the pure mother tongue of the species, the language given us before the heretics of Babel attempted to steal those of heaven's secrets that had not been bequeathed to men. It is a language that does not allow lying or confusion, but expresses clearly the will of the speaker inasmuch as it is known to him, and can be heard by the person to whom he is speaking."

Charles remembered the strange sound he had heard in Spiralla Plop's voice when they had first begun conversing, and now he realized it must have been the butterfrog's own language, echoing faintly behind what Pierre Marin called the mother tongue. And there was that faint whisper behind Pierre Marin's voice—not of a French accent, but actual French! Then something occurred to him. He quickly recapped Susan's escapade with the Time Pirates—the dressing up, the bird on her shoulder speaking French—and then he asked Pierre Marin:

"So are you saying Susan didn't have to pretend to speak French?"

Pierre Marin laughed for a long time before he could answer Charles, his face reddening well into his thinning hair.

"Oh, Abu!" he said (Charles guessed this must be the name of one of the pirates, probably the captain). "To fall for such a trick! I'm afraid he's grown too trusting after all these years. I must say, you come from a very clever family, Charles Oakenfeld. But no: Susan and President Wilson needn't have gone to such trouble. Indeed, it probably confused poor Abu to hear me speaking French rather than Moorish, which is his own language." Before Charles could ask Pierre Marin to explain, the former master of Drift House continued, "You see, the language of the Sea of Time works not in words, but in intention. It expresses the will of the speaker, rather than simply his thoughts. So if the speaker desired to say something *in French*, his words would indeed come out in that language. This is why the mermaids are able to speak their own tongue, though it costs them a great effort of concentration. Here, hold on." Pierre Marin knitted his brows as if he were thinking very hard, and then he said, "Am I made myself clearly?"

At first Charles didn't understand. There had been a slightly hollow sound to Pierre Marin's voice, and Charles realized it was because there was no echo behind it. He blinked in astonishment. "Were you just speaking *English*?"

Pierre Marin's brow relaxed, and he chuckled. "I hope so. Though, judging from your expression, my effort was only half successful."

To a scientific mind like Charles's, the last few minutes had been the most exciting in his life. His three days on the Sea of Time had been full of mysteries and puzzles—but here was a man who could analyze, dissect, and explain those mysteries as easily as his mother did the Monday crossword! "So when the mermaids lied," Charles said now, eagerly, "that was part of the same thing? Their will? Overcoming something natural?"

"What the mermaids have done goes beyond these paltry experiments," Pierre Marin said vehemently. "They have attempted to separate themselves from time's singular flow, and in so doing have brought falsehood to temporality itself. They have made themselves abomination." He shuddered, as if shaking the feeling off; when he resumed speaking, his voice was calm again. "I confess, I do not understand it fully. Perhaps truth requires falseness as beauty requires plainness against which to define itself." He put his pipe in his mouth so he could take his glasses off and polish them on the flounce of his cuff—Charles couldn't help but notice it was the exact same gesture he had just performed. "Time ees eternal," Marin said between clenched teeth, and then he put his glasses back on and took his pipe back in his hand. "The Sea of Time encompasses all of it, past, present, future. Eventually I will understand even the mermaids' secrets."

At this mention of time, Charles looked up. There was no sun in the sky to measure the advance of day, of course, but he still had the acute sense that the moment for saving Susan was slipping away. He got up and walked to the edge of the carpet to talk to Pierre Marin more closely. But when he jumped to the ground, something funny happened.

He bounced.

Or, not bounced really, but was thrown back up in the air. Not by Pierre Marin or anything Charles could see. Rather, it was like some kind of magnetic repulsion had forced Charles away. A moment later he was lying on his back on the flying carpet and looking up at the underside of the Trojan Horse's belly far above him, the sound of Pierre Marin's thin laughter reaching his ears with an equally faint tang of smoke.

"Come now, Charles. You're not thinking, or you would

have realized the Island of the Past permits no duplicates on its soil. This isn't Noah's ark—no pairs allowed here to repopulate the world. Only one representative of each species living in hollow hills, to commemorate its time on earth."

Charles sat up, a little dizzy. He pushed his glasses up his nose. Pierre Marin had come over to the carpet and was leaning on it with his elbows, his pipe smoking slightly in one hand, and as Charles looked into the older man's detached bespectacled face he felt his heart sink in his chest. For Pierre Marin's visage had an otherworldly quality to it—something so far from the normal sphere of wants and needs or hopes and fears that he looked almost like a statue, or a robot. He wasn't someone who *did* things, but who wanted to *see* what would be done. All at once Charles realized that the curiosity of the scientific temperament is not the same thing as the curiosity of the explorer: the latter must take himself out into the world to make his discoveries, while the former retreats to a place where the world can come to him. But Charles remembered that Pierre Marin was also the man who had overcome his own seasickness to reach new lands, new worlds, and it was this man he appealed to now. He crawled forward on his hands and knees, being careful not to press on any of the carpet's controls, and said,

"Please, sir. I understand that you'll want to say no to my request, and I know that you're here to observe what happens and not play a part in it. But Susan is my sister, and it's not her fault the mermaids captured her. And, and I love her, and I don't want her to die. Please, sir," Charles repeated, "will you help me save her?"

Behind his gold-rimmed spectacles Pierre Marin's eyes were as pale as the cloudless sky, but Charles could see thoughts racing through them as if a huge wind blew inside the old Frenchman's brain. For a moment his face seemed if anything even more distant and detached from what was going on around him. But there was an undercurrent of animation as well, and Charles knew he was reliving past battles—conquests, defeats, discoveries, and escapes Charles could only fantasize about. A small proud smile flickered on his mouth and his tongue darted out as if tasting the sea spray, and for a moment Charles's hopes surged.

And then the moment passed. Pierre Marin stuck his pipe between his teeth and said, "It ees not possible."

"Nothing is impossible," Charles insisted.

"No, Charles," Pierre Marin replied. "Some zings are eem-possible. Some are merely eemprobable." He took his pipe out of his mouth. "And some things I just won't do. I left the Time Pirates long ago, and when I did I made a pledge not to interfere in the course of history any longer, but only to observe its doings. Should your sister die, that would be regrettable, but you should not think of her as permanently removed from you. Now that you have access to the Sea of Time, you can travel back to the past and visit her anytime you want."

"You mean the past won't be destroyed as well?"

Pierre Marin's eyes narrowed. "Tell me what you mean, boy," he said, and Charles heard a trace of authority that hadn't been present before.

"The mermaids have all three Keys to the Great Drain."

"What? Impos—" Pierre Marin stopped himself. "The key from Drift House?"

"Yes. Murray, I mean, Mario found it, and I—" Now Charles broke off, but then honesty compelled him to continue. "We were misbehaving. We didn't know what it was, but we knew we shouldn't have been playing with it."

"And Quoin? Surely Abu didn't surrender his?"

There was an edge to Pierre Marin's voice now. An indignation, and an excitement too, and Charles wanted to make sure he built on it.

"They tricked him. Tricked us all really. We thought they had taken a mermaid prisoner, but actually she'd let herself be captured, and then Susan went on board and helped her get the key. I'm afraid we played into their hands."

"Pah!" Pierre Marin said, using Cosmo's favorite epithet. "Fooled by a parrot and a mermaid and a little girl, all in the same day! Quoin is a good man, but not the most imaginative, or farsighted. Who captains Drift House now?" he said unexpectedly, and his freshly sharpened eyes bored into Charles.

"Um, captains?" Charles stuttered. "I guess that would be Uncle Farley."

"'*Uncle Farley*'?" Pierre Marin repeated incredulously. "My Drift House is being commanded by an 'Uncle Farley'? Oh, *mon Dieu*, how has this come to pass?" He smacked the hand holding his pipe so violently against his other hand that ashes spilled all over the edge of the flying carpet. Charles could sense the ancient sailor wavering between resignation and resolution, and at length the onetime leader of the Time Pirates began speaking again.

"I have studied your world for a long while," he said. "Secondhand, of course, by means of the new arrivals on the island. The tales these creatures tell are terrifying—of being hunted down by the thousands, the hundreds of thousands, and then being picked off one by one until no one is left save you. Or watching your world disappear, forests burned off, waters filled with chemicals that kill your offspring before they're even born. Terrifying and heartbreaking—if also, alas, nothing new. Extinction is as old as the planet, and who knows what countless species my pirates will never even hear of, let alone track down and document, especially with the amount of time they spend in your era, where species are disappearing at an incredibly rapid clip. The information I receive from these latest arrivals is troubling in a different way, though, as if they indicated some larger problem I cannot yet make out. It is as if the problem were with time itself."

Charles listened to Pierre Marin speak with a mixture of fascination and impatience. With each word Susan was falling further and further into peril. But he could also sense that Marin was not merely talking, but talking himself into helping, and so, as calmly as he could, he said, "How can there be a problem with time?"

"Mmmm, yes," Pierre Marin said, half to himself. "It must seem strange to a boy of your world, in which science has done so much to make time seem like nothing but a more complex aspect of the physical world. But time—real time, Charles, and not the time measured by watches and calendars—real time is less a movement of atoms than of imagination. To be alive is to want, to desire, and no creature desires more than man. It is this collective urge that fuels the temporal flow. Think of it this way: one person in a large room doesn't affect the temperature, but a hundred will overheat it—all the more so if those hundred are large adults, rather than small children. As the number of human beings in the world has grown into the billions, their collective imagination has quickened time's pace to a breakneck speed. And I've heard stories that many of the streams and tributaries of the Sea of Time have become blocked so one can no longer access them. It is as if time has narrowed like a river on its way to a waterfall. I fear, Charles, that there is some terrible tragedy man is rushing toward at a speed too great to stop. And in this regard I wonder if Queen Octavia's plan isn't the right one. To close the drain, stop time forever, save what's here. No change, no new gains, but no loss either—save for your sister, of course. But perhaps that is a justifiable sacrifice."

Charles's jaw fell open. He had thought—had hoped, prayed—that Pierre Marin was talking himself into saving Susan, but instead he wanted to sacrifice her. It was—it—Charles didn't know what it was. It was too much. He snapped his jaw closed with an audible click.

At the sound, Pierre Marin glanced sharply at him, as if he'd forgotten Charles was there. A moment of electric contact passed between the two bespectacled faces, and then it was Pierre Marin's jaw that fell open, as if in wonder. He nodded his head.

"Yes," he said to himself. "Yes, I think I understand." He looked up at Charles. "You see, time has a will of its own. I don't know, perhaps it too is an aspect of our will, but it is one that runs counter to our conscious desires. For every action, an opposite reaction, yes? At any rate, this deeper current, if you will, is much subtler than what flows on the surface, much more mysterious and hard to chart. What I'm trying to say is that I suspect you and Susan and Mario have been brought here not so much by the mermaids or the pirates, but by time itself. Such a remarkable trio! A Returner, with the ability to travel to the future and come back! A girl who has managed to gather all three drain keys into her hand! And you, Charles!"

"Me?" Charles said warily. "I'm just the middle child."

"Pah," Pierre Marin said again, but softly, chiding. "You were able to find me, weren't you? Sometimes it is difficult for a man of my temperament to listen to his heart rather than his head. But you have reminded me that sometimes learning is an irrational act of faith as well as a more dogmatic process of investigation." Charles blushed, even as Pierre Marin rushed on. "I have sensed this problem in time's flow for a long time, and now I think—I believe, though I have little evidence to support it—I believe your job is not to stop time, but to save it, and perhaps save us all."

Charles's fingers were scratching into the flying carpet, just inches away from the rising sun that would have sent it soaring into the sky. He could contain himself no longer.

"Please, sir. Does this mean you'll help me rescue Susan?"

A flicker of movement in the sky caused Charles and Pierre Marin to look up. The last of the California condors was back, gliding to the head of the Trojan Horse, and with a single flap of its great wings it alighted between the ears.

Pierre Marin was looking up at the condor when he spoke.

"He still goes hunting. Though there is no such thing as hunger on the Sea of Time, still, hunting is all he knows, and every day he sets off across the Island of the Past on a fruitless quest."

"Perhaps he's not looking for food," Charles said. "Perhaps he's looking for a friend."

Pierre Marin looked down. He looked Charles straight in the eyes.

"Just let me get my sword," he said.

TWENTY-TWO

Belly of the Beast

SUSAN BRACED HERSELF WHEN FREJO'S lips closed. For some reason she was afraid the roof of his mouth was going to collapse on her. But even when she stretched her fingers up all the way, she couldn't touch anything.

She let her arms fall back to her sides.

The inside of Frejo's mouth was warm and moist, and the perfumed odor she'd smelled in the abandoned warehouse was almost overpowering, underlain slightly by a fishy odor. It was also completely black.

"Well, what am I going to do now?" Susan said out loud, then nearly jumped out of her skin when a giggle answered her through the darkness.

"What? Who's there?"

The giggle came again, but this time it was accompanied by a glow. At first it looked as if a hole had opened in the darkness, but then something shadowy reached into the hole, and when it emerged Susan could see that the something shadowy was an arm, and that the hole was actually the open end of a large sack. The hand at the end of the arm clutched a glowing wad that it lay on Frejo's tongue, and in the growing light Susan could see the figure the arm was connected to.

"Diaphone!"

Diaphone sat on Frejo's tongue, her tail splayed out before her, and pulled more and more of the glowing stuff out of the bag—it was the same phosphorescent seaweed that coated the hills of the mermaids' city, and soon the pile was so bright that Susan couldn't look directly at it. She looked at Diaphone's face instead, but Diaphone refused to meet her gaze until she'd pulled the last of the seaweed out of the bag and then, when she looked up, Susan could see that she was crying.

"I am sorry, Susan Oakenfeld. I betrayed you, and I am ashamed of my actions."

Whatever anger Susan might have felt evaporated instantly, and she ran to her friend and threw her arms around her. Diaphone returned her embrace tightly—but not too tightly, for mermaid arms are much stronger than human arms, and if she'd squeezed Susan with all her might (as Susan was squeezing her) she might have broken her ribs.

"Diaphone!" Susan cried again. "What happened? Why did Qu-Queen—why did *she* make you come with me?"

"Queen Octavia did not make me come," Diaphone said. "Indeed, she does not know I am here, and when she finds out she will be very angry." Diaphone shivered slightly. "The queen's anger is a fearsome spectacle. I am almost relieved that I will not be alive to experience it." She laughed slightly then, her mermaid's love of mischief overcoming her fear of Queen Octavia. "Or who knows, perhaps I *will* be around."

"I don't understand, Diaphone," Susan said. "What are you doing here?"

"I've come to try to save you. What Her Majesty wants you to do is wrong, and making you do it at the expense of your life is even more wrong. It is time"—and here she took Susan's hand lightly in hers—"it is time that a mermaid finally grew up."

Susan threw her arms around Diaphone again.

"Oh, thank you!" she said when she'd pulled away. "Even if we don't surv—" She shook her head. "No matter what happens, having you around will make it easier. Thank you." She frowned then, as the enormity of the situation impressed itself on her. "But how are we going to thwart her? Are we going to make a run—a swim for it? How will we get out of Frejo? And when are we going to leave anyway? I feel like I'm stuck at the airport waiting for the plane to pull away from the gate."

At this, a deep rumble shook the tongue, and indeed the

whole body of the whale. The rumble was enough to knock Susan down, although the tongue itself was soft and slightly springy, so the fall didn't hurt a bit. A moment later, she realized the rumble was Frejo's laughter.

I told you Sssusssan. The ride would be sssmooth asss sssilk.

The voice seemed to come from everywhere—from inside her as well as all around her body, the sibilant *s*'s arriving on rich perfumed air.

Susan looked at Diaphone.

"Is that Frejo?"

Diaphone nodded.

"It's very peculiar, but you'll get used to it." She scratched Frejo's tongue with her fingertips, the way you might scratch a dog, or a seal. "Thank you again, Frejo, for carrying us in safety and comfort."

Mmmm. . . . The sound of pleasure filled Susan's ears and brain. *You are welcome, Diaphone.*

"Frejo has agreed to help us," Diaphone said now.

"Are we going to make a break for it?"

Diaphone shook her head. "No," she said. "Frejo is the most powerful swimmer in the Sea of Time, but we mermaids are much faster. If they saw us change direction they would head directly for your brothers and, well, it is not worth saying aloud what they would do. And, as well, Frejo is determined to follow the rest of his kind through the Great Drain. But if there is any way for him to help us get back, he will do it."

"But everyone says there's no coming back from the Great Drain."

Here Diaphone touched the three keys dangling from the length of seaweed cord in Susan's hand.

"Someone managed to bring these back." The mermaid squeezed Susan's hand. "You must have faith. It is our only hope."

Then a long time passed during which the girls sat mostly in silence. Susan could just feel the deep, almost mechanical throb of Frejo's muscles moving them through the water with immeasurable, inexorable power. Diaphone mentioned that Frejo could build up to a great speed, faster than anything else that moved through the Sea of Time—with the exception of the mermaids, and of course he wasn't very *agile* either—to which Frejo replied with a deep rumble that rolled Diaphone and Susan around on his tongue as though they were in an earthquake. To the degree that she could forget about the seriousness of the moment Susan thought the bouncing around was pretty exciting, but Diaphone, who hated to look undignified, called out, "I apologize, Frejo, I apologize" until he stopped shaking them around. At one point Susan tied the cord containing the three keys firmly around her neck, grabbed up a handful of the glowing seaweed, and set off toward the back of Frejo's mouth. The aperture grew smaller and smaller until eventually she could reach out and touch its sides and the ceiling with her hands. But when she did the rumbling set in again, more pronounced than last time.

T-t-ticklesss, she heard then. *Sss-sss-ssstop. Sss-Sss-Sssusan!*

Susan apologized and made her way back to the big open cavern at the front of Frejo's mouth, where Diaphone was still sitting with her tail folded demurely beneath her. She had neatened her hair and wore a tense but resolved smile, and as Susan looked at her she realized the enormity of what Diaphone had sacrificed to help her, and she had to swallow the lump out of her throat. Trying to keep her voice light, she said,

"Don't you get curious? I mean, we're inside a *whale*."

"It is not exactly a novel experience for me," Diaphone answered, "though it has been a long time. When there were many whales and many mermaids, we used to spend much more time together."

Not all of it friendly, Frejo's voice rumbled through them.

"No," Diaphone admitted, "not all of it was friendly. Since meeting you, Susan Oakenfeld, I have looked back on many of the things done by me and my sisters and realized that not all of them were nice. Immortality," she said, and hesitated. "An endless life changes the way you look at things, and not necessarily for the better."

Susan was about to answer, but what she was going to say was cut off by a sudden jerk. The lurch was hard enough to knock her down and roll both her and Diaphone over the soft, slightly damp surface of Frejo's tongue, though neither was hurt.

"What was that?" Susan said.

Diaphone, sitting up and immediately combing her coppery hair with her fingers, didn't answer right away. Before she could say something Frejo's body lurched again, more violently than before. Susan and Diaphone rolled about wildly on his tongue.

"I-it's like turbulence," Susan stuttered, for Frejo's body was vibrating with a bone-shaking rattle that made her teeth clack together.

Diaphone was once again trying to straighten her hair, which, being very thick and very, very long, stuck out in a thousand different directions.

"I am beginning to see the virtue in your shorn tresses," she said, a miserable expression on her face. "What is turbulence?"

"It's when an airplane—"

But Susan was cut off again. This time Frejo seemed to bounce. Susan felt herself airborne for a moment, and then she plopped down on Frejo's tongue hard enough to lose her breath. Then:

Braccce yourssselvesss passsssssengerssss. We are in the drain.

"Diaphone!" Susan called out as she flopped around uncontrollably. "I'm scared!"

And all at once she felt Diaphone's arms wrap around her, cool, lithe, strong, like a suit of armor.

"Don't worry, Susan. Frejo will carry us safely."

Susan could feel the firm muscles in Diaphone's tail brace against the wild, rollicking ride, the flukes fanned wide for stability and pressed flat against Frejo's pink tongue. The pile of glowing seaweed scattered around Frejo's mouth, bouncing like fireflies and lending a thin, greenish glow to the vast pink cavern. The glow was eerie—scary, but pretty too, and Susan, a little more secure in Diaphone's tight grasp, found herself hypnotized by the sight. The lights flashed and flickered, left electronic trails through the air. Susan thought that if it was the last sight she was to see in her life then she'd better enjoy it. And just about the time she was thinking this, there was a shudder and thud bigger than all the other ones, and the light show disappeared.

At first she thought the lights had gone out. But then she realized that Frejo had stopped moving, and the bits of seaweed had stopped bouncing about. Frejo hadn't just stopped lurching and jerking: he'd *stopped.* The vaguely mechanical, muscular throb had stilled. The phosphorescent seaweed lay scattered all over Frejo's tongue and stuck to the roof of his mouth like glowing rivulets of water.

Diaphone squeezed Susan tightly once, then released her.

Sssusssan? Diaphone?

"Yes, Frejo?"

I think I am going to sssn-sssn-sssn—

"Susan!" Diaphone yelled then. "Run!"

And even as she spoke a great seam of light opened at the end of Frejo's mouth, and Susan, not quite sure what was happening, ran toward it. But even faster went Diaphone, in an awkward but incredibly quick three-pronged hop—arm, arm, tail—that had her bouncing past Frejo's lips several feet ahead of Susan.

As Susan emerged from the whale's mouth, she barely had time to notice the wide open vastness she found herself in. An iridescent plain stretched out before her, an immense lightless expanse soared overhead. Sand crunched beneath her bare feet. Vaguely she wondered what had happened to the water.

"*AAH—AAH—AAH-CHOO!*"

Frejo's sneeze was like an explosion. Susan was knocked flat on her stomach and rolled over several times until she came to a stop on her back, just in time to see a thousand tickling bits of glowing seaweed flash out of Frejo's open mouth through the air. A second light show shot out of what Susan assumed was his blowhole, little shredded bits of color that wafted through the air for a few moments before coming to rest on Frejo's back.

There was a moment of silence, and then Susan said,

"I'm sorry, Frejo, I don't have a tissue."

Frejo's laughter vibrated the sand beneath Susan's body. He was still chuckling when he said, "I'm sorry if I scared you, Susan. Sometimes you just have to sneeze."

Susan was about to tell him it was okay when Diaphone spoke behind her.

"Winds and waves!"

Susan turned and looked at Diaphone. She reclined on her elbows, her head arched back and eyes looking upward. Her hair was tangled all over her face and neck and arms, but she took no notice of it at all.

Gulping slightly, Susan turned and looked where Diaphone was looking.

Not a dozen feet away, a great column of water shot up into the air. It was only a little thicker than Frejo's body, but it shot up into the sky for what looked like miles and miles and miles before gradually opening out like the underside of an umbrella or a parachute, arching out for not-quite-forever before tenting down on all sides, encircling the whale and the mermaid and the human in a dome that was just a little bit larger than their imaginations.

The dome of water glowed slightly, and the sand beneath Susan's feet twinkled, but Susan wasn't sure which of them gave off the light and which reflected it, or if the light came from somewhere else. She had the strangest sense that the light came from inside *her*.

Susan had to skirt the mountain of Frejo's head to get closer to the column. She could feel a warmth coming off it—the air in that place was chilly but not quite cold—and she could also hear it humming with the force of its flight. She wanted to feel that hum and that warmth on her fingertips. She was so intent on reaching the column and sticking her hand in it that she didn't hear the swishing movement on the sand behind her. When she was only a few feet from the column a hand closed around her ankle.

"Susan."

Susan didn't know where the impulse came from. She turned and began kicking at Diaphone's arm with her free leg.

"Let me go, let me go!"

But Diaphone held on, her face set, implacable, her hair wild, until Susan's strength was spent.

Slapping once more at Diaphone's arm, Susan sank down on the sand in exhaustion.

"I just wanted to touch it."

"I had heard that the fountain's call is loudest in children." Diaphone said in reply. "I see now that the legends spoke the truth on this as on all things."

"I'm not a *baby*—"

"Susan," Diaphone cut her off. "If you had touched the fountain you would have been swept up in it. You would have become a part of it." With her free hand she gestured at the vast dome of water. She kept her other hand firmly around Susan's ankle.

Susan looked up at the dome. It was just so . . . so big. It was so big it didn't seem like good or bad mattered. It was . . . was just so . . . so . . . *big*. She just wanted to touch it.

She looked back down at Diaphone.

"What is it?"

"It is the Union of All That Is Separate. It is the Life Beyond Death. It is the End of Everything, and the Destination That Is Also a Departure."

The voice enveloped her in waves of ambergris. It was Frejo's.

Susan looked at him. The great whale lay flat on the sand, his iron-gray flippers stretched out beside him, the glowing black portal of the one eye she could see looking up at the column of water and the dome far above. Susan felt that eyes as big as Frejo's might be able to take in the whole of the dome in a way hers could not, but then she realized that, measured against the dome's vastness, she and Frejo were just the same.

Now Frejo's eye rolled down and looked at Susan.

"In the most basic sense, it is the water of the Sea of Time, compressed and coming out through the bottom of the temporal universe. It is also what will be lost," he said, "if you do Queen Octavia's bidding, and close those gates."

Susan turned.

There, just visible around the curve of the fountain, as plain as an old farm fence, stood three wooden doorframes, each defined by three pieces of wood—two upright posts and a single flat board laid atop them. In front of each doorway stood a low table. On each table stood a thick book, its cover laid open.

One book was black.

One was silver.

The last one was gold.

"They contain the Names of the Dead," Frejo said. "Past, Present, and Future. Lock them, and not another soul will join them."

TWENTY-THREE

Mal de Mer

In fact, Pierre Marin went for more than his sword. Three times he went into his shack, and each time he came out with his arms full, loaded down with bags, suitcases, and one large wooden trunk. It was only on his third exit that he had emerged with his sword belt buckled around his waist, and somewhat awkwardly—in fact Charles had to haul him up like a fish—climbed onto the flying carpet.

Charles pointed the carpet in the direction he thought he had come from and pressed on the star until it seemed the carpet would go no faster, and then they raced across the broad green field to Susan's rescue.

"It is a good thing we both wear these," Pierre Marin

shouted over the whipping wind of their passage. He tapped his glasses. "They protect our eyes from the speed of your wondrous craft."

Charles pushed his glasses up his nose. He was trying to brace himself without pressing any buttons that would swerve the carpet in one direction or another. They were going *awfully* fast.

Sooner than he would have expected, the Sea of Time came into view. The coastline was wide open and bare of anything even remotely human, let alone Drift House.

"I think we were a little to the left," Charles said, trying to keep the worry out of his voice. "Hold on." He pressed on the left arrows a little heavily, and the carpet jerked in that direction.

"*Mon Dieu,*" Pierre Marin said, "I have sailed in boats less stroppy than this."

Charles looked at Pierre Marin's face then, and he saw that it had gone a bit greenish.

"Oh no! Are you, that is, I mean, can you get seasick on a flying carpet?"

"Seasick? Me? Nonsense. I have sailed thousands of miles—thousands of years." But his voice was quavering, and Charles was afraid his passenger was going to be sick.

Still, he had other things to worry about. The carpet was racing up the coast of the Island of the Past, and still there was no sign of Drift House.

"Maybe I went the other—but I thought—oh, blast it! I think I've gotten lost."

"Nonsense," Pierre Marin said in as steady a voice as he could muster. "You strike me as a boy with an unerring sense of direction. It must be *les sirênes*."

Charles didn't mention the time he'd taken a shortcut across Central Park and gotten lost. "The mermaids?"

"They must have towed the house out into the water. Perhaps they meant to block the very possibility of our meeting." He looked unsteadily about the wide open sea. "I should have brought my spyglass."

Charles thought he should have brought Susan's. Then an idea occurred to him.

"Hold on," he called.

"To wha—yiy!!!"

For Charles had already pressed down on the rising sun, and the carpet was shooting upward with a stomach-turning arc. Pierre Marin fell backward, and his legs stuck up in the air where his head had been a moment before.

"Mon Dieu!"

Charles slammed on the brakes then, and Pierre Marin rolled back upright. His hat had fallen off his head, and his wispy hair flying around his face seemed like a manifestation of his thoughts, batting around behind the skin of his face, which had gone nearly as green as the grass of the Island of the Past, far below.

"Sands and stars, child! Never do that again."

But Charles had walked to the edge of the carpet and was looking out over the bare surface of the Sea of Time.

"I thought we could see better from up here," he said, pacing back and forth. He shaded his eyes with his hand, even though there was no sun to cast a shadow. But the sea's surface was polished clean as a marble. The only thing visible, far in the distance, was the low wispy froth of the mountain of mist rising about the Great Drain. From this vantage it looked tiny, harmless.

"Monsieur Marin," Charles called, "can you press on the star please?"

"Eh?"

"The star. Press on it lightly."

Pierre Marin lay a shaking hand on the star and the carpet began moving forward.

"Fascinating," Charles heard him say, and thought the man's voice sounded a little perkier. He felt the carpet speed up in little fits, and realized Pierre Marin must be tapping the star. "Ooh-hoo!" Pierre Marin chortled, his scientist's glee momentarily overshadowing his seasick stomach. "Yes, this is great fun!"

"The right arrows now."

"The, uh—oh yes," Pierre Marin said. The carpet lurched right, nearly tumbling Charles over the side.

"*Lightly,*" Charles said.

"Er, yes. Sorry, my dear b—"

"There! I see them!"

Pierre Marin inched forward on his hands and knees toward the edge of the carpet. "Where?"

"There." Charles pointed at a minuscule dark spot on the surface of the sea. Pierre Marin was squinting at the spot even as Charles ran back to the tree that branched up the center of the carpet.

"Hold on," Charles called, and with one hand he pressed the setting moon and with the other—he had to stretch to reach them both at the same time—he pressed the star. The carpet shot down like a roller coaster hurtling off its highest hill.

"My dear boooooooooyyyyyyyyy—"

The carpet shot out of the sky like a dive-bombing eagle. The dark spot on the sea grew with startling rapidity—now a spot, now a rectangle, now a thick dark box, now details visible, portholes, a parrot's head, the big square sail sticking up amidships. Too late, Charles realized he was plunging straight into the *Chronos*!

It was upon him sooner than he'd have believed. He slammed on the brakes, and if the carpet had been a car its wheels would have left long screeching streaks of rubber on the asphalt. The carpet stopped, but Charles and Pierre Marin didn't—for every action, Charles reminded himself, there is an equal and opposite reaction—and he and Pierre Marin and all his luggage flew onto the main deck of the *Chronos.* Charles bounced roughly against the wall of a cabin, but Pierre Marin rolled straight into the legs of Captain Quoin, who stood with his sword drawn against the airborne attackers.

Captain Quoin held his sword up in the air, looking down at the sprawling disheveled creature at his feet. Pierre Marin's face was so green that Captain Quoin must have wondered if it was a sea creature that had invaded him. Then a look of recognition crossed his face.

"Winds and waves! I wouldn't have believed it possible."

Charles sat up woozily. The Italian twins, Pierro and Pietro, eyed him warily, their swords in their hands.

"My old friend!" Captain Quoin said then. "Is it truly you?" And he hauled Pierre Marin up onto his wobbly legs.

"Wh-what's left of me," Pierre Marin said. He looked over at Charles. "You, sir, are *mad.*" But there was pride in his voice and a twinkle in his eye, beneath his green skin.

Soon enough, Pierre Marin and Charles had explained everything to Captain Quoin and the Time Pirates. Captain Quoin was quick to forgive Charles for his earlier double cross, though Pierro and Pietro—the former wearing a bandage on his head and the latter what looked like a padded diaper to protect his stuck bottom—still looked at Charles suspiciously.

"The fish-girls have taken the houseboat to the Great

Drain," Captain Quoin said now. "No doubt they are waiting to see if Susan will make good on her promise."

"Then we must fly!" Pierre Marin said. "If we can liberate the house, the mermaids will have nothing to hold over Susan."

"Yes," Charles said. "But how do we get that message to Susan?"

"Never fear," Captain Quoin said. "A boy with your pluck and ingenuity will surely find a way." He smiled when he spoke, a gleam of gold in his teeth. But his voice didn't sound nearly as confident.

"But there's no *wind*," Charles said, on the verge of hopelessness. Beat the mermaids? Find Susan at the bottom of the Great Drain? Who were they trying to kid?

"Sails and scales," Zhi Wo said then. "What kind of vessel do you think you're upon? What kind of *sea*?" And, so saying, he made his way to the *Chronos*'s poop deck. A big rudder wheel was mounted there, as well as a device that looked like the horn of an old-fashioned phonograph.

Zhi Wo put his lips to the mouth of the horn and blew. A blast of air emerged from the flared end of the horn—which was aimed, Charles saw now, straight at the sail. The sail puffed out and strained at its lashing, and the *Chronos* jumped into motion like a horse bolting from the starting gates.

"Oh. Oh dear." Pierre Marin said down on the main deck.

"There, there," Captain Quoin said in a voice that was half

amused, half sympathetic. "Let's just sit you down and, ah, Konstantin Dimitrivitch, fetch the bucket if you would."

Zhi Wo blew three more times through the horn, until the *Chronos* was skimming the surface of the sea in its flight. The speed was breathtaking.

Zhi Wo smiled mischievously at Charles. "My *Chronos* shall leave your tattered carpet in its mist!"

"We'll see about that!" Charles said, and he raced down to the carpet and leapt aboard.

It was hardly the time for fun and games, but at such desperate moments people will do anything to let off steam. And so Charles raced the *Chronos* three times, and three times he won. The third time he even rode in a full circle around the speeding ship. He would have raised his fist in victory, but he was too busy holding on for dear life at that breakneck velocity.

He was just coming around the port stern when he happened to glance ahead and saw a sight that chilled his heart: the huge column of mist that rose from the vast canyon in the Sea of Time: the Great Drain was upon them! The mist stretched into the air like a single solitary mountain, so high that Charles couldn't see the top of it. And now he heard the drain's roar as well, a thick gurgling noise, like the rumble of a hungry stomach. And then a cry from the deck of the *Chronos* sounded over the roar:

"Ahoy! Drift House, dead ahead!"

Charles peered down at the surface of the water, and there

it was, right on the edge of the drain: Drift House! How like a boat it looked on the Sea of Time, rocking in the frothy water speeding toward the drain. Not a racing vessel like the *Chronos*, but a great, invincible galleon. Spews of water burst from the prow of the solarium, and the crest flew proudly atop the glass pyramid's tip. Roiling waves splashed against the clinkered siding and sprayed off harmlessly, and the cannons gleamed proudly on the roof, their snouts aimed at all enemies.

One of the windows of the upper story flew open, and a familiar grizzled face poked from the portal.

"Ahoy *Chronos*! Lieutenant Cosmo holds the vessel Drift House in your name, and welcomes you aboard!"

As *if*, Charles thought.

But Cosmo wasn't Drift House's only occupant—there, emerging from the shadow of the cannon like a child peeking from behind its mother's skirts, was Murray.

"Charles!" he said, waving frantically.

"Murray! Hey, Murray!" Charles waved back.

"Charles, Charles!" Murray continued to yell. He waved even more frantically. *"Charles!"*

No, not waved, Charles realized then—*pointed*. Charles wasn't sure how he saw the shadow in the foaming water, nor what reflex prompted him to smack his hand down on the rising sun, but he did see the shadow, and he did smack his hand down, and the carpet shot into the air even as a flame-haired mermaid shot from the water. Charles barely cleared her. Her fist punched the underside of the carpet, knocking Charles to his stomach, and even as he swooped away he could hear her long bloodthirsty cry:

"Ree-yah!"

When Charles sat up he felt a little dizzy and out of focus. It took him a moment to realize that his glasses had fallen off. He squinted around the carpet's surface but couldn't make them out anywhere—they must have fallen into the sea. He shot up into the ragged edge of the mists coming out of the Great Drain, then looped around hard to the left (he wondered if he should use "port" on a flying carpet, or if that only applied to ships). When he emerged back into the open air he had to squint to survey the battle scene. There was a flurry of activity on the deck of the *Chronos* as the Time Pirates took up their battle stations with spears—harpoons, Charles supposed. Long, heavy-looking poles with barbed blades. Charles wouldn't like to be on the receiving end of one of *those.*

The churning sea waters were also busy. Mermaids leapt from the waves and hurtled their own harpoons when they were airborne, only to disappear beneath the water before retaliatory shots could be fired. There seemed to be dozens of them. It looked as though the humans were hopelessly outnumbered. The hull of the *Chronos* was studded with spears like a pincushion.

Charles squinted back at Murray to make sure he was okay. He could see now that the mermaids were forming a phalanx to keep the Time Pirates from approaching Drift House, but weren't menacing the house itself—which was, Charles saw now (or thought he saw; it was hard to tell without his glasses), anchored to the sea floor by a half dozen vine ropes at stem and stern. At least the house wasn't about to get sucked down the drain!

Convinced that Murray wasn't in any immediate danger, Charles swooped down to the *Chronos*'s deck. Captain Quoin rode the masthead as though it were a winged mount, his sword drawn and pointing out mermaids to his spearmen.

"Hard to starboard, Pierro! Close one, Jonno! Put a little muscle into it, this isn't badminton! Zhi Wo, aft port, aft port. Aft *port* you dyslexic scallywag you! Left! *Left!*" There was a clang then, as the captain used his sword deftly to deflect one of the mermaids' missiles. "Pietro, you call that a throw? Pah, you act as if the sea were your enemy and not the fish-girls!"

He noticed Charles then, hovering next to him.

"Well, laddy," he said, deflecting another hurled spear. "Think you can use that floatin' rug of yours to liberate the prisoners of yon houseboat?"

For a moment, Charles's heart thrilled. That was it! He would rescue them all—even the dwarf, who would be forced to admit Charles's superiority. But then his heart sank as he remembered:

"If you please, Captain. My uncle Farley."

"The stout one with the half-growed beard?"

"Yes sir. If you please, sir, he was injured in the last battle. I don't think he can walk, and Murray and Cosmo and I could never lift him."

"Aye, he's a big un is that one." A spear clanged off Captain Quoin's sword and whizzed by Charles's ear.

"Is that all you got, little missies? Come out of the water for a real fight, and I'll show you the strength of *my* arm!" He turned back to Charles. "It seems there's nothing for it but to shoot them fish-girls out the water one by one. Blast! It'll take a while, and we're racing against time here."

"Eh, Charles? Abu?" a weak voice called then.

Charles turned to see a strange face poking from the stairs leading below deck. At first he thought it was a cousin of the butterfrog's, so green were the cheeks and protuberant the eyes, but then he realized it was Pierre Marin.

"Ah, Pierre, my old friend. You and the sea was never mates." Captain Quoin laughed heartily, if sympathetically, at Pierre Marin's pathetic appearance.

"Unfortunately, Abu, you speak the truth," Pierre Marin said, smiling weakly at the captain's jibe. "Er, Charles. The, er, that is, the crate."

Charles looked behind him to the crate and baggage Pierre Marin had loaded on the carpet from his shack on the Island of the Past.

"Yes, sir?"

"If you would, ah, that is—"

"Open it, laddy, afore our good friend loses the will to speak!"

Charles, caught up in the spirit of the moment, used his sword to strike the lid from the crate. He hit the box so hard it tipped over on its side and a pile of colorless fabric spilled over the surface of the carpet.

"Ah, Pierre, my old friend. The *maladie* has addled your wits worse'n usual. It's too long since you've braved the sea."

It seemed to Charles that an annoyed look passed over Pierre Marin's face—or, rather, under the green pallor of his skin.

"If you'll just, ah—" He made some kind of weak gesture with his hands. "Spread them . . ."

"Pierre, Pierre. This is hardly the time for disguising ourselves in costume," Captain Quoin said somewhat gruffly—perhaps he thought his old friend was making fun of how easily Susan had deceived him. "The fish-girls is already spotted—"

"They're *nets*!" Charles yelled then. And he hauled one out to show. "We can snare them!"

"Winds and waves!" Captain Quoin said; then: "Ahoy!" he shouted. "Get up here, you miserable excuses for spearsmen, and equip yourselves more fittingly. The Time Pirates is going fishin', and them mermaids'll be wishin' they'd never *tangled* with us, har-har. My old friend," he added in a lower voice, "I'm sorry I ever doubted you."

But Pierre Marin made no reply. He seemed, as Captain Quoin had predicted, to have lost the will to speak.

In a moment each member of the crew of the *Chronos* was equipped with one of Pierre Marin's nets, including Charles. They were incredibly light and seemed to stretch out endlessly, and they had tether ropes that the Time Pirates fastened to the *Chronos*'s gunwales. Charles, whose carpet lacked anything so structural, tied his tether to one of the tasseled ends.

"Now Charles," Captain Quoin said to him, "hold 'er here, and whirl 'er above your head like so." The captain whirled his net as though it were a lasso. "Smoothly now," the captain said, "else you'll entangle your own self in your web. And then, when you sight your quarry—hya!" And, lunging, Captain Quoin let loose his net.

The bunched fabric shot through the air like a dark shadow, and then, when it reached the end of its tether rope, it burst open like a firework until it seemed that a huge spiderweb hung in the air over the sea's water. A thin shadow darkened the net, and a moment later a writhing tangled form fell toward the water—but not all the way. With a head-cracking *clunk!* the snared mermaid thumped against the *Chronos*'s hull, narrowly missing being impaled on one of her sisters' spears sticking out of the ship.

"Aiyee!"

The snared mermaid's ear-splitting shriek split the air as she twitched and wiggled in her net.

"Steady, boys!" the captain called, already outfitting himself with a second net. "Let's see what fish rise to our bait."

Charles wasn't sure what he meant until he saw a pair of dark shapes in the water and realized that two of the mermaids were coming to their sister's aid.

It was Pierro and Pietro who made the catch. They threw, the mermaids leapt, writhed, screamed, and a moment later two more twitching bundles thumped against the Time Pirates' ship.

"Whoo-hah!"

The voice surprised Charles—it was his own.

"That's the spirit, boy!" Captain Quoin said. "Now go and get yourself a fish!"

Charles pressed on the right arrows and the sun, whirling away from the *Chronos*'s deck. But it was to be another Oakenfeld child who made the next strike.

KA-BOOM!

The blast was so loud that time, if you will forgive the expression, seemed to stand still. There was a split second of silence that seemed to last an eternity. Charles could see Murray's face behind the downward-tilting cannon, solemn and focused. He could see the sailors on the *Chronos*, Zhi Wo leaning hard into the rudder to keep the ship from being sucked down the drain, the others' nets frozen midwhirl above their heads. He saw Cosmo in the front door of Drift House, his sword brandished at a mermaid caught in midleap from the water and Captain Quoin back on the *Chronos*'s masthead, sword in one hand, net in the other, his mouth open but no sound emerging. Even the roar of the Great Drain seemed to have been silenced.

And then a hollow geyser erupted a hundred feet into the sky in a whoosh of air replacing water and water replacing air. At precisely the moment Diaphone was preventing Susan from being sucked up into the fountain that is the End of Everything, a miniature version of that column of water flew up all around Charles, shooting up and surrounding him in its hollow, safe center and then arching away in showers of mist. When the last of the mist had cleared away, Charles saw no fewer than four dark shapes lying in the water, unmoving save for the occasional twitch of fluke or finger.

"Way to go, Murray!" Charles shouted—but he made a mental note to tell Murray later that he wasn't supposed to play with matches.

"Zhi Wo, bring 'er around hard to starboard. Starboard, you turned-about helmsman, *starboard*. STARBOARD!"

Even as the *Chronos* seemed practically to be spinning on its axis toward the stunned mermaids, an ear-splitting—no, bloodcurdling—yell erupted from Drift House.

"ARGH!"

Charles whirled around, squinted hard. A mermaid had pinned Cosmo with her spear to the side of the house. At first Charles thought the dwarf was stabbed, but then realized it was just his thick fur shirt. The mermaid's blade held him in place like a pin through a mounted butterfly, and in her other hand she held a giant rock, raised to dash his brains out.

Cosmo's face was unreadable behind his thick beard, but his voice was clear enough:

"This house 'n' its occupants is under the protection of Lieutenant Cosmo, first mate of the *Chronos*, fish-girl. It's more'n a spear 'n' a rock as'll let the likes of *you* into it!"

Even before the mermaid could let loose her stone, Cosmo hurled his sword. It was an awkward position to throw from, and it was a sword too, and not made for throwing. But Cosmo's aim was true. A length of steel appeared out of the mermaid's back, and without a sound she fell backward into the water. Charles was kind of glad he was missing his glasses then, because he wouldn't have liked to see her face clearly. A moment later a thin red cloud obscured her sinking form.

And then: silence.

No cries from the Time Pirates, nor from Drift House. No mermaids appearing from the depths, no *thwat* of hurled nets or *thunk* of spears striking the hull of the *Chronos* or *boom* from the cannon. And:

No sound from the Great Drain.

A voice came from the roof of Drift House, echoing oddly over the silent water.

"Charles!" Murray pointed. "Charles!"

Charles turned away from the two stilled vessels. The great

column of mist had nearly vanished—only a few wisps still snaked through the air, like lonely ghosts looking for their friends, and then, one by one, they too faded away. The impossibly enormous hole in the Sea of Time had been reduced to a little dimple in the swell, even now smoothing out into an expanse of water as flat as the carpet Charles rode upon.

The Great Drain had been closed.

"Charles!" Murray was still calling. "Charles, look out!"

It seemed to Charles then that his carpet was shrinking. It took a moment for him to realize that something, hands, no, hooks—why did he have to go and lose his glasses!—no, *tentacles* had grabbed it at all its edges and was wadding it up out from under him. The carpet was half its full size, a third, a quarter. Only then could Charles see the woman's torso grafted onto the immense octopus body. She hung upside down off the shrinking wad of flying carpet, so her sharp smile was simultaneously a wicked frown.

"The game is over, boy," the strange hybrid creature said haughtily. "The drain is closed forever. Susan is gone. And you, Charles Oakenfeld, shall be my personal slave."

The tip of one of Queen Octavia's tentacles wrapped around Charles's ankle, and Charles could feel the terrible grip of the suckers, pulling him into an unimaginable present that had no future and no past, and no end.

TWENTY-FOUR

Susan's Choice

SUSAN STOOD BEFORE THE THREE Books of the Dead. As she got closer she saw that their pages were turning—impossibly thin pages turning impossibly fast. The pages were so thin they seemed to add nothing to the left-hand side of each book, nor take away anything from the right, but they made a rustling sound as they turned, and stirred up the faintest breeze. The black Book of the Past was nearly finished, she saw, while the silver Book of the Present was open to its middle, and the gold Book of the Future had only just cracked its binding.

The pages whispered by, the names inscribed on them too blurred to read.

Susan felt something hard in her hand. She looked down, and was surprised to see that she'd pulled the keys from her neck—she had no memory of doing so. She noticed that they were warm, and vibrating the tiniest amount. Pulling almost, in the direction of the books. When Susan was a very little girl, Mr. and Mrs. Oakenfeld had taken her to visit a farm in the country, where she had held a baby piglet. The little pink ball of flesh had been hot and wiggly in her hands with its desire to get back to its mother's teat, and Susan thought the keys felt a little like that as well. They seemed to yearn for the keyholes in the book covers before them, and it was all she could do not to slam them shut and turn them locked. Still looking at the rippling pages, she called over her shoulder, "Frejo? What are they?"

In the vastness of that wide-open space beneath the fountain's canopy, Frejo's enormous voice seemed as small as hers.

"It is said that they are the only things in this world that are neither *made* nor *born*, but merely *are*. That all thinking beings have been cast in their likeness. That they are the true record of time—the time that was, the time that is, the time that will be, all coexistent, co-equal, and no moment more real than any other."

"But they're just names. How can names be time?"

"Each name represents a being, Susan. A thinking, feeling, wanting *being*. And, more than that, each name represents that being's desire to be an individual, distinct from everyone else, filled with dreams as well as the longing to realize them. Time is nothing more than the accumulation of all this emotion. This desire to be alive."

Desire. It was a word Susan had always associated with grownup situations, but now she realized that it meant much more than that. It was also the word Pierre Marin had used in his speech about time to Charles, though Susan didn't know that. But she understood something else.

"Frejo, if I close the books, will people stop . . . feeling?"

"Not exactly. But their emotions will lose their focus, their strength, their goal, their meaning. Eventually their lives will become a blur to them, without fear or joy or passion—most important, without end."

"Zombies," Susan said. "They'll become zombies." In her mind's eye she saw her mother and father, walking through their apartment in New York without seeing each other, without missing their children, walking aimlessly, endlessly. She shuddered. To shake the image from her mind, she asked Frejo, "Why are they called the Books of the Dead? Why not the Books of the Living? The Books of Life?"

"Because life is only one part of existence. The visible part. The measurable part. The tiniest part."

"Like the tip of an iceberg?"

"Yes, although that description is too simple. The universe stretches out on all sides from this life: left and right, forward and backward, around and through, before and after and in directions that we cannot perceive with our mortal senses. That is why all travelers upon the Sea of Time are able to travel back in time, and some very lucky ones, like your brother Murray, are able to go forward as well."

"Is it because he drank the water?"

"Not exactly. It is more apt to say that he drank the water because he is special. The sea—those books behind you—knew that he would drink the water long before it entered his mind to do so. Just as they knew you would be here, standing before them."

All this time Susan had been staring at the books. But Frejo's words made it seem that they were staring back at her, and she turned to look at the whale instead. She was looking at him head-on, and it was a bit like talking to the side of a barn.

"I'm not sure I understand, Frejo."

A flicker of an expression crossed Frejo's enormous mouth. If you have ever snapped a garden hose and watched the ripple travel from one end of it to the other, you have seen something like what Susan saw on Frejo's face. What I mean is, if it was possible for lips as long as a boxcar to grin, they did.

"You do understand, Susan. Of the three of us here, I think you are the only one who does understand. You are the only one of us to have lived in a temporal universe. The only one to have contemplated mortality." Frejo sighed in a wash of intoxicating breath that made Susan slightly lightheaded, and then he said, "Why else would the fountain have chosen you for this moment?"

"But I thought the mermaids—" She glanced at Diaphone, who looked at her with an expression of helplessness that mirrored her own.

"The mermaids are indeed wondrous creatures. But they realize neither their full power, nor, more important, the limits of that power. Though they called Drift House onto the Sea of Time, and compelled you to their aid by shameful means, yet even they were the pawns of forces they couldn't understand, any more than you could understand, at first, how you ended up on the sea. But now you and you alone stand before the books with the keys in your hand."

Susan's brain ached with trying to understand everything that Frejo had said.

"The Sea of Time contains every moment of existence," the whale continued. "If you know how to navigate its waters, you can emerge anywhere, anywhen. You can come and go through a bathtub's leaky faucet, or a puddle that only stands for a few hours after a rainstorm, or the rainstorm itself. Through a cataract gushing from a rent in the earth or the condensation that forms on a window when it is cold outside and hot inside. It doesn't matter how big or small the body of water is, whether it can contain a form your size, or mine: time is the most malleable of all things, and can shape itself to any container. But," Frejo continued, and here his voice grew even deeper, "as one moves closer to the time you and your brothers inhabit, many of the rivers and bays of the sea have been dammed or drained or otherwise partially or wholly blocked. It is as if time's stream has been narrowed, quickened, the paths to the future reduced and so the many possible futures reduced as well. It is a problem I have spoken to Pierre Marin about, though neither of us understands it fully."

For the first time in the conversation, Susan found a tangible fact she could hold on to, and she grabbed at it.

"You know Pierre Marin?"

"All who live in and on the Sea of Time know the intrepid seasick sailor. I have not seen him in many, many years, but I think he would agree with me that your arrival here cannot be accidental, or incidental. I believe you and your brothers have been accorded nothing less than the task of restoring time to its proper flow—its proper perspective. The sea summoned you, and the sea does not act without a reason."

"But *why*? Or . . . or how? How can I—we—Charles and Murray and I—restore *time* to its proper flow?" Susan found it hard enough to say, let alone contemplate.

"That I do not know, Susan. But I think I have figured out a way to start you on your journey. It will be difficult and dangerous, but I can think of no alternative method. But first, as the bearer of the keys, you must make your decision."

"My decision?"

"Whether or not to lock the Books of the Dead, and close the great fountain forever."

"But I thought we had decided—"

"*We* can decide nothing, Susan." Frejo's voice was more gruff now. "Though other beings have been in possession of all three keys at one time or another, you are the first who has ever stood before the books with all of them in her hand, save perhaps whoever fashioned them in the first place. There can be no mistake in that. Like all keys, the keys in your hand desire to be fitted into their locks. Like all books, these want only to be finished. It is up to you to decide whether or not to grant their wish. To end the story now, or allow it to continue for a few more pages."

At first Susan didn't understand. Why would she want to close the drain? If she did that, she would never be able to return to Charles and Murray, to Uncle Farley and President Wilson and the surface and the Bay of Eternity and her parents back in New York City. Why on earth would she do that? But as she looked down at the three keys in her hand they seemed to twinkle at her. They were warmer than they had been before, and practically twitched in the direction of the books. She could feel how much they wanted to be fitted into the locks, and she could feel something else, something in her—something maternal—that wanted to grant their wish.

With an effort, she looked up at Diaphone. The mermaid

returned her gaze silently, inscrutably, but the outer tips of her flukes twitched once, twice, as though there were some huge energy she were just barely suppressing within her.

Susan looked at Frejo then, who regarded her with one enormous eye. The eye was black and reflective and shiny, though maybe not quite as shiny as it had been before, as if it were beginning to dry out in this arid place. Indeed his entire body had lightened somewhat, from a dark chalkboard gray to the color of old asphalt as the water evaporated from his skin. For the first time since Susan had met him, he did not seem impregnable. He seemed immensely, immeasurably old.

"I only ask that if you decide to close the books," Frejo said in a weary voice, "that first you allow me to move into the fountain, so that my name too can be recorded, and my death be not empty."

Susan looked down at the keys. Her fingers had curled around them at some point, and she wasn't sure if she had curled them or if the keys had done it. When she spoke, it was to the three pieces of metal clutched in her hand as much as to Frejo or Diaphone.

"I. Don't. Understand."

"Then you must act without understanding," Frejo said. "You must listen not to me and not to the keys, but to yourself."

Susan tried to listen, but she couldn't tell one voice from another—not when they were all silent and all speaking inside her head. She closed her eyes to concentrate better, and thoughts began whizzing through her brain as quickly as the turning pages in the books. She saw Frejo's eye, first of all, and Diaphone's twitching tail, and the keys in her hand, the after-images of what lay before her. And then her mind flashed backward: she saw the mermaids' glowing city and the lonely green mounds of the Island of the Past and the vast endless blueness of the Sea of Time swirling into the spiral of the Great Drain. The glowing parrot's head at the front of the *Chronos* looming out of the darkness. The roots of the great banyan tree filling up the solarium like a jungle gym. The gentle sway back and forth as Drift House rocked free of the present into the anytime/no time of the sea, and the rain that had fallen down on her and Charles and Murray and Uncle Farley like a net on their first day at Drift House, and the smell of her mother's neck when she had kissed Susan goodbye, and the inside of the car during the long drive from New York to the Bay of Eternity, and—and the horrible event that had made them leave the city in the first place. The smoke billowing up in the sky and settling back down on the city in an ash of fear. And she imagined some things that hadn't happened either: she imagined sailing in Drift House to all the places she had read about in history. She had gotten an A in history not just because she studied hard but because she *loved* history, she wanted to learn all about it, and what better way to learn about it than by *going* there? But even as she thought that, she knew the answer: she would not have to visit history, she could *be* it.

She could plant the keys in the Books of the Dead like seedlings and then step into the fountain that was the End of Everything before its flow was cut off forever (for some reason she knew she would have time to do this) and soar upward and disperse into a billion pieces of time itself, as if her body were only temporarily solid and would finally be returned to its true, its rightful state, like the cubes of bouillon her mother dissolved in hot water to make soup. She could see this last image so clearly it seemed she felt it. It seemed that she was already dissolving—that the most solid parts of her were the keys in her hand. They glowed before her closed eyes like three bones of her skeleton that had been laid bare and wanted to be covered again—covered in the skin of the books behind her.

There was a sound then. A thin little sound like:

phwat

Susan waited for the sound to be repeated, but it didn't come again. With an effort she opened her eyes. She saw the keys first. They surprised her a little, sitting on the flesh of her palm rather than growing from it. They had come to seem like such a part of her. But they were not, after all. They were separate.

And then her eyes were caught by a glint of light. There, not three feet in front of her, lay a pair of glasses that had obviously fallen on the sand, producing the *phwat* she had heard.

Not just any pair of glasses: *Charles's* glasses.

A cry burst painfully from Susan's lungs. It felt as though she were being ripped in half like a sheet of paper, and the sound that emerged from her mouth was produced not by her vocal cords but by the tearing of one half of her from the other.

"Oh, Frejo, please! Get me out of here! Get me back to the surface, please! I've got to help Charles!"

Another strange, magnified expression rippled across Frejo's gigantic face, mirrored in miniature on Diaphone's: whale and mermaid were both sighing in relief.

"Then quickly, Susan. Back in my mouth."

Susan replaced the keys around her neck. She could still feel their longing for the books behind her, but she refused to turn around and look at them. She grabbed Charles's glasses off the sand and stuffed them in the pocket of her pajamas and ran to Frejo's open mouth and climbed inside. Diaphone was already inside waiting for her and she took Susan's hand and helped her over the slippery part of Frejo's bottom lip, and then the top lip curled down and sealed itself to the bottom and Frejo's mouth went black. There was no bag of seaweed to brighten the damp, warm, perfumed cave this time.

Back. Farther back.

The voice inside the mouth vibrated all through Susan.

"But we can't see, Frejo. Which way is back?"

Back! the voice rumbled through her. *Hurry!*

It was hard going. Diaphone had to let go of Susan's hand to drag herself along, and the slithering sound her body made on the wet tongue sounded like a huge snake crawling behind her. The tongue, which had seemed cushiony before, now felt wet and treacherous. Time and time again Susan slipped and fell.

"Frejo, please. I can't see any—"

Back!

Through some sixth sense Susan could feel the passage narrowing around her, the walls and roof closing in, the heat of the deep innards of Frejo's body growing stronger, and the scent of perfume as well, so powerful that Susan felt dizzy. Eventually the passage grew so narrow that she was able to feel the sides of what must have been Frejo's throat, and soon after that she had to duck and then drop down on all fours because of the lowering roof. She could feel Frejo's tongue twitch and quiver beneath her with the urge to sneeze, but he held it in. She was dragging herself along on her elbows, the sound of her ragged breath so loud in her ears that she could no longer tell if Diaphone was following along behind her, and she was just about to beg Frejo to let them return to the front of his mouth when she felt her head burst into a more open space. In a moment her body was through, and she felt herself in a narrow vertical shaft. A moment later she felt Diaphone's head against her ankles and, twisting some, and pulling on her friend, she helped her into the chamber, where the two stood pressed against each other like a pair of campers sharing a single sleeping bag.

"Susan!" Diaphone said. "I think I understand!"

Susan, who could barely breathe in the tight space, didn't understand, and was too miserable to ask. The scales of Diaphone's tail were surprisingly sharp and abrading the skin of her leg through her pajamas, and right at the bony part of her hip Charles's glasses were pressing sharply into her skin, and she was afraid that at any moment one of them—Charles's glasses, or her skin—was going to break. And over and above this she felt an ache in her chest directly beneath the keys, a longing for the peace they had offered, and that she had rejected.

"We're in the tube that leads to Frejo's blowhole! Do you understand?"

Susan didn't. She was trapped at the bottom of the world and the end of time inside a whale—inside a whale's *blowhole* apparently—with a mermaid rubbing up against her like a giant piece of sandpaper, and nothing, absolutely nothing made sense at that moment besides the jab of Charles's glasses against her hip. She was going to return those glasses to Charles. He was so *helpless* without them. With an effort, she managed to shift ever so slightly, so that Diaphone's hip—if mermaids have hips—no longer threatened her brother's vision.

Take a firm hold of her, Diaphone.

Frejo's voice had thinned considerably. It no longer seemed like a voice at all, no longer carried any hint of perfume. It was as if the whale were already fading away, replaced by Diaphone's arms, which encircled her as they had on the incoming journey. But this time Susan could take no solace in their grasp. They were just one more thing squeezing her chest, stealing her breath.

Listen closely now. I don't think the fountain can take me as long as the keys are inside me. It cannot dissolve that which dissolves itself. I am going to heave myself across the portal between the Great Drain and the fountain and attempt to wedge myself in place, and then I am going to shoot you from my blowhole into the vortex. My body should slow the water pressure considerably, and the force of your expulsion should give you some extra momentum as well, so that if you swim very, very hard you may be able to push through the drain's current to the surface. But you will have to swim as you have never swum before.

Susan felt as if she were listening to the sound of a television being played in another room. As if Charles were watching some horrid silly unbelievable monster movie about sea creatures. The only problem was, *she* was the star of the movie.

Susan Oakenfeld?

It took all her effort to speak.

"Y-yes, Frejo."

It has been a privilege knowing you. You have already done great things and made difficult choices in your life. You have no reason to be afraid now.

"But I *am* afraid, Frejo. I'm terrified."

Do you remember the story your uncle told you? About the time he first took Drift House onto the Sea of Time and floated helplessly toward the Great Drain until something stopped his progress and pushed him back to safety?

Susan nodded. Then, realizing the whale couldn't see her, she said, "Yes, Frejo."

That was me, Susan. I saved your uncle, and I will save you.

Susan felt a little of the weight lift from her chest. "Thank you, Frejo."

You have come this far, Susan. You must complete the journey. When I give the sign, take a deep breath, and don't let it out until you hear Charles say your name. Now brace yourselves. I shall have to jump and then turn on my back. It will not be as smooth as the ride down.

There was an earthquake then, as Frejo jumped, and then another, as he jumped again. Susan felt like the great body all around her was ripping apart and falling in on itself, crushing her.

"Diaphone! I'm scared!"

"Be brave, Susan! Be brave!"

If what had just happened had been like an earthquake, then there are no words to describe what happened next. Frejo's body began to vibrate with such force that Susan felt like the bones were being rattled out of her skin. She realized they must be in the fountain! At the very nexus of the place where all the water from the Great Drain—all the water from the entire Sea of Time!—had to squeeze through its narrowest aperture and come bursting out the other side.

And then a lurch twisted her sense of gravity, so that she felt she was lying on her side. And then another, so that she felt she was upside down. She *was* upside down, she realized. But what was upside down on this side of the Great Drain was right side up on the other: Frejo was aiming them toward the surface of the sea! And through her terror she felt the first glimmer of hope: Frejo was going to shoot them to the surface! He was going to shoot them to the stars even, like a rocket bearing tidings from another world!

Now, Susan—BREATHE!

The walls of the blowhole opened up about her and her lungs expanded and the air rushed into them even as water plunged down on her head and Diaphone's arms tightened around her and the water was all around them and then the walls contracted again and with an upward burst that drained the blood from her head Susan felt her body and Diaphone's shoot from the blowhole of the last talking whale the world would ever know and into the lightless soundless motionless waters at the bottom of the Sea of Time.

And there they stayed, unmoving.

At some point Susan had closed her eyes and she didn't open them now. She could feel the weight of the sea's water over every square millimeter of her body save for the loop of flesh under her arms and across her chest where Diaphone's arms encircled her, and her shoulders and upper back, which pressed into Diaphone's chest behind her. The water seemed to bear down on them from every direction, but most strong on her head, holding her as firmly at the bottom of the sea as stars are held in the sky. It took a long time for her to feel the undulations of Diaphone's body through the arms that held her, to realize that the water was pressing most firmly on her head because they were in fact shooting up against its current at great speed. Her lungs burned with the desire to take another breath but Diaphone's arms were holding her so tightly that she didn't think she could have sucked in air even if she'd tried. A great fire built up in her chest, so hot that it felt as though it would scorch the life out of her. She felt her head begin lolling back and forth in the current of their passage, felt it bang against Diaphone's face behind her, told herself to hold it upright, couldn't remember how to do that. Couldn't remember what she'd forgotten. Couldn't remember what her head was. Couldn't remember . . . couldn't . . .

Oh Charles! Susan thought. I've got your glasses. You'll never be able to see again!

With a whooshing sound that seemed strange to Susan's ears—sound? what was sound?—Diaphone shot out of the water and into the air. As if the water itself had been holding Susan's eyes closed they fluttered open, but the bright blurs moving all around her refused to resolve into something she could understand. Sight? What was sight?

Then:

"Susan!"

With a gasp, Susan opened her mouth and sucked a huge cooling gulp of air into her burning lungs, but let it out almost immediately in a single cry.

"Charles!"

It all happened in a split second, but Susan felt like she was moving in slow motion. She and Diaphone were still airborne, still ascending even, so great had been Diaphone's speed, and like a snapshot the scene on the sea flashed evenly across her field of vision: Drift House, the *Chronos*, the seven mummy-shaped figures writhing in the water and the eighth shape that swam languidly toward them, one of its tentacles sticking out of the water and wrapped around the neck of a squint-faced boy rolled into what looked like an old rug floating a few feet above the water—water that was bubbling now, as if some great beast had let out a sigh deep beneath the surface.

As I said, it all happened in a split second—but it was such a long second that Susan all but forgot how she'd come to be airborne, and when Diaphone spoke in her ear Susan had almost forgotten her, despite the arms wrapped tightly around her torso.

"Grab for the carpet, Susan!"

And with no further warning Diaphone launched Susan into the air even as her own body fell toward the surface of the sea, which itself seemed to be falling away as, somewhere, far below, the waters of time rushed through to the world beyond this one.

"Diaphone!" Susan yelled as the mermaid fell into the rapidly spinning spiral. "No!"

But even as she yelled Susan found herself smashing into the tube of the flying carpet. Her reflexes took over and she grabbed the tubed carpet as though it were a log and clung to it with all her might. She held on to the carpet as, until just a moment ago, her friend had held on to her, and as she saw Diaphone's body swirling down the ever-deepening drain, her magnificent emerald tail, her coppery hair, her small mouth set not in a pout but in a tiny, determined smile, Susan realized that the last act of the mermaid's life had been to save Susan's.

"Oh, Diaphone," Susan whispered. "Oh no."

The water had receded far below by then, and Susan could see no sign of her friend in the rising mist. Instead, one by one, like boxcars or a caravan of camels, the netted mermaids spiraled after their renegade sister in the freshly opened torrent, and Susan could hear their cries of anger and fear over the awakening roar of the Great Drain. Susan counted five of them before the mists made it impossible to see anymore.

"Goodbye," Susan said, and imagined the word spiraling down the water after her friend. "Goodbye, Diaphone. Thank you."

The voice that spoke next was the only voice that might have distracted her from her sadness at that moment.

"Susan!" Charles called from the other end of the rolled-up carpet. "Help, Susan!"

"Charles!" She began to inch forward on the carpet. "Don't worry, Charles, I'm coming. I've got your glasses, don't worry."

But her brother made no reply. She could feel his body writhing inside the carpet, but he didn't say a word. And then, when a long black tentacle curled itself around the carpet in front of her, she understood why.

Susan had to crane her neck to see directly under the flying carpet. There was Queen Octavia, looking up at her with a grimace of such hatred that Susan practically leapt from the carpet into the Great Drain.

"You have betrayed me, Susan Oakenfeld! I shall take great pleasure in ripping your limbs off your body one by one!"

One of Queen Octavia's tentacles was curled about the body of the carpet just in front of Susan, and another snaked farther forward—around Charles's neck, Susan guessed, which must be why he hadn't answered her. The other six tentacles held the last two netted mermaids and were busily untangling the bonds that held them.

"Save some for me, Your Majesty. I have detested the brat girl since the moment I laid eyes on her."

The voice was Ula lu la lu's.

"You don't scare me," Susan said now. "I've been through the bottom of the world and back again. Anything you could do pales by comparison."

Queen Octavia's mouth opened to answer, but before she

could a loud screeching cut her off, and a green and red blur seemed to fill the air around her face.

"Rak! Rak! Rak!"

"President Wilson!" Susan exclaimed.

The parrot didn't answer. He divebombed Queen Octavia's face over and over again, slashing at her with his sharp claws and beak. The queen beat at him with her hands, which seemed clumsy instruments compared to her tentacles, but the latter were still busy clinging to the two mermaids who hadn't gone down the drain, and to Charles's neck, and the carpet.

Inching forward again, Susan grabbed at the cold tip of the tentacle wrapped around the carpet. It took all her muscle to budge it, but as formidable as Queen Octavia's strength was, it was already overtaxed by the burden of holding up her own thrashing body and the two netted mermaids. With a sudden snap, the tentacle twisted free.

But now the queen's weight hung from the single tentacle wrapped around Charles's throat. As Susan crept forward she could see his face: his mouth open, gasping for a breath it couldn't take, his red cheeks already deepening into purple as the oxygen seeped from his blood.

Below her, Queen Octavia was trying to whip President Wilson out of the air with her newly freed tentacle. A cruel smile had appeared on her face: she didn't look worried at all.

"Watch your brother's face as he dies, Susan, and know that the same death will be coming for you."

All at once a red and blue form fluttered onto the carpet between Susan's face and Charles's. It was Marie-Antoinette. She took the time to fix Susan in the eye, and then it seemed to Susan as if the Time Pirates' parrot winked at her.

"Boys!" she said. "They always need saving, don't they?" And she drove her beak into the tentacle wrapped around Charles's neck, and the tentacle snapped open as if it had been stung—as indeed it had.

Queen Octavia's scream of pain and fury turned just as quickly into a wail of fear and despair chorused by Ula lu la lu and the other netted mermaid as they dropped into the churning, mist-shrouded waters below them. In a moment their bodies and their cries had been swallowed up by the Great Drain, and the only sound Susan could hear was Charles's gasps for breath.

Susan crawled the rest of the way forward on the tube of carpet (Marie-Antoinette flew to the other end of the carpet, where she was joined by a panting but jubilant President Wilson) until her face was directly over her brother's. She gave him a big kiss on each of his cheeks, and she couldn't tell if the red in his cheeks was a blush or just the residual effects of Queen Octavia's attempt to choke him to death. His neck was ringed with a nasty polka-dotted rash from the tentacle.

"And now, Charles Oakenfeld," Susan said. "I think you

really *must* thank me, for I have brought you back your spectacles." And reaching into her pocket, she set them on his face.

Charles blinked at his sister as his eyes focused behind his glasses, as if he were making sure it really was her. Then:

"How many times do I have to tell you not to say 'spectacles'?" he said. "It's affected."

EPILOGUE

PERCHED PRECARIOUSLY ON THE LATTICE of the solarium, Pierre Marin struck the base of the weathervane with a hammer. As metal struck metal a bright ringing noise sang out into the air of the Sea of Time.

DING—DING—DING!

A few feet away, Uncle Farley and the three Oakenfeld children sat on Charles's flying carpet and watched with interest as, with a jerk, Pierre Marin twisted the weathervane on its hinge. It squeaked mightily but turned. He poured some oil from a can on the hinge and it turned more freely. With another application it spun under his hand as freely as a top.

"There you are," he said.

Uncle Farley squinted. "Really?" he said. "That's all it was? A little rust?"

Pierre Marin had given Uncle Farley a dose of some kind of medicine that had cleared up his head wound, although he still felt a little woozy and would continue to, Pierre Marin said, until he got back to Canada and put his body back on a real clock, and real food. His glasses had been broken in the fighting with the mermaids, and that would also have to wait until they got back to Canada. Which, it seemed, would be soon.

Pierre Marin laughed now. "Sometimes an objective hidden in plain sight is the hardest to find," he said. Taking Uncle Farley's hand, he stepped nimbly from the solarium to the carpet. "I'm sorry it didn't require some magical incantations or potions, but yes, I'm afraid it was just a little rust. You should have no trouble piloting the house back to the Bay of Eternity now." He laughed again. "It must have been quite frustrating for you. With the antenna stuck in place, the homing stations are basically just big radios."

Now Charles scratched his neck, which was bruised red in places from the last grip of Queen Octavia's tentacle. "But they don't have any mechanical parts. How can they work as radios?"

"Why would they need mechanical parts?" Pierre Marin said with a twinkle in his eye. "They're made from the wood of the radio tree." He chuckled a little at his joke, then said, "You children have so many questions. I would have thought a week on the Sea of Time would have driven the curiosity from you."

"A week!" Susan and Charles said at the same time. "Is that all it's been?"

The old Susan would have stuck her tongue in her cheek, the old Charles would have pushed his glasses up his nose. But their experiences had changed them. They looked at each other a moment, then burst out laughing, and then the group boarded Charles's flying carpet to descend to the house's main levels.

"It has been a *full* week," Uncle Farley said. "And now I think it is time we headed back to Canada."

"Yes," Pierre Marin said, "I suppose that is true." He flashed a significant look at Charles. "One does not realize how much one misses the company of one's fellow creatures until one has been among them again."

"We'll be back," Charles said. He tapped expertly on the carpet and it began to move out over the edge of the solarium. Turning to his uncle, he said, "We will be back, won't we, Uncle Farley?"

Uncle Farley sighed heavily, his hands bracing himself on the carpet as if it were an amusement park ride. "I suspect I couldn't prevent it even if I wanted to."

A voice piped up from the back of the carpet.

"We'll be back," Murray said. "You two more than I, I think. But we'll all be back."

Susan and Charles looked at their younger brother, as if they'd forgotten he was with them. He sat without fidgeting on the carpet, looking more like a judge than a five-year-old. The golden locket gleamed around his throat in the sourceless light of the Sea of Time. The carpet had descended to the first floor now, and the rich blue water of the sea stretched out endlessly around them.

"It's such a shame Diaphone's not here," Susan said, looking out at the water. "She—she gave her life to save me."

Uncle Farley looked as if he were about to say something, but a voice coming through the open window of the dining room cut him off.

"Madam, please," said President Wilson. "Don't be coy with your vassal. I exist only to be of service."

"Be of service," Marie-Antoinette parroted, and Susan fancied that only she could hear the secret laugh in her voice.

"Ahoy, maties!"

The voice was Captain Quoin's. He and the rest of the Time Pirates stood at the edge of the Island of the Past—not on the island itself, but on a raft pulled up to the shore. About a hundred yards out the *Chronos* rocked gently in deeper water. There they all were: First Mate Lieutenant Cosmo, Helmsman Zhi Wo, the twins Pierro and Pietro, ship's boy Konstantin Dimitrivich, and Jonno, the *Chronos*'s transtemporal navigator. Captain Quoin stood at their head, wigless, swordless, but around his neck hung a shiny silver chain and from the chain hung the third of the three keys—the silver one—that had so recently traveled to the bottom of the world and back.

Tapping as delicately as a pianist, Charles steered the flying carpet over to the *Chronos*'s crew. When carpet and raft were edge to edge, hands were shook all around.

"Goodbye, Farley," Captain Quoin said. "Sorry about that nasty cut I gave you."

"Goodbye, Lieutenant Cosmo," Susan said. "I'm sorry I had to tie you up."

"Pierro, Pietro," Charles said. "Goodbye."

"Mario," Jonno said. "Nice shooting with the cannon. Goodbye."

"Pah, boy," Cosmo said to Charles. "And are you afraid to shake me hand after all this? Put 'er there, and let's let bygones be bygones." Charles reached to shake the dwarf's hand only to be pulled into a tight embrace. "You're a brave lad, and don't let anyone tell you different. You did your family and fellow sailors proud."

Susan saw then a familiar covered bundle at the end of the line of pirates. "Why, it's the dodo! Have you brought her to release her?"

"Aye, Susan," said Konstantin Dimitrivich, who besides being ship's boy was also the *Chronos*'s head of specimens. "She will live on here, the last of her kind, so that none may ever forget her gentle, trusting species."

Last of all, the Oakenfelds and Uncle Farley said goodbye

to Pierre Marin. The old Frenchman hugged them each long and hard, and seemed so loath to let them go that Susan found herself inviting him to come back with them.

"It's your house. Your ship. You're the captain."

Charles saw the same distant look come over Pierre Marin's face as when he had first asked for the Frenchman's help in rescuing Susan. His eyes glazed over and looked far away, and a brief smile revealed a flash of the gold tooth at the side of his mouth. His hand floated to his chest and rubbed at the gold key that hung there. Then, slowly, he shook his head.

"No, Susan. Thank you but no. My place is here now, my task. I am compiling a history of all the inhabitants of the Island of the Past, and I am far from finished. The island is larger than it appears, and loops back on itself many times, and for every creature I catalog Captain Quoin here seems to bring along one or two more." He nodded at the dodo, which Konstantin Dimitrivich was coaxing from her cage. "I imagine she has an interesting story to tell."

The dodo blinked in the bright sunless sky. Tentatively, she stepped off the Time Pirates' raft onto the sand. With a somewhat heavy step, Pierre Marin followed the last of the dodos onto the Island of the Past, the one and only human who would be allowed on its surface. Susan found her throat tight, and, swallowing the lump down, called after him.

"But won't we see you again? There's so much you can teach us."

Pierre Marin stared at the dodo for a moment, then turned back to Susan.

"There is even more that you can teach yourselves," he said. "But if you have questions, you'll always know where to find me."

He seemed about to set off then, when Murray suddenly spoke.

"Monsieur Marin?"

"Eh? Yes, Mario? I mean, Murray?"

Murray walked to the edge of the carpet, and motioned for Pierre Marin to come closer. The onetime sailor bent down—his good leg, Susan noted, instinctively coming forward to support the bulk of his weight, the golden key dangling safely from his neck, its glint matching that of the locket that hung from Murray's. Murray spoke directly into Pierre Marin's ear while the old man nodded and said "Yes" several times, and finally, "I do think you're right." He stood back up and shucked the bag from his back—all that remained of the load of luggage he'd set out with two days ago—and pulled from it a big, boxy-looking camera.

"This is a souvenir I picked up from the late nineteenth century," he said. "I should like a portrait if you don't mind. For, uh, posterity."

Susan wondered why Pierre Marin was stuttering slightly, but Charles understood immediately.

"It's for the locket! Murray's locket!"

"Hmmm, what's all this?" Pierre Marin said.

"Nothing," Murray said quietly, but Susan was sure she heard a slightly mischievous tone underneath the seriousness. "Susan and Charles are just poking their noses into things, as usual." And then, with his back to Pierre Marin, Murray mouthed the words *"He mustn't know."*

Pierre Marin looked confused for a moment, but then he shrugged and said, "There now, everyone gather close together. You too, Farley, it wouldn't be a family portrait without the kindly uncle." The four voyagers stood shoulder to shoulder (or shoulder to hip, in the case of Murray and Uncle Farley), smiling resolutely into the lens. There was a near-inaudible click—no flash—and then it was over.

"You'll have to come back now," Pierre Marin said, stowing the camera back in his bag and hoisting it onto his back. "For a print, if nothing else. Well, goodbye then. Safe journey home."

"Monsieur Marin," Susan said, "are you going to walk? Charles made it sound like it's an awfully long way."

Pierre Marin was looking at the dodo, who had found her land legs and started off at a brisk, slightly zigzagging pace across the immense green plain, as if in search of the hollow hill that would become her new home. He turned now, to look back to Susan and offer her one last smile.

"It seems to work for that sweet creature," he said, waving at the dodo's diminishing form. He hoisted his bag and took a step forward with his good leg, then set his pegleg in front of it, and suddenly he was a step away from her. Good leg, pegleg, good leg, pegleg: Susan could see how, eventually, Pierre Marin could reach any destination he set his mind to.

"And besides," the ancient mariner called over his shoulder when he was almost out of earshot and the dodo was just a gray blur in the distance, "I have all the time in the world."

Charles's Glossary of Affected Words

appellate: There's a whole other meaning to appellate having to do with courts and the legal system and all that, but that's not how Cosmo is using it. He's working from the French word "appellation," which means "name." So what he's saying is that it would be inadequate to *call* the Island of the Past a floater, since it moves through the water very rapidly. As you can see, it's not just the Oakenfelds who use affected language . . .

archipelago: An archipelago is a chain of islands, like Hawaii, or the Aleutian Islands off the coast of Alaska. It's an old

Latin word—and you know people are being extra affected when they use Latin.

avian: Avian means "birdlike." But doesn't "avian" sound cooler? And, when you get right down to it, isn't that the main reason to use affected words? To sound cool?

baleen: Another name for baleen is whalebone—i.e., yes, it's a bone from a whale, but a very specific kind of bone, which hangs from their upper jaw kind of like a big scrub brush, and is used to strain plankton out of the water, which they then eat. Since you probably came to this entry from *scrimshaw*, I won't make you look up plankton. They're really, really, really small fish, and none of them were harmed in the writing of this story.

bereft: This is what happens when your uncle is a scholar: he uses fancy words like "bereft," when all he has to say is "sad," which is what bereft means. Fortunately, Charles is too well-mannered to say "It's affected" to Uncle Farley.

besotted: This is an old-fashioned way of saying "in love with" or "having a crush on." Since President Wilson is an old-fashioned parrot, it suits him, don't you think?

blanched: Turned white. The opposite of blushed.

buck up: "Buck up" is kind of how the English say "Put your game face on!"

chiaroscuro: Yes, chiaroscuro. Well, it's a painting technique dating to the Italian Renaissance—that's, like, the 1500s, which is to say, a *really* long time ago. It uses light and shadow to make shapes look three-dimensional like a photograph, instead of flat like a cartoon.

crumpet: Actually, Charles is right: a crumpet really *is* an English muffin. Mrs. Applethwaite just happens to make especially good ones.

[I] daresay: This is a contraction of "I dare to say," which is basically the fancy way of saying "I guess" so that no one knows you're really guessing.

deigned: Deign means to do something that's a little bit beneath one's dignity. Here, a rather uppity parrot (I'm not giving away any secrets!) has decided to talk to some lowly humans.

demeanor: Literally, "the way in which someone behaves." See, sometimes affected words do save you a lot of space.

deuced: This is one of those words English people use when they want to use stronger language but there are children around. It's basically a polite insult—or, really, the least impolite insult. Americans would say "darn."

digestif: This is not just fancy English, but fancy French (a language that President Wilson knows quite well, as we'll see soon enough). A digestif is a food or beverage you take to aid digestion. Digestion–digestif: get it?

encroach: Encroaching is taking something that doesn't belong to you. This can be stealing, but that's not quite what Diaphone means here. What she's really telling Susan is not to take away Ula lu la lu's dignity by making her feel bad, since, presumably, Ula lu la lu feels pretty bad about being a prisoner. As we shall see, Ula lu la lu's got a thick skin, and it would've been pretty difficult for Susan to take her dignity from her, or anything else for that matter.

endive: A white leafy plant used in salads that just happens to be shaped a lot like a rowboat. It is the sort of vegetable Susan would eat, but not Charles.

execrable: Bad. Really bad. Really, really bad. Like, so bad you have to go the thesaurus and look up synonyms for "bad" just so you can say how bad it is.

exhorting: Isn't this a cool-sounding word? Ex-ORT-ing. I think it kind of sounds like something a talking pig might say. Of course there are no talking pigs in this book, so I should probably just give you the definition and let you get back to the story, huh? Exhorting is when you try to get someone to do what you want, like when you exhort your mom to give you the last piece of apple pie by telling her you'll clean your room or wash the dishes or something. That's exhorting.

fathom: A unit of measurement used for water depth, about six feet; thus Shakespeare's famous monologue that begins "full fathom five thy father lies" is a poetic way of saying your father is under about thirty feet of water. I'm guessing he probably doesn't have one of the mermaids' golden bubbles either. "What golden bubbles?" you say. You'll just have to keep reading, won't you?

fulcrum: A fulcrum is that thing underneath the middle of a seesaw, so it can go up and down, which you could probably figure out from context. In fact I bet you didn't look this word up at all: you're just seeing it because you've finished the book and don't want the reading experience to end, right?

gibberish: Nonsense. Noises that mean nothing. Duh.

gunwale: The side of the deck of a ship. It's called the gunwale because the guns were mounted in wales, or wells, cut into the railing. I don't know why they don't just call them "gunwells"—and to make it even more confusing the word is pronounced "gunnels." Language is so strange sometimes.

hypothesis: This is the ultra-fancy way of saying "theory," which is the semi-fancy way of saying "guess." See *I dare-say*, above.

imbecilic: If you really want to make someone feel dumb, then call them dumb by using a word they don't even know. Which is what imbecilic means: "dumb." Of course, neither this book nor its publisher advocates that you should ever call anyone dumb, or even imbecilic. You should always be nice to people, even if they're dumber than you.

insolent: Rude. Insulting. Trash talk. The kind of thing that's funny when you say it to your friends, but when you say it to your parents you get in trouble.

inverted commas: That's what the British call apostrophes, i.e., quotation marks. When you think it about it, they actually do look like upside-down commas, don't they?

iridescent: Shiny or reflective, but also full of all the colors of the rainbow. Besides fish scales (and mermaids' tails), oil slicks are commonly called iridescent.

ironic; irony: Irony is when you say one thing but mean the other. For example: "Sure, Mom, I'd *love* to do the laundry, fold the clothes, and put them away. I can't think of anything I'd rather do on a Saturday afternoon!" That's irony. You might know it by its less affected name: sarcasm.

litter: You're probably familiar with kitty litter, but this isn't that kind of litter (we didn't want you imagining a prince sitting in a litter box). No, this kind of litter is a chair mounted on a platform that servants carry around for you. If litter is an affected word, then sitting on a litter is just about the most affected kind of sitting there could ever be.

locution: This is a shorter way of saying "a particular manner of speaking." On second thought, I'm wondering if I should have just said "particular manner of speaking," since that's easier to understand, even if it takes more words.

longitudinal: You know those lines on a map or a globe? Some of them go up and down, and some of them left and right? The ones that go up and down are lines of longitude

(just in case you're wondering, the ones that go left and right are lines of latitude).

"Lu-u-u-cy": President Wilson, being much older than any of us, can remember television shows from, like, the Dark Ages. In this case, he's referring to a famous joke on a show called *I Love Lucy* starring Lucille Ball, who was a wacky housewife who always got in trouble. Whenever her husband caught her, he'd say, "Lu-u-u-cy, you got some 'splaining to do." You can still catch the show on reruns, if you parents let you stay up late, or record it for you.

mess: Here, Cosmo is using the seafarers' meaning of mess, which is their word for "kitchen." Since they call it a mess, you can imagine what the food that comes out of it tastes like.

minced: When the floor is cold and you walk with your toes curled up so as little of your feet as possible touch it, you're mincing.

mollified: Soothed, calmed—like when you give a baby a pacifier to stop it from crying.

muesli: Muesli is the on-steroids version of All-Bran or Grape Nuts or one of those cereals that has enormous amounts of fiber to keep you regular. It's so hard it actually hurts your mouth to chew it, but if you use good whole milk and slice some fresh fruit on it, it's actually kind of good. And it keeps you regular.

nasty turn: A nasty turn is a shock, and not the pleasant kind.

no recuerdo: Spanish for "I don't remember." How Murray knows Spanish is another story. In fact, it's another book; if I wasn't working on this glossary I'd be writing it right now.

old boy: "Old boy" is kind of the Victorian way of saying "dude."

palisades: Palisades are cliffs that go on for a long really long time—not up and down but right and left. Often they run along a coastline, as they do here, and along the Hudson River, as Susan is remembering.

penchant: This is a way of saving the few letters it would take to type "a definite liking" or "predilection," which is what penchant means. Of course, the time I saved in typing you use up in turning back to this glossary.

perilous: Dangerous. I'm sure there's a funnier way to explain it, but really, perilous just means "dangerous," and I wish Charles had said "dangerous" because then I wouldn't have had to type this.

preposterous: Something is preposterous when it's so dumb that it's like, "Dude, that's *dumb.*"

procedural: Doing things in their proper order. Like washing your hands before dinner and washing the dishes afterward, as opposed to washing the dishes before dinner and washing the dog after. That just wouldn't make sense, would it?

queer: Well, we've all heard of a certain television show in which a clutch of, how shall we say it, enthusiastic guys teach another guy how to dress right and spruce up his pad to make girls like him. This isn't that kind of queer. This kind of queer just means "weird." Personally, I think that show's a little weird.

rectify: Rectify means "fix." Yup, you're right. Uncle Farley probably should've just used the word fix.

replica: I think replica is less affected than cool-sounding. Kind of science fictiony. But in any case it just means "a copy of something else."

scrimshaw: Especially popular in the 1900s, scrimshaw is the carving of little figurines out of baleen. You probably want to look up *baleen* now, don't you?

spectacles: Eyeglasses. You probably already knew that, and are annoyed you had to turn four hundred pages for no new information.

spume: Literally, "foam or froth." You're more likely to see spume on top of a cappuccino than shooting out of a whale's blowhole, unless you happen to be an ocean-going fisherman.

squabbling: A squab is a baby pigeon. Apparently baby pigeons scream at each other all the time—fighting over those worms, I guess—because squabbling means "arguing noisily," something Charles and Susan are all too good at.

stroppy: Okay, so stroppy actually means that you're always in a bad mood and ready to yell at someone for no reason at all (we've all had an aunt or an uncle or a babysitter who's stroppy). So to call a boat stroppy, as Pierre Marin does here, means that it tosses violently on the water—which, if you're prone to seasickness as Pierre Marin is, can make *you* quite stroppy too.

tarry: Delay, waste time. Of course, there's no time to waste on the Sea of Time, so I guess we'll just go with delay.

tempered: Although it sounds like "tempered" is something that would make you mad, it's actually the opposite: it's something that calms you down. Go figure.

translucent: See-through, but not 100 percent see-through, like wax paper, or iced tea.

trisecting: To cut in thirds. You know, there's only one piece of pie left, but you and your dad and mom all want a piece of it, so you have to trisect it.

trompe l'oeil (TROMP LOY): More French. By now you're thinking, this book has English, French, Italian, and Spanish in it—is it a tour bus or is it a novel? Think about it this way: the next time you have to draw a picture in art class, you can tell the teacher it's an example of trope l'oeil painting and he or she will be quite impressed, I promise. Of course, you probably just turned back here to get the definition, so I suppose I should give it to you now. Trompe l'oeil means "to fool the eye"; basically it's painting that's so real-looking, you almost think you're looking at flesh and blood and trees and sky. For information on how this is done, see *chiaroscuro*.

unadulterated: Unadulterated means "pure." Which means that "pure unadulterated" is repetitive. My bad.

utterances: Things that come out of your mouth that aren't food. In other words: words. I've got to give this one to Charles. Susan was really being affected here.

vicissitudes: The only word harder to say than "Mississippi," and harder to spell as well. And you know what it means? Change. That's all. Just "change."

Victorian: Queen Victoria ruled England from 1837 until 1901, which is, first of all, a really long time (the longest anyone has ever ruled England, in fact), but it is also the time people refer to as "Victorian." Or, if you're being a little affected, "Vic*to*rian."

virago: A virago is either a strong and courageous woman, or else a bossy and noisy woman. I think you know which definition applies to Queen Octavia.

writhe, writhing: If you've got a nasty older brother or sister who holds you down and tickles you (of if you're the fun-loving older brother or sister who holds down their annoying younger sibling and tickles them) then you know what writhing is. It's that squirming you do when you *really* want to get away from someone.